Industrializing Organisms

HAGLEY PERSPECTIVES ON BUSINESS AND CULTURE

PHILIP SCRANTON & ROGER HOROWITZ, EDITORS

Beauty and Business
Commerce, Gender, and Culture in Modern America
Edited by Philip Scranton

Boys and Their Toys?
Masculinity, Technology, and Class in America
Edited by Roger Horowitz

Food Nations
Selling Taste in Consumer Societies
Edited by Warren Belasco and Philip Scranton

Commodifying Everything
Relationships of the Market
Edited by Susan Strasser

Industrializing Organisms
Introducing Evolutionary History
Edited by Susan R. Schrepfer and Philip Scranton

Industrializing Organisms
Introducing Evolutionary History

Hagley Perspectives on Business and Culture, Volume 5

Edited by
Susan R. Schrepfer
and Philip Scranton

ROUTLEDGE
NEW YORK AND LONDON

Published in 2004 by
Routledge
29 West 35th Street
New York, NY 10001
www.routledge-ny.com

Published in Great Britain by
Routledge
11 New Fetter Lane
London EC4P 4EE
www.routledge.co.uk

Copyright © 2004 by Taylor & Francis Books, Inc.

Routledge is an imprint of the Taylor & Francis Group.
Printed in the United States of America on acid-free paper.

All rights reserved. No part of this book may be reprinted or reproduced or utilized in any form or by any electronic, mechanical, or other means, now known or hereafter invented, including photocopying and recording, or in any information storage or retrieval system, without permission in writing from the publishers.

10 9 8 7 6 5 4 3 2 1

Library of Congress Cataloging-in-Publication Data

Industrializing organisms : introducing evolutionary history / edited by Philip Scranton and Susan Schrepfer.
       p. cm. — (Hagley perspectives on business and culture ; 5)
Includes bibliographical references (p.   ).
   ISBN 0-415-94547-X (hardcover : alk. paper) — ISBN 0-415-94548-8 (pbk. : alk. paper)
1. Agricultural biotechnology. 2. Agricultural biotechnology—Environmental aspects. 3. Transgenic organisms. 4. Agricultural innovations. I. Scranton, Philip. II. Schrepfer, Susan R. III. Series.
   S494.5.B563I53 2003
   631.5'233—dc21

2003011931

Contents

Preface ... vii
PHILIP SCRANTON

Introduction: The Garden in the Machine: ... 1
Toward an Evolutionary History of Technology
EDMUND RUSSELL

Part I: Plants, Profits, Politics, and Power

"For Profit and Pleasure": Peter Henderson and the ... 19
Commercialization of Horticulture in
Nineteenth-Century America
SUSAN WARREN LANMAN

Biological Innovation in American Wheat Production: ... 43
Science, Policy, and Environmental Adaptation
ALAN L. OLMSTEAD and PAUL W. RHODE

Creating an Industrial Plant: The Biotechnology ... 85
of Sugar Production in Cuba
MARK J. SMITH

Manufacturing Green Gold: Industrial Tree Improvement ... 107
and the Power of Heredity in the Postwar United States
WILLIAM BOYD and SCOTT PRUDHAM

Part II: Animals, Aggression, Arrogance, and Analysis

War Horses: Equine Technology in the American Civil War ... 143
ANN N. GREENE

Turbo-Cows: Producing a Competitive Animal in the ... 167
Nineteenth and Early Twentieth Centuries
BARBARA ORLAND

Canine Technologies, Model Patients: The Historical ... 191
Production of Hemophiliac Dogs in American Biomedicine
STEPHEN PEMBERTON

Making the Chicken of Tomorrow: Reworking Poultry as Commodities and as Creatures, 1945–1990 ROGER HOROWITZ	215
Hogs, Antibiotics, and the Industrial Environments of Postwar Agriculture MARK R. FINLAY	237
Afterword SUSAN R. SCHREPFER	261
Notes on Contributors	267
Index	269

Preface

PHILIP SCRANTON

Late in summer 2001, a mildly phrased but pointed query flitted across my computer screen, sent from University of Virginia historian Edmund Russell to the several hundred subscribers of Envirotech, an online discussion group appealing chiefly to scholars interested in the intersections of technological and environmental history. "Are animals technology?" Russell asked, sitting back to watch the fur fly (though while traveling, out of touch with his e-mail, he missed a fair portion of the vigorous exchanges he'd triggered). Scores of colleagues around the globe, including Rutgers' historian of technology Paul Israel, weighed in on various aspects of the issue, generating a spirited crosstalk that its initiator termed "all the things scholarly debate should be—wide-ranging, insightful, at times heated, stimulating, helpful."[1]

Though many "spoke up" in the "animals as technology" discussion, I confess that I lurked—actually, I lurked and pondered. A year earlier, a group of Rutgers historians, developing a new doctoral track in the history of technology, environment and health, had proposed to the Rutgers Center for Historical Analysis (RCHA) board that it devote the center's resources for 2001–3 to exploring themes close to the heart of their cross-disciplinary research and teaching field. Since the late 1980s, RCHA had sponsored a series of two-year "cutting-edge" projects, bringing in graduate students, post-docs, scholars within and beyond the university in many fields to examine consumer societies, the black Atlantic, utopias, and other terrains for innovative historical research. After our proposal was accepted and funded, this volume's coeditor, environmental historian Susan Schrepfer, and I, a historian of technology and industry, headed up the seventh RCHA project: Industrial Environments: Creativity and Consequences.

In the months before Ed Russell's Internet debate broke forth, Susan and I, with Paul Israel's amiable assistance, had arranged a schedule of weekly seminars for the first year, had selected fellows, and had sketched an idea for a spring 2002 conference. This gathering, we expected, would center on one of several ideas we'd been developing, but "animals as technology" rapidly supplanted all the others, for we recognized the Envirotech exchanges' intellectual vibrancy. Susan's research on plants and forests led us to expand the framework question to "Organisms as technology?" Soon after we contacted Ed Russell to draft a conference keynote and fashioned a call for papers, proposing a scholarly discussion of Industrializing Organisms: Plants, Animals and Technology. It read, in part:

While the breeding and manipulation of plants and animals for human purposes has been occurring for millenia, efforts to produce standardized and more efficient organisms suitable for large-scale agricultural and industrial processes are more recent phenomena tied to the rise of modern industrial societies. When, where, and how have industrializing humans sought to "improve" plants and animals in order to better integrate them into technological processes and systems? To what extent was the modification of organisms an essential element of modern technology? What kind of design considerations were unique to living organisms and what kind of obstacles did these present to human efforts at redesign? What have been the consequences of such modifications?[2]

We discovered, in our interactions with the scholars whose work follows in this volume, that these were provocative questions. Only some of them have been addressed in depth thus far, but each may provide an intellectual stimulus to support further research in the area Russell outlines in his introduction as constituting "evolutionary history." His conceptualizing essay leads into the collection's two main sections, focused conventionally on "Plants" and "Animals." Part 1, loosely proceeding forward from the mid-nineteenth century, provides windows into commercial horticulture (Susan Warren Lanman), the diversity and transformations of wheat production (Alan L. Olmstead and Paul W. Rhode), the sugar sector's version of "factories in the fields" (Mark J. Smith), and the construction of tree improvement as a modern business-science partnership (William Boyd and Scott Prudham). Turning to the animal kingdom and cycling back to antebellum and Civil War decades, we encounter first the mobilization and commodification of horses as instruments of war (Ann N. Greene), then the drive to refine and classify Swiss milk cows (Barbara Orland), the production and tending of hemophiliac dogs as vehicles for medical experimentation (Stephen Pemberton), the postwar project of creating broiler chickens (Roger Horowitz), and the parallel pharmacological restructuring of hog raising (Mark R. Finlay). Susan R. Schrepfer closes the volume with reflections on the essays and their relationship to environmental history.

Each of *Industrializing Organisms*'s contributors gained, individually and intellectually, from encountering one another in our April 2002 meetings. It is our hope that you, our readers, will also profit from their collective investment in bringing these essays to publication. If anything, these studies indicate that human intervention has telescoped/compressed evolutionary dynamics in the service of our economic priorities and in relation to political/cultural imperatives. The implications of such biological activism remain to be fully assessed, but careful evaluations will, we believe, be grounded in research that extends the concepts, issues, and approaches you will encounter in the essays of this book.

Acknowledgments are a customary and key element of any book's preface. At Rutgers, the editors are indebted to Deborah White, the New Brunswick history department chair and RCHA's general director, to Paul Israel, head of the

Edison Papers Project at Rutgers and our closest collaborator, and to Lynn Shanko, who flawlessly manages both the RCHA's operations and the faculty folks who try their hands at organizing successive projects. At the conference, commentators Phil Pauly and Keith Wailoo, of the Rutgers faculty, plus Lynn Swartley and Greg Hise, RCHA fellows, offered presenters critical responses to their work, along with avenues for extending its reach and refining its quality. At Routledge, Karen Wolny has been an advocate for the Hagley Perspectives series and the best of colleagues in helping series coeditor Roger Horowitz and me to learn what makes an essay collection publishable. Of course, Susan and I are grateful for the diligence and responsiveness of "our" authors to the arrays of questions and comments we offered on early drafts of the essays published here. Without their conviction that this volume could contribute to thinking creatively about historical patterns of human interventions into the natural world (in eras before gene-splicing and contemporary versions of biotechnology), the Industrializing Organisms conference's legacy would be little more than fond memories. Thanks to one and all.

Notes

1. Edmund Russell, "Animals as Technology," retrieved on November 14, 2001, from Envirotech@lists.Stanford.edu.
2. Call for Papers, "Industrializing Organisms," Rutgers Center for Historical Analysis, retrieved November 16, 2001, from http://rcha.rutgers.edu/spring-conf.html.

Introduction
The Garden in the Machine:
Toward an Evolutionary History of Technology

EDMUND RUSSELL

In the classic story of industrialization, machines replaced human and animal muscles. Waterwheels replaced the oxen that turned grindstones in mills. Steam engines replaced the mules that had pulled boats upstream. Gunpowder replaced the arms that had drawn bows or flung spears. Gasoline engines replaced the horses that had drawn plows, wagons, and carriages. Electricity replaced the hands that had beaten eggs and scrubbed clothes. Somewhere along the way, our understanding of "technology" followed suit. In the minds of many historians, technology has consisted of machines and, more recently, systems of machines and humans. Hiding behind this view is an assumption about the relationship between technology and nature: technology replaced or modified nature, but nature was not technology. But since machines are always made from metal, wood, rubber, and other products of nature, the assumption boils down to—put bluntly—nature having to be dead to be technology.

One of the most enduring metaphors in the history of technology expresses this idea neatly. Leo Marx famously argues that the locomotive epitomized the intrusion of technology into the American countryside in the nineteenth and twentieth centuries. "The machine in the garden," he called such intrusions. Even cattle, sheep, and horses recognized that technology radically changed nature, Marx suggests.[1] While disagreeing on the desirability of such change, most historians before and since Marx have seen the relationship between technology and nature (wild as well as pastoral) in much the same way: as technology intruding into nature.

But the smoke of industrialization (along with romanticism and the Cartesian dichotomy between nature and humans) has obscured our view. Yes, machines brought radical changes to nature. But no, cattle, sheep, and horses did not simply witness the intrusion of technology into nature. Their ancestors had walked among the vanguards of technology. They were not machines, but they were biological artifacts shaped by humans to serve human ends. They were technology and, in the root meaning of the word, biotechnology. To capture this reality, we need to reverse Marx's metaphor and see the garden (nature) in the machine (technology).

Moreover, industrialization never came close to replacing living organisms completely; industrialization was a biological as well as a mechanical process. Machines, plants, and animals coexisted, and industrialization needed living organisms to succeed. Machine shops and factories relied on belts to transfer motion from one wheel to another, and livestock produced the leather that became the belts. The waterwheels that powered mills would have been useless without crops to supply grains for grinding and trees to supply lumber for sawing. The growth of railroads, a paradigmatic example of industrialization, caused the number of draft horses in the United States to increase rather than decrease. As Ann Greene shows in this volume, it did so by opening new markets for farm products (justifying greater investment in livestock by farmers) and creating demand for wagons to transport goods to and from depots.

Evolutionary History

To understand the role of nature in industrialization, we need a new way of thinking about technology and nature. Reversing Marx's metaphor is a first step, but where do we go from there? This volume, and the conference from which it grew, gives several possibilities. Each essay offers a unique angle of approach. A theme tying them together, however, is the idea that people have shaped organisms to serve human ends. One way to look at that process is through the discipline that specializes in the plasticity of organisms—evolutionary biology. At the same time, we want to capitalize on history's expertise in analyzing the range of human social variables that have driven such change. My suggestion is that we develop a new field called evolutionary history: the field concerned with the role of evolution in the human past.[2]

Until now, evolutionary history has meant the topics or events studied by evolutionary biology. The usage suggested here is a new, additional meaning, envisioning evolutionary history as a field that links history with biology to create an evolutionary historiography. Its essential contentions are that humans have shaped the evolution of other species, and that such intervention has significantly changed both humans and other species.

Evolutionary history creates the opportunity to reach a fuller understanding of the past than would be available from either discipline alone by pairing strengths to overcome weaknesses. A virtue of history is its emphasis on human complexity and change; a virtue of biology is its emphasis on natural complexity and change. A weakness of history is its tendency to ignore, collapse, or black box the complexity and dynamics of nature; a weakness of biology is its tendency to ignore, collapse, or black box the complexity and dynamics of human society. Their synthesis, however, offers us a chance for new research questions and insights.

Let us look first at what evolutionary biology brings to the table. It inherited a good start on evolutionary history when, over a hundred years ago, Charles Darwin used domestic animals as evidence and inspiration for his theory of evolution through natural selection. He chose the term "natural selection" partly because it linked human-induced ("artificial") selection with selection in the wild better than other terms he considered.[3] *On the Origin of Species* opens with a chapter on domestication before turning to selection in the wild in the second chapter.[4] Moreover, Darwin described another of his books, *Variation of Animals and Plants under Domestication*, as providing the evidence upon which his theory of evolution by natural selection was based.[5]

The next major step in evolutionary theory, the neo-Darwinian (modern) synthesis, also developed partly through attention to anthropogenic evolution. The synthesizers coupled Darwinian evolutionary ideas with population genetics to explain the means of variation and inheritance. To the synthesizers, evolution took place when gene frequencies changed from one generation to the next. Given enough such changes, new species would arise. Unfortunately, the synthesizers—like Darwin—had no examples from the wild to illustrate their ideas. Then one of the leaders of the synthesis, Theodosius Dobzhansky, learned that insecticides seemed to have lost their ability to kill pests in California orchards. Dobzhansky seized on this pattern to illustrate his theory. Some individual insects happened to carry genes that enabled them to survive insecticide treatment. The "resistant" individuals survived and reproduced, while their "susceptible" brethren became evolutionary dead ends. Soon, individuals carrying genes for resistance dominated the population. The insecticide had not changed; the insects had.[6] Recently, several evolutionary biologists have emphasized the importance of anthropogenic evolution.[7]

What evolutionary biology brings, then, is a workable theory that explains how and why organisms change over time, including under domestication. Three ideas anchor the theory of Darwinian evolution today: variation, inheritance, and selection. Individuals must vary in some trait; they must pass their trait on to their children; and possessing that trait must increase or decrease an individual's chances of survival and reproduction. None of this happens in a vacuum, of course, which brings onstage a fourth consideration: the environment. Organisms do not evolve toward some Platonic ideal; they evolve in response to the particular circumstances in which they find themselves, and success in an evolutionary sense means simply that you contribute more offspring to the next generation than do other members of your species.[8]

Humans can influence this process at several points. We can increase variation (by increasing mutation, importing stock from elsewhere, and by inserting genes from other organisms). We can decrease variation through inbreeding and cloning. Traditionally, we have had to rely on an organism's own reproductive machinery to handle inheritance, but now genetic engineering enables us to see that an organism inherits a trait from an unrelated organism. We can control selection by choosing which individuals mate (bull and cow, stallion and mare, male and female flowers) and which do not. We can also shape selection by stacking the deck. Humans did not pick which individual insects mated in California orchards, but they did change the survival rate (thus reproductive rate) of resistant versus susceptible individuals.

This stage of analysis, in which one asks why humans and how humans did what they did, is where history enters with its particular expertise. All fields of history have something to add, but our main concern here—the history of technology—is particularly well equipped. Historians of technology have developed sophisticated ideas about (1) why and how humans have shaped machines the way they have, and (2) how those changes have in turn shaped human society. My suggestion here is that we apply those same insights to the ways humans have shaped organisms.[9]

Historians of technology have shown that many social factors shape the nature, development, and use of technology. Such factors include politics, labor-management relations, economics, warfare, science, institutional strategy, national identity and styles, culture, gender, race, and class.[10] Although historians of technology have focused most of their attention on human-machine interactions, we have seen a growing interest in human-machine-nature interactions.[11] (This interest has found an institutional home in Envirotech, a special interest group within the Society for the History of Technology.)[12] For the most part, though, even historians in Envirotech have maintained Marx's nature-technology dichotomy. Technology shapes nature, or nature shapes technology, but rarely do we see nature and technology merging.

A few scholars, though, have taken the next step. Largely interested in agriculture, these researchers have—explicitly or implicitly—analyzed organisms as technology. One group has focused on ways people have *intentionally* changed other species through breeding. Jared Diamond made domestication, and the social and biological changes that follow, central to the rise of urban civilizations, wars of conquest, spread of disease, development of technology, and written history. Joel Tarr and Clay McShane have described horses as essential technology in cities. William Boyd, Deborah Fitzgerald, Jack Kloppenburg, Harriet Ritvo, and John Perkins have traced the rise of plant and animal breeding from less formal to more formal systems. They have identified several factors as driving changes in breeding practices and effects, including craft knowledge, government sponsorship of research, the rise of the science of genetics, capital accumulation, commodification, national security, founding of agricultural research stations and universities, institutional ambitions, international trade, rural economics and politics, class anxiety, and concern about hunger.[13]

A second group has focused on ways people have *unintentionally* affected evolution. John Perkins, Mark Boyd, and I have drawn on Dobzhansky's ideas to explain the rise of resistance to pesticides and antibiotics.[14] (This is no small phenomenon. By 1986, scientists had identified some 450 species of insects and mites, 100 species of plant pathogens, and 48 species of weeds that had evolved resistance to pesticides.)[15] Stephen Mosley has noted that dark peppered moths became more common near Manchester as coal smoke on trees made their lighter brethren more visible to predaceous birds.[16] Joseph Taylor has argued that hatcheries changed salmon evolution by narrowing genetic diversity and selecting for fish that clumped rather than dispersed. By causing more damage to large than small fish, dams selected for smaller and faster-maturing salmon.[17] The importance of unintentional anthropogenic evolution can be measured, among other ways, in lives. By 1995, the U.S. Centers for Disease Control and Prevention had concluded that evolved resistance to antibiotics contributed heavily to sixty thousand annual deaths in the United States from hospital-acquired infection.[18]

Moving forcefully in this direction is essential if the history of technology is to contribute to debates about biotechnology. Histories of biotechnology are important because they can delineate factors driving biotechnological change. The Biotechnology Industry Organization's website implies that widely accepted goals drive development of biotechnology: feeding the world, improving health, cleaning up the environment, defending against biological warfare, and so forth.[19] Those goals are important, but are they the only ones? The virtue of evolutionary history is that it focuses on the "why" of human-induced evolution. Historians of technology are perfectly situated to address these questions because they have developed ways of understanding why humans have shaped machines and technological systems; now the task is to apply these insights to living technology.

The stakes are high. In 2002, with twelve million citizens on the brink of famine, Zimbabwe rejected emergency shipments of food from the United States because the shipments contained genetically engineered corn. Zimbabwean leaders feared that some of the corn might be planted, contaminate other corn plants with its pollen, and doom its future export trade to Europe. There, fears of damage to health and environment had led to sharp limits on imports of bioengineered food. Between the jaws of advocacy of biotechnology, on the one side, and criticism, on the other, twelve million hungry Zimbabweans found themselves squeezed in a vise not of their creation.[20] If modern biotechnology does not convince us that organisms sometimes are technology, with fully as much potential for good or ill as any other technology, it surely will convince our successors.

Studying organisms as technology will likely meet objections. In an Envirotech online discussion during the summer of 2001, a number of scholars objected to the suggestion that organisms sometimes were technology. Some participants objected that technology acted on nature, but nature was not technology. Others argued that we demean animals (and ourselves) by thinking of them as machines, which opened them and us to exploitation. Still, this research agenda will also find support. Several participants in that debate used horses bred for particular jobs as examples of animals used as technology. Another suggested that if something can be patented, as some life forms now are, it is technology. Overall, supporters seemed to outnumber objectors.[21]

Taking on this task strengthens the growing ties between the history of technology and environmental history.[22] The latter field focuses on the interaction between humans and nature in the historical past. Environmental historians have identified a range of social factors that influence the way people interact with nature. Many are the same as those studied by historians of technology—economics, politics, gender, race, class, war, scientific and technological institutions, and corporate strategy.[23] Recently, and with great controversy, some environmental historians have argued for displacing wilderness preservation as the ideal way for humans to interact with nature. This shift means paying attention to working landscapes, urban as well as rural, and recognizing that human intervention can result in progress as well as decline.[24] It implies a much greater focus on the natural elements with which people commonly interact, such as lawns, urban forests, weeds, and farm animals.

Unlike the history of technology, which has a long tradition of unpacking technological development, early work in environmental history often black boxed technology.[25] One common narrative line runs: People developed the technology they wanted, which changed nature, which usually backfired to harm people and nature.[26] Recent research is helping to change that framework. Most of this literature, though, still portrays nature and technology as separate.[27] Marx's "machine in the garden" metaphor continues to prevail. Evolutionary history, which stresses the malleability and transformations of organisms to suit human purposes, helps guide the attention of historians of technology and environmental historians alike to the central role that organismal technology has played in human history.

Research Tactics

Evolutionary history stands out from other fields of history by using evolution strategically. From its perspective, humans and nature have been molding each other for millennia. Given this strategy, what tactics might an evolutionary historian deploy in research? The essays in this volume suggest several approaches that might be especially useful to historians of technology interested in industrialization.

Suggestion 1: Replace the biotechnology industry's terms "biotechnology" and "new biotechnology" with "macrobiotechnology" and "microbiotechnology." The Biotechnology Industry Organization (BIO) defines biotechnology as the "use of biological processes to solve problems or make useful products." It describes any use of organisms, starting ten thousand years ago with domestication, as biotechnology, and it describes recent technologies using cellular and molecular processes as "new biotechnologies."[28] These terms are confusing. "Biotechnology" refers both to all biotechnology and to a subset, any biotechnology before the arrival of new biotechnology. On their face, the terms do not clarify what distinguishes new from other biotechnology. We need terms that link traditional and new biotechnology in ways that enable us to see both continuity and change.

I suggest we use *biotechnology* to refer to all living technology (its root meaning) and to subdivide biotechnology into *macrobiotechnology* and *microbiotechnology*.[29] "Macrobiotechnology" refers to technology in which humans intervene on the level of the whole organism (such as traditional breeding). "Microbiotechnology" refers to technology in which humans intervene on the cellular or subcellular level (such as genetic engineering). It is easy to find boundary cases, so these divisions are heuristic rather than absolute.[30] Their virtues are that "biotechnology" links past and present technologies while "macrobiotechnology" and "microbiotechnology" point to key differences. Such differences include contrasting techniques (breeding whole animals and plants versus using techniques from molecular biology), contrasting rates of genetic change (slower versus faster), and contrasting means of inheritance (inheriting genes only from parents versus inheriting genes from potentially any other organism). We need not change our research topics to microbiotechnologies to demonstrate the importance of our work for today's problems. But by using the terms micro- and macrobiotechnology in introductions and conclusions to books and articles, we can make the connection between past and present clear while also clarifying their differences.

Ann Greene's essay on Civil War horses, to pick one example from this volume, is classic macrobiotechnology. Through breeding, humans created horses that varied in size, strength, and speed. These differences suited some horses to cavalry, some to artillery, and some to pulling wagons. Scaling up to thinking of these horses in the overall category of biotechnology, Greene's story suggests social forces that might shape micro- as well as macrobiotechnologies. Although traditional breeding took too long to make significant changes in horses during the Civil War, we see the same press for standardization and interchangeable parts familiar from other histories of military technology. To decrease the need for tack suited to horses of multiple sizes, and to speed replacement of horse fallen in harness, Union Quartermaster General Montgomery Meigs standardized the sizes of horses for each task. We know from other studies that technology developed for military use often finds its way into civilian use, even if qualities desirable in a military setting were suboptimal for civilians.[31] Might we expect, then, to see standardized microbiotechnology developed for the armed forces transferred to civilian use even if it is suboptimal?

William Boyd and Scott Prudham's essay about trees illustrates a transition from macrobiotechnology to microbiotechnology in an unusually short time. American foresters turned to selective breeding only in the twentieth century. By the end of the same century, researchers had created transgenic poplars and spruces. Driving this change, Boyd and Prudham argue, has been the effort of capital to convert reproductive biology into a means of accumulating wealth.

Unfortunately, biotechnology that has seemed ideal for this project might be less than ideal from an ecological or evolutionary point of view. As macro-biotechnology, cultivated tree plantations radically simplified the species composition of forests, and thus the species that depend on forests, over large areas. As microbiotechnology, genetically modified trees might spread genes to wild trees, which would develop traits, such as herbicide resistance, desired only in the cultivars. Although technological means may have changed, the biotechnological goals shared by government and private researchers remained much the same.

Suggestion 2: Conceptualize biotechnologies as factories. In analyzing the relationship between farming and industrialization, historians of technology have focused most closely on machinery and other tools produced by factories for farming. Such tools include tractors, plows, disks, harrows, combines, fertilizers, and pesticides. But farms are not just consumers of industrial products. As the 1916 textbook quoted in Mark Finlay's essay notes, farmers were manufacturers; they too transformed raw materials into useful products. Seeds, fertilizer, pesticides, calves, and feed were the inputs, and food and fiber were the products. A hog farmer referred to his operation as a "hog farrowing assembly line." As the twentieth century progressed, "factory farming" became a familiar phrase. It usually referred to farms relying on sheds to raise animals under controlled (and often crowded) conditions to maximize production and profit, but the concept was capacious enough to encompass whole farms.

We can also, however, view a farm not as one factory but as a complex of factories functioning on a variety of scales. Some of the factories are under roofs, but others are not. Smaller factories play essential roles in transforming raw materials into products. Mark Finlay quotes in his essay one scientist as saying that hogs needed to be modified because one should not use "poor machinery to put the raw product [feed] through." Mark Smith in his essay emphasizes the centrality of sugarcane's key characteristics to industrializing sugar production. The biology of sugar production meant that sugar refineries had to go to sugar fields, not the reverse. Moreover, organismal factories carry out complex and difficult tasks. Fritz Haber and Karl Bosch received Nobel prizes for figuring out how to fix atmospheric nitrogen. We readily regard as factories the machinery that use the Haber-Bosch process to make fertilizer. Why not legumes as well? Legumes are not machinery, but they do transform atmospheric nitrogen from one form to another, more useful form for humans.

Moreover, biological factories are essential in industrialized agriculture. No one has yet figured out how to transform sunlight, carbon dioxide, and a few nutrients into grain—except by subcontracting the job to plants. The same goes for meat production and animals. Roger Horowitz's essay describing chicken production, including development of the Chicken of Tomorrow, captures the need for keeping chickens at the center of production while at the same time fitting these animals to more technological settings. Industrialization on the farm has meant building ever more efficient factories partly by improving machinery and partly by improving biology.

Alan Olmstead and Paul Rhode illustrate in their essay how such a perspective can overturn established interpretations of history. Economic and agricultural historians have long attributed increased farm productivity (per worker) in the nineteenth and early twentieth centuries to machinery. Implicitly, they assumed that the biology of farming remained constant. This assumption is understandable if we think of crops as fixed. But organisms are not fixed. They constantly evolve in response to their environment, either on their own or through human efforts. Olmstead and Rhode show clearly that, without breeding, wheat harvests would have plummeted, all the machinery in the world notwithstanding, as insects, weeds, and diseases evolved to storm the factory gates. Biological development was roughly as important as mechanical innovation in boosting productivity.

The failure of economic and technological historians to recognize the importance of biology suggests a striking asymmetry in attention to biological and mechanical aspects of farming. We would not dream of seeing increased productivity over decades if farmers bought a fixed number of tractors, never maintained them, never replaced them, and never bought other implements for the next one hundred plus years. Historians would intuitively recognize the need for change in machinery. Evolutionary history encourages us to look for the same in organisms. Thinking of organisms as factories (or as machine equivalents) might help this process.

Pushed another step, we might even reverse our assumption about the type of technology best suited to industrialization. Usually, we have seen machinery as compatible with (and driving) industrialization. But biotechnology might in fact be better suited than machinery. One way to increase efficiency in a factory is to reduce the number of steps required to make a product. Imagine if we collapsed steps so much that the assembly line was also the product itself. So each time General Motors made an automobile, it had to ship out the assembly line as part of the car. General Motors would have to build a new assembly line for each car, an impossible proposition. But biotechnology has made this feat possible repeatedly. Organisms convert raw materials into products—feed into meat, for example—and then leave the factory as the product itself. Moreover, they leave behind new, self-organizing assembly lines that also will become products, ad infinitum. Perhaps, then, the future of industrialization lies in becoming ever more biological rather than less.

Suggestion 3: Think of biotechnologies as workers. One of the fascinating aspects of biotechnologies is their ability to perform a variety of roles. They resemble factories, but they also resemble workers in factories. Like human laborers, they cannot work all the time (witness the running down of horses in the Civil War); need to eat and drink during the workday; require direction from managers; work well only if temperature, humidity, and light fall within certain ranges; have a limited life span; wear out with age; require special shelter; demand more resources to work harder; reproduce sexually (occasionally asexually); and even stop working all at once.

Historians of technology have learned much about workers in factories and workshops: who they are, where they come from, how they interact with one another and with management, what conditions they encounter, and what impact they have on products and companies, to name just some of the dimensions. It would be fruitful to take these insights and apply them to organisms used as technologies. How do they interact with one another and management? How do they convince managers to alter the conditions of work? How are they compensated for their work? What techniques do managers use to get them to do more work? How have human and organismal labor proved fungible?

For example, the essay by Stephen Pemberton illustrates ways in which human and animal laborers resembled and substituted for each other. When Kenneth Brinkhous began studying hemophilia, he relied on an itinerant labor supply. If a hemophiliac happened to turn up in the hospital where Brinkhous worked, he would draw the blood he needed to carry out his studies. To overcome the unreliability of such labor, Brinkhous hired hemophiliac Jimmy Laughlin as a regular employee who washed equipment as well as supplied blood on demand. But even Laughlin was not ideal, for using him as a test subject threatened to kill him. Hemophilic dogs proved to be better workers than Laughlin. They lived in the laboratory around the clock, supplied hemophilic blood on demand, in ever-growing quantities, drew no salary, and could die in tests without raising moral qualms. At the same time, the biological similarities of human and canine laborers placed some of the same demands on their employer. The life-threatening conditions for both types of labor demanded surveillance, surgery, and ready blood transfusions. Industrialization made canine labor possible by producing effective treatments to keep hemophilic dogs alive. The combination of machinery and biology increased the supply of a valuable product, hemophilic blood, in a way neither could have accomplished alone.

Pigs, too, became more productive workers when encased in a technological environment. As Mark Finlay shows, the introduction of feeds fortified with antibiotics and vitamins reduced the length of time sows needed to nurse their piglets. As a result, meatpacker Jay Hormel noted, the sow could "immediately be *put back to work* producing another litter instead of performing no other service than milking her litter" (emphasis added). Like human workers on assembly lines, sows found themselves in a "speed up" designed to boost productivity.

Suggestion 4: See biotechnologies as products.   Factory, worker, and product—biotechnologies have been them all. Historians of technology have studied ways in which industrialization encouraged product standardization, mass production, brand marketing, and shared cultures of consumption spread over large areas. Some products have found their way onto the market as branded products, such as Fords and Apple computers, while others have been commodities, such as screws and nails. Highly processed products have lent themselves most easily to such analyses, perhaps because they are most recognizably the products of industry. Biotechnologies too have become commodities and branded goods.

One of the aims or results (or both) of industrialization has been product standardization. This feature has been especially important for branded goods, for which quality control is crucial to maintaining the brand's reputation. Organisms also have undergone standardization while posing particular challenges. One of the most obvious is sexual reproduction, which rejiggers genetic endowments of offspring every time. A standard technique for producing certain characteristics more reliably is inbreeding, which reduces genetic variation. (This process has created new problems in the form of genetic diseases from expression of double recessive genes, illustrating again how challenging organisms can be.) As Barbara Orland reveals in her essay, Switzerland set out in the nineteenth century to create a standard national cow. Out of countless varieties of cows, Swiss authorities decided on the measurements and appearance of the Swiss cow. Herd books helped this process by recording pedigrees. A national brand, if you will, resulted. Similarly, Susan Lanman's essay on seed producer Peter Henderson links commodification of seeds with commodification of leisure. Through expert selection, seed trials, and dependable growers, Henderson managed to produce reliable products from varied organisms.

Roger Horowitz demonstrates in his essay the speed with which the poultry segment of industrialized agriculture could change animals and markets. Concerted efforts by breeders and producers led to the postwar creation of a Chicken of Tomorrow, a meaty breed well suited to mass production. As this breed replaced older varieties on farms, marketers changed the bird's public identity as well. Rhode Island Reds and other names of varieties disappeared from grocery store labels as broilers, fryers, breasts, and thighs arrived to take their place. One kind of product differentiation (based on chicken parts) superseded an earlier kind of product differentiation (based on breeds).

Last, as Gerard Fitzgerald reminded those who attended the Rutgers Center for Historical Analysis' April 2002 conference "Industrializing Organisms," some biotechnology products are valuable precisely because they are alive. Inert tularemia would be of no use. Tularemia pathogens need to infect, reproduce themselves in, and overcome enemy soldiers. Industrializing this biotechnology demanded that researchers develop methods that would strictly govern when and where tularemia would be allowed to go about its business of reproduction, thereby advancing military and national interests.

Suggestion 5: Deemphasize the plant-animal dichotomy as a primary way of organizing ideas. The original title of the conference on which this introduction is based was "An Anatomy of Animal Technology." The conference organizers asked if I could widen it to include plants. I agreed. Yet it was hard doing so because I had structured my ideas around animal anatomy, and eventually that plan went by the wayside. Eventually it became clear to me that, ironically, I was recapitulating the evolution of organizing ideas in biology. Fifty years ago, most universities had a department of zoology and a separate department of botany. In the 1960s and 1970s, many merged into a single department of biology. Now they are fissioning again into departments of cellular and molecular biology on the one hand, and departments of organismal, ecological, and evolutionary biology on the other. As a result, people who study plant cells feel a greater kinship with people who study animal cells than with people who study plant taxonomy or ecology.

In the macrobiotechnology era, the plant-animal dichotomy worked fine. As we move into the microbiotechnology era, though, it will work less well. Now that we can move genes across taxa, their origins in plants or animals matter far less. Tobacco plants that glow in the dark, thanks to a firefly gene, exemplify the kingdom-spanning potential of microbiotechnology. Is it plant or animal? The answer is both.

Suggestion 6: Expand evolutionary history. A common call among historians of technology is to link scholarship more closely to other fields, thereby demonstrating the importance of technology in history. Evolutionary history has the potential to suggest a number of such links and encourage integration. To understand why people have shaped other species as technology, we might well turn to history of science, cultural history, economic history, political history, and social history. The list could continue; it quickly becomes apparent that every field of history intersects this new approach, has something important to contribute, and might well gain in the transaction.

If evolutionary history grows, we will have to tread carefully if we extend evolutionary analyses to human behavior. We know the damage wrought by social Darwinism and eugenics. It seems clear, at any event, that we will never be able to reduce human behavior to genes alone. Environment plays a central role in determining which genes get expressed. We do not have enough genes to govern all behaviors, but humans have enormous behavioral plasticity, and some combination of individual choice and social variables interact to shape behavior.[32]

Ideally, evolution would help focus our attention on human social problems in need of addressing. Much as we like to think of ourselves as separate from nature, selection still wields its scythe. In some parts of the world, diseases such as tuberculosis, malaria, and infant diarrhea still kill millions of people, selecting in the process for individuals who, through genetic endowment *or* social advantages, can fend off infection. In fighting these and other diseases, humans are locked in a coevolutionary arms race against pathogens.[33] Humans develop drugs, pathogens evolve resistance against them, humans develop new drugs, and the cycle continues.

Once we think coevolutionarily, we encourage ourselves to examine not just how humans shape organisms, but how organisms shape humans. This perspective does not imply a return to the older tradition of technological determinism. Rather, because of its emphasis on change, coevolution provides us with an unusually flexible way of thinking about how humans, nature, and technology have and will continue to shape one another.

Conclusion

As we incorporate biology more fully into our understanding of industrialization, we shed light on the present as well as the past. Industries today continue to depend on plants, animals, and microorganisms. Industrialized agriculture would disappear without plants to capture and transform the sun's energy into sugars and proteins. Bakeries and breweries would have to close their doors without yeast to transform sugars into carbon dioxide and alcohol. The construction industry would have to undergo massive transformation without trees to change carbon dioxide and water into cellulose in lengths and densities useful for homes and offices. Pharmaceutical companies rely on plants to invent molecules with pharmacological properties that the companies can then copy. Genetic engineering, one of the most high-tech industries in the world, would shut down without organisms to supply and receive genes that do certain kinds of work. By the end of the twenty-first century, under the continued influence of the biotechnology industry, historians will likely take organisms for granted as forms of technology. The essays in this collection suggest fascinating ways to start developing the concepts and approaches that will generate a broader understanding of these relationships.

Notes

The participants in the Industrializing Organisms conference have my gratitude for helpful comments on an earlier version of this essay. Philip Scranton was especially helpful with ideas and editorial insight. A sabbatical leave from the University of Virginia and a grant from the National Science Foundation (SES-0220764) supported this work.

1. Leo Marx, *The Machine in the Garden: Technology and the Pastoral Ideal in America* (New York: Oxford University Press, 1964), 195.
2. This definition, and some other ideas in this essay, appear in Edmund Russell, "Evolutionary History: Prospectus for a New Field," *Environmental History* (forthcoming). While this essay focuses on the role of evolution in the history of technology, the journal article focuses on its role in environmental history. Evolutionary history could be a field, subfield, or a research program; this essay uses "field" for the sake of simplicity.
3. Charles Darwin, *Variation of Animals and Plants under Domestication*, Vol. I (New York: New York University Press, 1988 [1875]), 5.
4. Charles Darwin, *On the Origin of Species* (New York: New York University Press, 1988 [1859]).
5. Darwin, *Variation*, I:1 (note 1).
6. Theodosius Dobzhansky, *Genetics and the Origin of Species* (New York: Columbia University Press, 1937), 161.

7. Juliet Clutton-Brock, *A Natural History of Domesticated Animals* (Austin: University of Texas Press, 1989); Juliet Clutton-Brock, *A Review of the Family Canidae, with a Classification by Numerical Methods* (London: British Museum [Natural History], 1976); Juliet Clutton-Brock, *Horse Power: A History of the Horse and the Donkey in Human Societies* (Cambridge, Mass.: Harvard University Press, 1992); Steve Jones, *Darwin's Ghost: The Origin of Species Updated* (New York: Random House, 2000 [1999]); Stephen R. Palumbi, *The Evolution Explosion: How Humans Cause Rapid Evolutionary Change* (New York: Norton, 2001); R. K. Wayne and E. A. Ostrander, "Origin, Genetic Diversity, and Genome Structure of the Domestic Dog," *Bioessays* 21 (March 1999): 247–257.
8. Douglas J. Futuyma, *Evolutionary Biology*, 3rd ed. (Sunderland, Mass.: Sinauer Associates, 1998). For applications of evolutionary ideas to efforts to conserve biological diversity, see Richard B. Primack, *Essentials of Conservation Biology*, 2nd ed. (Sunderland, Mass.: Sinauer Associates, 1998); Gary K. Meffe and Ronald C. Carroll, *Principles of Conservation Biology*, 2nd ed. (Sunderland, Mass.: Sinauer Associates, 1997).
9. For applications of evolutionary ideas to design and nonliving technologies, Walter Vincenti, *What Engineers Know and How They Know It* (Baltimore: Johns Hopkins University Press, 1990); George Basalla, *The Evolution of Technology* (Cambridge, England: Cambridge University Press, 1988).
10. Examples of a huge literature include John K. Brown, "Design Plans, Working Drawings, National Styles: Engineering Practice in Great Britain and the United States, 1775–1945," *Technology and Culture* 41 (2000): 195–238; Ruth Schwartz Cowan, *More Work for Mother: The Ironies of Household Technology from the Open Hearth to the Microwave* (New York: Basic Books, 1983); Claude S. Fischer, *America Calling: A Social History of the Telephone to 1940* (Berkeley: University of California Press, 1992); Donna Haraway, *Simians, Cyborgs, and Women: The Reinvention of Nature* (New York: Routledge, 1991); Gabrielle Hecht, *The Radiance of France: Nuclear Power and National Identity after World War II* (Cambridge, Mass.: MIT Press, 1998); David A. Hounshell and John Kenly Smith, Jr., *Science and Corporate Strategy: Du Pont R&D, 1902–1980* (New York: Cambridge University Press, 1988); Thomas Parke Hughes, *Networks of Power: Electrification in Western Society, 1880–1930* (Baltimore: Johns Hopkins University Press, 1983); Sheila Jasanoff, Gerald E. Markle, James C. Petersen, and Trevor Pinch, eds., *Handbook of Science and Technology Studies* (Thousand Oaks, Calif.: Sage Publications, 1995); Nina E. Lerman, Arwen Palmer Mohun, and Ruth Oldenziel, "The Shoulders We Stand On and the View from Here: Historiography and Directions for Research," *Technology and Culture* 38 (1997): 9–30; Donald A. Mackenzie, *Inventing Accuracy: An Historical Sociology of Nuclear Missile Guidance* (Cambridge, Mass.: MIT Press, 1990); Philip Scranton, ed., *Beauty and Business: Commerce, Gender, and Culture in Modern America* (New York: Routledge, 2001); Philip Scranton, *Endless Novelty: Specialty Production and American Industrialization, 1865–1925* (Princeton, N.J.: Princeton University Press, 1997); Bruce E. Seely, *Building the American Highway System: Engineers as Policy Makers* (Philadelphia: Temple University Press, 1987).
11. Jeffrey K. Stine and Joel A. Tarr, "At the Intersection of Histories: Technology and the Environment," *Technology and Culture* 39 (October 1998): 601–640; Special Issue on Technology and Environment, *Technology and Culture*, July 1997; Brian Black, *Petrolia: The Landscape of America's First Oil Boom* (Baltimore: Johns Hopkins University Press, 2000); Adam Rome, *The Bulldozer in the Countryside: Suburban Sprawl and the Rise of American Environmentalism* (New York: Cambridge University Press, 2001); Edmund Russell, "'Speaking of Annihilation': Mobilizing for War against Human and Insect Enemies, 1914–1945," *Journal of American History* 82 (March 1996): 1505–1529; Edmund Russell, "'Lost Among the Parts Per Billion': Ecological Protection at the United States Environmental Protection Agency, 1970–1993," *Environmental History* 2 (January 1997): 29–51; Edmund Russell, "The Strange Career of DDT: Experts, Federal Capacity, and 'Environmentalism' in World War II," *Technology and Culture* 40 (1999): 770–796; Edmund Russell, *War and Nature: Fighting Humans and Insects with Chemicals from World War I to Silent Spring* (Cambridge: Cambridge University Press, 2001); Jeffrey K. Stine, *Mixing the Waters: Environment, Politics, and the Building of the Tennessee-Tombigbee Waterway* (Akron, Ohio: University of Akron Press, 1993); Joel A. Tarr, *The Search for the Ultimate Sink: Urban Pollution in Historical Perspective* (Akron, Ohio: University of Akron Press, 1996); Richard White, *The Organic Machine* (New York: Hill and Wang, 1995); James C. Williams, *Energy and the Making of Modern California* (Akron, Ohio: University of Akron Press, 1997). Rut-

gers recently began a new Ph.D. program in the history of technology, environment, and health. Paul Israel, "New Ph.D. Program Announcement," *Envirotech Newsletter* 1 (September 2001): 1. The University of Virginia has created a committee on the history of environment and technology, which is overseeing a new graduate field in the history of environment and technology.

12. Jim Williams, "Envirotech an Official SIG," *Envirotech Newsletter* 1 (September 2001): 1.
13. Jared Diamond, *Guns, Germs, and Steel: The Fates of Human Societies* (New York: W. W. Norton, 1998); Clay McShane, "The Urban Horse as Cyborg," June 12, 2000 (manuscript in author's possession); Joel A. Tarr, "A Note on the Horse as an Urban Power Source," *Journal of Urban History* 25 (March 1999): 434–448; William Boyd, "Making Meat: Science, Technology, and American Poultry Production," *Technology and Culture* 42 (2001): 631–664; Deborah Fitzgerald, *The Business of Breeding: Hybrid Corn in Illinois, 1890–1940* (Ithaca, N.Y.: Cornell University Press, 1990); Jack Ralph Kloppenburg, Jr., *First the Seed: The Political Economy of Plant Biotechnology, 1492–2000* (New York: Cambridge University Press, 1988); John H. Perkins, *Geopolitics and the Green Revolution: Wheat, Genes, and the Cold War* (New York: Oxford University Press, 1997); Harriet Ritvo, *The Animal Estate: The English and Other Creatures in the Victorian Era* (Cambridge, Mass.: Harvard University Press, 1987); Donna Haraway, "Universal Donors in a Vampire Culture: It's All in the Family: Biological Kinship Categories in the Twentieth Century United States," in *Uncommon Ground: Rethinking the Human Place in Nature*, ed. William Cronon (New York: W. W. Norton & Company, 1996 [1995]), 321–366; Robert E. Kohler, *Lords of the Fly: Drosophila Genetics and the Experimental Life* (Chicago: University of Chicago Press, 1994); Nicholas Russell, *Like Engend'ring Like: Heredity and Animal Breeding in Early Modern England* (Cambridge, England: Cambridge University Press, 1986).
14. John H. Perkins, *Insects, Experts, and the Insecticide Crisis: The Quest for New Pest Management Strategies* (New York: Plenum Press, 1982); William Boyd, "Making Meat: Science, Technology, and American Poultry Production," *Technology and Culture* 42 (2001): 631–664; Russell, *War and Nature*.
15. National Research Council, *Pesticide Resistance: Strategies and Tactics for Management* (Washington, D.C.: National Academy Press, 1986), 16–17.
16. Stephen Mosley, *Chimney of the World: A History of Smoke Pollution in Victorian and Edwardian Manchester* (Cambridge, England: White Horse Press, 2001), 45.
17. Joseph E. Taylor, *Making Salmon: An Environmental History of the Northwest Fisheries Crisis* (Seattle: University of Washington Press, 1999), 203–204, 206, 233.
18. David Brown, "'Wonder Drugs' Losing Healing Aura," *Washington Post*, June 26, 1995, p. A1.
19. Biotechnology Industry Organization, "Bio," retrieved January 14, 2002, from http://www.bio.org/.
20. Rick Weiss, "Starved for Food, Zimbabwe Rejects US Biotech Corn," *Washington Post*, July 31, 2002, A12.
21. Envirotech, "Are Animals Technology?" July 20–27, 2001, retrieved January 28, 2002, from http://www.udel.edu/History/gpetrick/envirotech, *Envirotech*, listserve, "More Animals as Technology," July 28–August 1, 2001, retrieved January 28, 2002, from http://www.udel.edu/History/gpetrick/envirotech.
22. The American Society for Environmental History does not have interest groups, but Envirotech meets at its conferences. *Environmental History* ran a special issue on technology and the environment in the spring of 1994.
23. Donald Worster, ed., *The Ends of the Earth: Perspectives on Modern Environmental History* (New York: Cambridge University Press, 1988); Donald Worster, *Dust Bowl: The Southern Plains in the 1930s* (Oxford: Oxford University Press, 1979); Cronon, *Uncommon Ground*; William Cronon, *Changes in the Land: Indians, Colonists, and the Ecology of New England* (New York: Hill and Wang, 1983); William Cronon, *Nature's Metropolis: Chicago and the Great West* (New York: W. W. Norton, 1991); Carolyn Merchant, ed., *Major Problems in American Environmental History: Documents and Essays* (Lexington, Mass.: D. C. Heath, 1993); Andrew Hurley, *Environmental Inequalities: Class, Race, and Industrial Pollution in Gary, Indiana, 1945–1980* (Chapel Hill: University of North Carolina Press, 1995); Martin V. Melosi, *The Sanitary City: Urban Infrastructure in America from Colonial Times to the Present* (Baltimore: Johns Hopkins University Press, 2000); Russell, *War and Nature*.
24. William Cronon, "The Trouble with Wilderness; or, Getting Back to the Wrong Nature," in Cronon, *Uncommon Ground*, 69–90; Richard White, "Are You an Environmentalist or Do You Work for a Living?" in Cronon, *Uncommon Ground*, 171–185.

25. Stine and Tarr, "At the Intersection of Histories"; but see Martin V. Melosi, *Garbage in the Cities: Refuse, Reform, and the Environment 1880–1980* (College Station: Texas A&M University Press, 1981); Jeffrey K. Stine, *Mixing the Waters*; Joel A. Tarr, *The Search for the Ultimate Sink*; James C. Williams, *Energy and the Making of Modern California* (Akron, Ohio: University of Akron Press, 1997).
26. Roderick Nash, *Wilderness and the American Mind*, 3rd ed. (New Haven and London: Yale University Press, 1967); Cronon, *Changes in the Land*; Carolyn Merchant, *The Death of Nature: Women, Ecology, and the Scientific Revolution* (San Francisco: Harper and Row, 1980); Worster, *Dust Bowl*.
27. But see White, *The Organic Machine*.
28. Biotechnology Industry Organization, "Bio."
29. Philip Scranton coined "microbiotechnology" and I "macrobiotechnology."
30. One example of a boundary case comes from Gerard Fitzgerald. See Gerard Fitzgerald, "The Biography of a 'Purely American Disease': *Francisella tularensis* and the Industrialization of a United States Biological Weapon, 1911–1960." Paper presented April 5, 2002, at the conference Industrializing Organisms: Plants, Animals, and Technology sponsored by the Rutgers Center for Historical Analysis, New Brunswick, New Jersey. The pathogens in Fitzgerald's paper are microorganisms, but their modification and use would be considered macrobiotechnology under the definitions presented here. I am less attached to these particular terms than I am to the goal of developing terminology that expresses both continuity and change between older and newer biotechnologies.
31. Russell, "The Strange Career of DDT."
32. Paul Ehrlich, *Human Natures: Genes, Cultures and the Human Prospect* (Washington, D.C.: Island Press, 2000); Stephen Jay Gould, *The Mismeasure of Man* (New York: Norton, 1996); Richard C. Lewontin, Steven Rose, and Leon J. Kamin, *Not in Our Genes: Biology, Ideology, and Human Nature* (New York: Pantheon Books, 1984).
33. Douglas J. Futuyma and Montgomery Slatkin, eds., *Coevolution* (Sunderland, Mass.: Sinauer, 1983).

I
Plants, Profits, Politics, and Power

"For Profit and Pleasure"
Peter Henderson and the Commercialization of Horticulture in Nineteenth-Century America

SUSAN WARREN LANMAN

Visitors arriving in nineteenth-century American cities frequently focused on the thriving industries, impressive architecture, and many amenities that marked these bustling metropolises, but often failed to appreciate the complex web of resources supporting these urban spaces. Few observers understood the "industrial gardens" accompanying urban development and the changing consumption habits of city dwellers. Knowledgeable gardeners, however, carefully observed the intensively cultivated plots and acres of greenhouses surrounding teeming urban areas, as well as market wagons mired in the traffic of city streets, riots of voluptuous produce offered up at market stalls, and blossoms in fashionable florists' plate glass windows displayed like precious gems. In an ongoing symbiotic relationship, the city supported complex horticultural innovations that, in turn, facilitated ever denser urban spaces. Few gardeners understood this labyrinthine relationship better than Peter Henderson, a gifted and shrewd Scottish immigrant gardener. In the course of his long career, his multiple roles as private gardener, market gardener, seed dealer, author, and florist proved crucial to his financial success and influence. Attuned to the urbanization and industrialization that marked New York City and its neighbor, Jersey City, he intuitively grasped the underlying structures and efforts needed to succeed in an industrializing America.

Specifically, Henderson's marketing techniques gradually spread his "products" over broad geographic areas. He shrewdly chose plants, as well as the methods for growing them, based on the resources required, subsequent yield ratios, and their compatibility with labor rationalization. He then promoted these same seed and plant choices to an emerging middle-class market open to the commodification of their leisure. Henderson effectively employed public sector resources to support selective marketing and distribution of his wares. While keenly aware of new technologies, he advocated and adopted them only when they proved financially expedient.

Henderson's accomplishments highlight emerging technologies in horticulture, but also suggest some key points about the cultural, social, and economic forces fostering them. Capitalism played a crucial role in dispersion and adoption of these gardening innovations, but at the same time markets reflected government expenditures, subsidies, and regulations, rather than standing as autonomous and self-regulating. Federal government support for railways, canals, and postal services, along with municipal services such as fire protection, piped water, organized produce markets, streetcar lines, docking facilities, and enforcement of public health measures, all contributed to the feasibility of denser population concentrations. Focusing on late-nineteenth-century economic and social structures as they absorbed and shaped the impact of new technologies provides a complex and nuanced understanding of developments within gardening. In other words, changes *surrounding* horticulture clarify alterations *within* it. Some of these departures from established practice originated abroad.

Major changes occurring in Great Britain's seed industry, for example, place the American experience in context. As the post, telegraph, and railroads facilitated communication and commercial transactions during Queen Victoria's reign, purchased seed gradually began supplanting saved seed. By the 1880s consumers could buy seeds in pharmacies, grocery shops, hardware stores, and even at book dealers, and the continuing expansion of seed firms further encouraged packet purchases. By the 1890s British merchants imported huge quantities of seed from Europe, Australia, New Zealand, Canada, and the United States, where climate and labor costs led to cheaper production. London market gardeners, who originally supplied many of the seedmen, experienced continental competition that discouraged the continuation of specific British seed strains.[1]

In America Henderson thus followed trends that began earlier abroad, but his combined roles as market gardener, nurseryman, seed dealer, florist, and horticultural author provide a privileged opportunity to examine the complexity of those changes. Britain and other European countries experienced the economic effects of globalization throughout the nineteenth century. In consequence, commercial horticulturalists on both sides of the Atlantic developed effective strategies for overcoming their challenges, although globalization by its very nature frequently obfuscated the origins of innovations. Henderson's life offers the chance to consider this immigrant's career and his impact in considerable detail. We begin with Henderson's traditional horticultural training in Scotland, examining the myriad factors contributing to his financial success and influence, and ultimately his impact on U.S. horticulture.

Scottish Networks: From Tradition to Innovation

Both Peter Henderson and his family embraced the myth of the self-made man, but his actual upbringing and early career reveal a far more interesting story, as well as the power of social networks based on ethnicity. Henderson's background included an excellent basic formal education and practical training in all aspects of horticulture. Born as the second son of an estate agent in the village of Pathhead, near Edinburgh, Henderson benefited from local parish schooling until age fifteen, because his father could afford the school fees. He subsequently clerked in Edinburgh, and then apprenticed as a gardener at Melville Castle, supposedly at the suggestion of a friend already working in the gardens. The actual decision to train for the gardening profession probably derived from a number of other factors as well. His maternal grandfather, Peter Gilchrist (1740–1810), had enjoyed a successful career as a nurseryman and florist, and Peter Henderson's older brother James also followed the occupation. Furthermore, Henderson's parents could pay the substantial premium (usually about the sum of £20) for the indenture articles and subsequent training required by the formidable and deeply respected George Sterling (c. 1806–1885), head gardener at Melville Castle.

Sterling's knowledge and the extensive collection of over 15,000 plant species at Melville Castle, including huge numbers of exotics, ensured that his staff gained a thorough grasp of botany. A strict disciplinarian and a stickler for botanical nomenclature, each week he assigned each of his apprentices and undergardeners a group of plants to be identified and labeled during their "free time" in the evening.[2] While at grammar school, Henderson complained bitterly to his older sister that he saw little practical value in the Latin he was obliged to memorize, but this earlier schooling quickly allowed him to become proficient in plant identification. In acknowledgment of Henderson's ability, Sterling sent him to Ballantyne's Nursery in nearby Dalkeith after only a few months of apprenticeship to identify and label a collection of hardy herbaceous plants, which resulted in the reward of a sovereign (a coin worth £1). Upon learning that Henderson spent his off-duty hours collecting specimens for his herbarium to familiarize himself with plant names, Ballantyne offered him a silver medal for the completed project. The following year Henderson won the Royal Botanical Society of Edinburgh's medal for the best herbarium of native and exotic plants.[3] During his four years of apprenticeship, he also studied mathematics at night classes in Edinburgh.

Although proud of Henderson's intellectual accomplishments, which reflected positively on his apprenticeship at Melville Castle, Sterling railed at him for any practical task undertaken in a slipshod fashion, especially if the head gardener detected something incorrectly executed. Sterling expected even his youngest staff members to be competent and diligent. The young Peter endured many lectures on the difference between appropriately dry and "killing dry." On one particularly heinous occasion, the head gardener dragged his apprentice up by the ear to view a deceased victim of his underling's ineptitude, a lesson Henderson never forgot.[4] Under the "Scottish system" apprentices rotated through all the gardening tasks. They usually began in the kitchen garden, then moved to the flower garden, pleasure grounds, frame yards, and finished with the most valuable plants raised under glass.[5] In sum, Henderson's four years constituted some of the best and broadest horticultural education available at the time.

Upon completion of his apprenticeship, Henderson secured passage on the sailing ship *Roscius* and arrived in New York harbor by spring 1843. He found employment in George Thorburn's nursery and floral business at Astoria, Long Island. Grant Thorburn, the first seed dealer in New York and the firm's founder, had emigrated from Dalkeith, Scotland, so Henderson's apprenticeship at Melville Castle provided him with a sound reference. According to Henderson's own account, he spent some of his time as an assistant in the firm's large floral establishment in New York City learning bouquet design as well as wreath and cross production. The work also gave him an introduction to commercial horticulture and New York City's business environment.[6]

Beginning gardeners in the 1840s followed the British tradition of shifting employment frequently in order to broaden their skills. Having gained some experience with the American climate, Henderson obtained employment with fellow Scotsman Robert Buist (1805–1880) of Philadelphia in 1844. Born at Cupar, Fife, near Edinburgh, and trained at the Edinburgh Botanic Gardens, Buist had emigrated in 1828. By the 1830s his roses, camellias, and verbenas had attracted attention, and he had written several articles in addition to his books *The American Flower-Garden Directory* (1832) and *The Rose Manual* (1844).[7]

Henderson joined the staff of Buist's Exotic Nursery as its proprietor began expanding his grounds and buildings. The main glasshouse range extended to almost three hundred feet, and Buist added a new greenhouse using experimental tank heating. He had personally imported plant material in the 1830s from Great Britain and acquired other material through plant hunting expeditions; thus Henderson worked with the latest imports and discoveries. His new job included opportunities to observe plant breeding and development. Buist constantly introduced new roses and tested immense quantities of seedling camellias.[8] Henderson's excellent Scottish training and his complete dedication to his job soon earned him the trust and friendship of his employer. During cold periods Henderson voluntarily spent his winter nights at the greenhouses, tending the multiple fires to ensure even temperatures and to guard against the problem of overheated flues setting fire to the wooden structures. Intermittently snatched sleep on the floor of a warm glasshouse proved preferable to a cold boarding-house room and long treks in the snow to check fires.[9]

About 1846 Henderson left Buist's nursery to design and install a garden for Charles F. Sprang of Pittsburgh. Customarily in this period, Henderson's new job would have been obtained upon recommendation from Buist or by virtue of the nurseryman's direct selection. Throughout the nineteenth century many large nurseries, in both Britain and the United States, provided job referral and placement for competent men in their ranks. During this time Henderson also wrote his first article for C. M. Hovey's *Magazine of Horticulture*, probably inspired and assisted by Buist's connections with the publication. Throughout his early years, Henderson benefited from a superior education and his membership in a far-reaching Scottish horticultural network. In turn, he wrote a number of articles on emigration for the British press in the 1880s and helped recent arrivals find employment in America. Such efforts allowed him to continue traditional practices and simultaneously enlarge his pool of skilled labor because many of the immigrants found employment at his nursery.[10]

Market Gardening: Balancing Land, Labor, and Capital

At the conclusion of his Pittsburgh assignment in 1847, Henderson possessed $500 in capital and headed east to New Jersey to begin market gardening on rented land in partnership with his older brother James, who had emigrated to America at an earlier date. Their entry into market gardening coincided with price reductions in glass occasioned by the cutting of heavy tariffs dating from 1824. In 1846 the United States reduced window glass tariffs to a uniform 20 percent rate and then further cut the rate to 15 percent in 1857. The reductions resulted in large glass importations from producers in England, Belgium, and France. Commercial gardeners near ports particularly benefited from a cheaper product made possible by the European workers' lower wages. Less-expensive glass prices enabled the Henderson brothers, who had limited capital, to afford more of the portable glazed wooden frames needed for protecting early crops from the cold.[11]

Jersey City proved an ideal venue for market gardening because the rapidly expanding population of nearby New York City provided a ready market for fresh vegetables and fruits as well as luxury items such as early strawberries and cut flowers. Limited ferry service across the Hudson River in the early years of the nineteenth century and the vast areas of marshy shoreline limited development in some sections of Jersey City. Thus prices for renting or purchasing land suitable for market gardening remained within reason. Additionally, the refuse and industrial waste from the area's developing industries provided inexpensive fertilizers for knowledgeable gardeners.

Henderson understood the need to manure and enrich his soil regularly in order to obtain prolific crops. In his first few years he applied large quantities of "night soil" obtained from Jersey City scavengers, mixed it with stable manure, charcoal, sawdust, or any other easily obtained absorbent, then applied it at a rate of about thirty tons per acre. As the effectiveness of any one single fertilizer tended to decrease over time, Henderson experimented with nearly every imaginable source of industrial waste in the vicinity. Hops refuse from local sources such as Cox's Brewery, Limbech & Betz's Brewery, and the Palisade Brewery proved an excellent free fertilizer, although the brewers eventually realized its value and then charged market gardeners the cost of the best stable manure for their waste. Henderson valued the hops as much for its mulching and pulverizing properties as its fertilizing value. He also successfully experimented with a number of other options. "Sugar house scum" from the New Jersey Sugar Refining Company, the Havemeyer Sugar Company, and Nathaniel Tooker's Molasses House, thoroughly mixed with soil or muck and composted, constituted an excellent additive when applied at about twenty tons per acre.[12]

The establishment of a facility for slaughtering cattle brought on hoof by the railroads called the *Abattoir*, at Communipaw in 1866, provided Henderson with valuable new fertilizer sources. Not only did the animals produce the usual manure, but their bones became a valuable source of phosphorous in the form of bonemeal. In 1874 the stockyard moved to larger quarters in Jersey City and by 1884 an average of 1,000 beef cattle, 1,300 sheep, and 2,000 hogs entered the yards daily, providing lavish amounts of fertilizer at acceptable costs to the market gardeners, who heaped ever greater quantities on their increasingly valuable land.[13]

Jersey City industries met other horticultural needs as well. The huge P. Lorillard's Snuff and Tobacco Factory, established in 1760, employed hundreds of women and girls for stripping leaves from stems of cured tobacco ribs. Gardeners, who purchased the discarded stems, either soaked them in water to create an infusion that became a potent liquid insecticide or burned them in a vaporizer to fumigate the glasshouses. In the late nineteenth century, the raw material for this nicotine insecticide cost London gardeners about thirty times more than their New Jersey counterparts.[14]

The industrial expansion that drove Jersey City's growth provided byproducts for horticultural use at advantageous rates so that, even as land values rose, market gardening remained a viable land use, provided that entrepreneurs applied more manure. In this instance, New Jersey gardeners followed the lead of London market gardeners, who obtained their manure from rail cars carrying excrement out of the city. Those with expensive land in close proximity to London applied as much as two hundred tons annually per acre to support the intensive, multiple crops they harvested including lettuce, strawberries, and cucumbers.[15] In a rather unique symbiotic relationship, the city fed off produce grown in its own excrement and enriched the market gardeners facilitating the process. Jersey City market gardeners also earned substantial profits by using waste from urban and industrial development.

As early as 1860, Peter Henderson declared his real estate worth $50,000 and his personal estate at $20,000 in the U.S. census.[16] He purchased property in Jersey City and eventually accumulated roughly a ten-acre tract as additional parcels became available. He then gave up his old rented land in Jersey City when he moved a mile away to his new property at Jersey City Heights.[17] Henderson's financial success derived from his careful scrutiny of capital investment and labor costs as well as his ability to shrewdly evaluate the smallest detail related to either. The savings realized through this strategy, multiplied over an enormous number of operations, yielded substantial profits. Cost effectiveness drove the construction of his production facilities, the selection of his plant material, and his choice of workers. Furthermore, decisions regarding each of the above remained interdependent.

Operating within the classic triangle of land, labor, and capital, market gardeners constantly endeavored to find a balance. As urban New York and New Jersey land prices rose, rents for market garden plots in the 1880s reached as much as $100 an acre yearly. Yet proximity to market remained crucial. As Henderson explained by way of example, a single wagon load of his cabbages might bring in $50 on Saturday and only $30 the following Monday morning. By being within a half hour of the great wholesale Washington Market in Manhattan, he could arrange for several of his wagons to make three trips in the course of a single day when prices reached high levels. High rents also pressed market gardeners to invest in as much as fifty to seventy-five tons of manure per acre to realize large crop yields, although they never reached the two-hundred-ton amounts used by their London counterparts who had access to cheap and almost endless supplies of human and animal excrement. For American market gardeners, by contrast, driving down labor costs remained crucial in balancing the equation because U.S. wages remained relatively high.[18]

As early as 1869, Henderson explained to his readers the lower cost of plants in the United States and the higher profit margins realized in his adopted country. After pointing out the labor saving schemes and innovative growing techniques used in America, he noted that two of the largest London nurseries employed an average of fifty workers per year and then contrasted those numbers with American requirements. "The same quantity of glass would be worked here in a style quite equal to theirs, as far as the quality of the plants goes, with less than one-third the number [of employees]."[19]

In the revised 1887 edition of the same book Henderson made the point more succinctly.

> While the price of labor is from one-third to one-half more in this country than in Europe, nearly all the products of the nursery, greenhouse or garden are sold lower here than there—not merely lower, but in a majority of cases at less than half the price.[20]

Moreover, "the high cost of labor has long ago forced us to use our ingenuity in simplifying our work."[21] As an example, he noted that his men could propagate two thousand cuttings per day on growing benches, while propagators in the leading London nurseries averaged about five hundred daily. In potting off rooted cuttings he noted that his "crack workmen" could complete ten thousand in a single day, while men in the London establishments might not match half that number. Henderson specifically attributed the increased productivity to task simplification and the selection of workers with a propensity for "rapid movement" to whom he paid relatively high wages.[22] The "Champion," James Marvey, was a case in point.

> He had been in my employment for nearly twenty years and had ever distinguished himself for rapid and neat workmanship, for, some years before his death, he had repeatedly potted 10,000 cuttings, in two and a half inch pots, in

ten consecutive hours, and had attained on one occasion the extraordinary number of 11,000 in ten consecutive hours. I paid him for years $5.00 per day, and always considered him one of our cheapest workmen, because, not only did he earn all he got, but his example fostered a spirit of emulation among our other employees, valuable alike to themselves and to us.[23]

Such productivity depended on standardizing as many aspects of each operation as possible, including the flower pots the men handled. As early as 1839, C. M. Hovey, editor of the *Magazine of Horticulture*, called for standardized pot sizes, but potters continued to produce goods according to their own proclivities.[24] Short of ordering from only one supplier, finding uniform pots proved to be a challenge.[25]

Further, hand production by small potters exacerbated the proliferation of shapes. While the British did not use machinery for pot production until after World War II, Americans began mechanized production of small-size pots as early as the 1870s, and both retail and wholesale catalogs listed machine-made pots up to four-and-one-half inches in size.[26] These pots offered the possibility of standardized sizes beyond the control of individual potters. Pot-making machinery required a substantial initial investment, but drastically reduced labor costs. One unskilled man and boy, by placing clay balls in revolving molds and pressing down on a heavy lever, produced five to ten thousand pots a day, depending on size. Their combined wages remained lower than one skilled potter turning out an average of one thousand pots by hand throwing on a traditional wheel.[27]

At the 1888 Society of American Florists' convention held in New York City, members voted to adopt standard sizes for pots; and the *American Florist* subsequently published a template depicting the configuration for each pot size.[28] At the 1890 convention the Whilldin Pottery Company of Philadelphia received a certificate of merit for their standard pots, a facsimile of which appeared in the firm's subsequent advertisements.[29] The new standard pot caused so much interest that the *Gardeners' Chronicle* described it in detail for British readers.[30] Although many potters initially refused to accept the standardized dimensions because the heavy rim of the new pots could not profitably be thrown by hand, they eventually acquiesced to the standard sizes, given their popularity with growers.[31]

Standard sizes less liable to breakage when stacked made sense for market gardeners and florists, such as Henderson, who ordered their pots in lots of ten thousand or more. Always aware that carelessly piled pots suffered significant breakage rates, Henderson's foremen ordered the garden boys to pick up all used pots promptly. The lads washed and placed all pots in large soap packing boxes, sorted so that only a single size nestled neatly in any one box. Strips tacked to the end of each box ensured that when stacked, the box rather than the pots bore the weight. When needed, the entire box could then be taken to the work area. For rapid potting, the lad passed each pot as needed to the better-paid senior gardener.[32]

Efficient pot storage and "rapid movement" in plant production made possible the plant potting assembly line. Further, plants grown under like conditions in identical pots facilitated the empirical selection of specimens offering better performance. Henderson, the plantsman and botanist, always possessed an eye for spotting superior plants. His definition of "superior," of course, remained strongly entwined with market considerations. At the same time, such considerations often meant that plants meeting his approval enjoyed wide dispersion.

As in the case of terracotta pots, Henderson made other capital investments with close attention to plant requirements and labor costs. His selection of watering equipment illustrates this point. Given his superb training under George Sterling, Henderson understood the moisture requirements of all his plants. Woe betide an employee if the "boss" inspected the glasshouses under a worker's care and detected a "killing dry" sound as he tapped the pots. In the 1860s Henderson's plants depended upon long water tanks under the benches. Two men using West's Force Pumps, which connected to 125-foot hoses of one-and-a-half-inch diameter, could water about seven thousand square feet of glasshouse space in an hour. The job required at least six hours when the men worked with traditional watering cans dipped into the tanks. As Henderson noted, "of all the labor-saving apparatuses introduced into our gardens, the force pump is the most valued."[33]

Because Henderson initially rented his market gardening land, his major glass structures consisted primarily of three-by-six-foot sashes resting on wooden frames heated by hot beds composed of manure. The existing forcing pits and glasshouses with heat running through tile flues provided superior structures for growing plants, although in many instances the frame yard yielded substantial crops. He could, for example, realize $1,000 of profit per acre of land covered in sash by July and still bring in a second crop in the autumn.[34]

However, when Henderson moved to his new property in 1864, he invested heavily in the latest technology for new forcing pits and greenhouses. Given the higher prices of glazing during the Civil War and the new technologies incorporated into the glasshouses, they cost about $10 per lineal foot, which was twice as expensive as houses with flues. He built his structures in eleven-by-one-hundred-foot modules and attached them to one another to create thirty-three-by-one-hundred-foot complexes heated by a single Hitching's Patent Boiler connected to a four-inch pipe running under each of the benches along the entire length of the structure. Although he considered the old-style flue heating to be both economical to build and a satisfactory heat source for the plants, Henderson preferred hot water pipes for his new greenhouses. One boiler fire heated an acre of glass that would have required nine separate fires with flue heating. With fuel costs similar for both systems, hot-water labor costs only amounted to one-third of the previous level under the flue system. The hot water heating nearly doubled the initial cost of Henderson's greenhouses, but in five to six years the additional profits garnered through superior technology covered the cost of his investment.[35]

Henderson applied the same considerations to plant devices, soil mixtures, and plant shipping. After satisfying the basic cultivation requirements of the plants, capital investment and most especially labor costs remained his key concerns. His foremen's daily logs and his own close observations as he worked on the premises guaranteed that he could figure the cost of almost any horticultural task.[36]

Given his efficiency in controlling costs and selecting plant material that yielded large crops, Henderson substantially benefited from price inflation during the Civil War. Jersey City market gardeners filled federal contracts for produce used at the West Hoboken army camp. They also provided fruits and vegetables for huge numbers of troops from New York and New England waiting in Jersey City for trains to take them into the battle areas.[37] As Henderson readily stated, "war prices" meant wholesalers paid double rates for market garden crops between 1861 and 1866.[38]

Henderson benefited from market conditions in the late 1860s because he remained vigilant about the shifting costs of land, labor, and capital. His substantial investments in property, as well as far superior glass and heating technology prior to 1864, resulted in handsome profits during the war despite labor shortages. He also skillfully employed the byproducts from other industries in the area to defray or reduce his fertilizer and insecticide expenditures. By embracing the industrial activities of his neighbors, he minimized his own products' costs.

The Seed Business: Standardization, Transportation, and Distribution

Starting in 1871, Henderson capitalized on the knowledge he had gained as a market gardener to develop his seed business. He understood the need for reliable seed sources and the characteristics his peers valued, such as compact growth, high yields, good germination, and early cropping. In the late 1840s, Henderson had grown his own seed or exchanged and purchased seed from neighbors in preference to buying it from merchants. Even reputable seed dealers then found it difficult to control seed quality and purity, and imports from distant sites exacerbated the problem of nontransparent exchanges. At least market gardeners purchasing seed from neighbors knew the quality and characteristics of the original crop, any possible cross-pollinators grown in the vicinity, and the date the crop went to market. Saving one's own seed involved letting a portion of the crop ripen to seed and forgoing the profits of a harvest on that acreage. As land prices near the urban center escalated, it became increasingly desirable to obtain seed grown on less valuable land in other areas. Because growers competed with one another to get their crops into the lucrative New York City markets at the earliest possible date, seed that produced early crops commanded a premium price. Many gardeners eagerly sought it, appreciating the potential profit. For example, the first melons into a summer market could realize $1,500 per acre, whereas those following might only be sold at $500 to $1,000 per acre.[39]

Henderson also understood the level of frustration experienced when a grower could not procure desirable seed at any viable price. His own experience with the Jersey Wakefield cabbage drove that point home. Early cabbage constituted one of the most profitable market gardening crops, because its price, even after the Civil War, averaged $600 per acre. The best and earliest variety originated with Francis Brill of New Jersey, who in 1842 planted seed obtained in England under the name "Early Wakefield." This highly successful variety became a staple of New Jersey growers. As a variety producing small amounts of seed, however, its seed often sold at $20 per pound, making it about five times as expensive as other varieties. Obviously the New Jersey growers tried importing the "Early Wakefield" seed from English growers again, but varieties sold under that name subsequently failed to exhibit the characteristics they so valued.[40] Our earlier brief look at the British seed trade may well explain why.

Henderson's first edition of *Gardening for Profit* (1867) spread the cabbages' fame and made it a mainstay of his seed business. As he gradually turned large portions of his Jersey City property into a seed testing grounds, some of his most intensive efforts went into selecting the earliest bearing cabbages for seed and then locating reliable seed growers. A German immigrant by the name of Old Carl, who would not sell seed to him, habitually marketed the first "Jersey Wakefield" every year—to Henderson's chagrin. Eventually he learned that his rival's secret consisted of removing young shoots produced from cut stumps and treating them as cuttings to be grown on and then ripened for seed. While not a viable process for raising seed in quantity, this approach helped Henderson's seed growers to produce improved and pure stock for bulk seed raising.[41]

Profits often followed from recognizing superior varieties of seed and bringing them to the marketplace, rather than in hybridizing and selecting seed. For example, "Henderson's Early Summer" cabbage originated with Abraham Van Sicklen of Jamaica, Long Island. Henderson purchased twenty pounds of Van Sicklen's seed at a wholesale price of $600 the previous year and introduced it to the market in 1874 under its new moniker. Although maturing ten days later than "Jersey Wakefield," it weighed about twice as much, making it a highly profitable variety.[42] As a knowledgeable plantsman, Henderson followed trends in Europe and noted the 1861 introduction of double zinnias by the famous Paris seed firm Vilmorin. He subsequently imported their seed and raised his own flowers, noting that the plants often failed to produce double blooms until the end of summer. He first commercially marketed them in 1865.[43]

Henderson's commercial success initially rested on his reputation as a savvy market gardener as well as his willingness to test seeds to ensure quality and purity of strain. He also benefited from the high prices market gardeners realized during the war years, motivating them to purchase top quality seed. Further, the publication of *Gardening for Profit* in 1867 provided a manual for veterans interested in launching their own businesses, and, of course, these readers became his natural customers. In a profession with long apprenticeships and closely guarded "secrets of the trade," Henderson exploded into print with simple and clear directions, plus extremely practical advice. He not only sold books; he created customers as well.

Henderson's seed growing remained integrated with his other horticultural activities. His original partnership with his brother dissolved after the first few years when James obtained land in the hill section of Jersey City and concentrated on vegetable growing. Peter became increasingly interested in raising ornamental plants, and he also opened an office in Mellvain & Orr's plant store in New York City (1853). At this point he marketed ornamental plants on a daily basis and by noon each day offered any unsold stock at auction. By 1862 he moved into the seed store of fellow Scotsmen James Fleming and William J. Davidson at 67 Nassau Street and began selling his goods through annual catalogs and newspaper advertisements. He also raised plants expressly for wholesale auctioning.[44]

In 1871 Henderson set up his own seed business with his son Alfred and William H. Carson, at Cortland Street in New York City. Within the first year of operation, he began cross-marketing his products with the slogan, "everything for the garden." Some of his business practices simply echoed those of other early gardeners and seedmen. However, Henderson's knowledge of market gardening in a period of vast urban expansion made him particularly aware of the market gardeners' need for pure, high-quality seed strains to ensure good crops of marketable varieties. Of course, any knowledgeable market gardener sprouted his seed in damp cotton or flannel to ensure viable seed before committing large fields to a single seed source, but Henderson grew his trial seed through the entire season. He invited readers and any potential clients to visit his test areas at the Jersey City nursery. By 1890, his ten-acre complex contained nearly four acres of glass flanking either side of Randolph Avenue, which intersected the site. A cobbled driveway led to the one-story brick building housing Henderson's office.[45] Since a visit to the complex only entailed a short ferry ride across the Hudson River and a pleasant walk from the boat landing, many individuals visited his impressive grounds. They made their seed purchases with the knowledge that he planted those strains on his own acreage.

In 1876 Henderson formed a new partnership with his son Alfred and his son-in-law James Reid, under the title Peter Henderson & Company, at Cortland Street in New York City. Their advertising and catalogs appealed to amateur and recreational gardeners as well as to professionals, and they sought a national as well as a regional market. Their physical location proved to be one of their firm's major strengths, given that Hudson County, and particularly Jersey City, served as a major transportation point. The Morris Canal connected the Delaware to the Passaic and Hudson Rivers as early as 1836, and railroad construction began simultaneously in the area. By the 1870s, the Erie Railroad and the Pennsylvania Railroad provided the terminus for at least a dozen different rail lines.[46] The Lehigh Valley, the Philadelphia and Reading, the Susquehanna, and the Baltimore and Ohio Railroads all established major facilities in Hudson County by the end of the century.[47] Moreover, the seed firm's motto, "every post office a Henderson Seed Store," emphasized the importance of the nation's expanding postal system. Henderson could easily import or export merchandise by both rail and water. Ocean steamers from abroad, including those from the Cunard, Hamburg, Bremen, White Star, and Cardiff Lines, all docked on the west side of the Hudson. The Cunard Line docked in Jersey City proper, and Henderson therefore used ready transportation when he visited Britain in 1872 and imported choice plants on his return home.[48] Company warehouses on the Morris Canal and the frequent ferry service between Jersey City and Manhattan meant that Henderson could split his workday between the nursery grounds adjacent to his house and the seed establishment, which sat at 35 and 37 Cortland Street in Manhattan (the site of the now destroyed World Trade Center).

Henderson also benefited from U.S. government postal rates for shipping small goods. The 1869 cost, two cents per four ounces, for any parcel not exceeding four pounds, made seed readily available to almost all potential customers.[49] Even twenty years later, doubled to sixteen cents a pound, postal rates robustly supported the seed business.[50] The Society of American Florists, an early trade group focused on the business interests of professional horticulturalists, was active in postal rate politics. It included what we would today call retail florists, nurserymen, and wholesale growers. Henderson valued the group's ability to lobby for favorable postal rates for plant shipments, because the issue dovetailed with so many of his own interests.

In developing his business, Henderson worked to achieve good quality seed, offering consistent performance. With expert selection, regular seed trials, dependable growers, and rigorous inspection of his suppliers' fields, Henderson & Company sought to standardize their products. Having a central location for the trial areas and greenhouses, the warehouse facilities, and the seed store facilitated both importation and quality control of the product. After packaging by a low-paid female labor force, seed orders quickly reached their destinations via the extensive railway network that converged on Jersey City.

Unlike many easily consumed items, seed required horticultural skills if consumers were to realize the full benefits of the product. As cities expanded in the late nineteenth century, market gardeners proliferated around the parameters and along the railway lines leading to urban hubs. Further, as disposable income increased for the middle class, even small cities and larger towns usually supported at least one florist. Amateur gardeners, including many women, found themselves attracted to the floral business. Unfortunately, many of these new horticultural entrepreneurs lacked the skills necessary for economic survival, yet Henderson's seed business depended upon their success. Furthermore, recreational gardening demanded that an entirely new segment of the population acquire gardening skills. In short, mass marketing required that Henderson educate potential consumers of his products.

Garden Writing: Products for the Masses

Henderson's efforts followed a pattern established by earlier horticultural writers. By the middle of the nineteenth century, head gardeners and nurserymen frequently wrote for the horticultural press in both Great Britain and America. Thus Henderson's early articles followed established precedent. He covered a wide range of gardening topics in Hovey's *Magazine of Horticulture*, *Gardeners' Monthly*, *Rural New Yorker*, *Country Gentleman*, and most frequently the *American Agriculturalist*.[51]

When *Gardening for Profit* first appeared in 1867, it echoed earlier publications by such plantsmen as Bernard McMahon, Joseph Breck, and Robert Buist. These authors sought to provide readers with garden instruction adapted to the U.S. climate, while simultaneously helping them understand the broad range of plant material for sale at the writer's stores. Henderson's book reached a different audience, however. Orange Judd, the highly successful owner and editor of the *American Agriculturalist*, appears to have actively solicited Henderson's manuscript, a speculation that seems to fit with the thrust of the book. Judd came from a rural family in upper New York state and worked his way through Wesleyan and Yale as he studied science and agriculture. He fervently championed clear, concise writing directed at working people with limited education. Thus the straightforward, lucid instructions on market gardening offered in Henderson's book reflected the preferences of its editor. Orange Judd and Company subsequently published *Practical Floriculture* (1869) and *Gardening for Pleasure* (1875). A fourth book, *Henderson's Handbook of Plants*, first appeared in 1881 as a Henderson & Company publication. So too did *Garden and Farm Topics*, a compilation of previously published material in shortened form. Henderson & Company also published *How the Farm Pays*, primarily written by William Cozier with market-gardening-related material contributed by Henderson.

These three books, with their clear step-by-step instructions, went through several printings. Henderson eventually authored revised editions of the first two in the 1880s. He used these to educate both potential professional and amateur gardeners about plant cultivation, in order to increase seed and plant sales, yet his writings also contained subsidiary themes. These included attention to fraudulent business practices, a realistic understanding of production costs, and an appreciation of various plant characteristics and qualities.

Fraudulent practices among seed dealers certainly constituted a long-standing problem, and shrewd professional gardeners prided themselves upon recognizing the tricks of the trade. The most succinct statement of these practices appeared in Henderson's essay, "Humbugs in Horticulture." While Henderson avoided divulging the legal names of culpable individuals, he offered his readers stories about scams and identified the perpetrators by their street names such as the "Blue Rose Man" and "Dutch Peggy."[52] Henderson drove home the fact that fraudulent dealers in seeds and plants considered the neophyte as their traditional prey and that even professional horticulturalists found themselves duped on occasion.

Henderson sought to provide his readers with a realistic understanding of horticulture costs. He readily quoted typical construction, land, supply, and labor costs for production of marketable items and urged readers to make their planting decisions based on final net profits. Freely sharing his advice with individual visitors, he dutifully answered hundreds of readers' letters on specific problems.

Henderson also emphasized identifying plant characteristics that enhanced profitability; and he particularly valued early or predictable maturity dates, size, color, and, last, taste, though only in a limited number of items. Even in *Gardening for Pleasure*, a volume specifically directed to the amateur, noncommercial gardener, he retained these preferences. By placing seed selection within this context, Henderson helped define the guidelines for a "valuable" plant. In selecting and selling vegetable seed, for example, market exigencies shaped his decisions. He needed reliable early and late varieties so seed purchasers could get their product to market when seasonal prices peaked. Plants displaying characteristics that made for simpler and faster harvesting (e.g., bush beans) became desirable. When selling seed to the amateur flower grower, he valued characteristics such as ease of germination, novelty, and visual appeal. In each case, the market determined the plant considered valuable.

By helping to define reliable and beneficial plant characteristics, as well as increasing reader awareness of fraudulent practices and production costs, Henderson increased the desirability of his own products. He astutely realized the premium price a standardized product of predictable quality could command in the marketplace. Market gardeners by hard experience or shared tales recognized the games of hucksters and delighted in outwitting them, but newly returned veterans hoping to earn a living or recreational middle-class gardeners had little taste for such ruses. Henderson sought to sell an image of integrity and confidence as well as horticultural products.

His books, of course, veered into blatant advertising at times. The chapter on gardening implements in *Gardening for Pleasure* touted the merits of specific products such as Henderson Lawn Mowers, "Hale's Perfect Mole Trap," French patterned watering pots, and White's trellis (also plugged in the pea growing section).[53] A red-printed page in *How the Farm Pays* offered a free copy of Henderson's 132-page "finely illustrated" 1885 catalog with color plates to book purchasers, who could detach and mail the perforated coupon sheet with their names and addresses. Of course, all the later publications regularly advertised Henderson's other books in their end pages, just as his annual catalogs in turn promoted the books.

Over the decades Henderson's unerring instinct for promotion led him to support and write about contests for prize produce, a plant potting competition in Madison Square Garden, and just about any other event that garnered positive publicity for his businesses. He also wrote all of his own catalog and advertising copy until 1880. One color lithograph in a circa 1880 seed catalog featured a view of Henderson's four acres of greenhouses and six acres of growing area, but also deliberately included the extensive industrial development surrounding his complex as well as the view across the Hudson to New York City. This image placed Henderson's horticultural complex within the landscape of industrial and urban America, and it gave him the satisfaction of asserting the legitimacy of his "industrial garden" to his customers.

The Millionaire Florist: Consumer Culture in the Age of Conspicuous Consumption

Peter Henderson, the market gardener and seedman, stated his occupation as "florist" in the 1880 census.[54] The term carried far broader implications than it does today, but his choice of that title allows us to understand some very significant changes occurring in consumption patterns during the decades of his career in America. In the 1840s, the market for highly perishable luxury goods such as flowers usually remained quite limited, because the wealthy often maintained country greenhouses with private gardeners, whereas the emerging middle class lacked the income for such luxuries. If the latter did indulge in flowers, they typically purchased them as potted plants. Funeral pieces and simple flat or one-sided bouquets for special occasions composed of cut flowers sold only in large urban areas such as Boston, Philadelphia, and New York City. Speaking before the New York Horticultural Society in 1880, Henderson clearly remembered that city-wide flower sales for New Year's day in 1844 (the biggest sales day of the year at that time) usually did not exceed $1,000 in New York City. In 1880 he estimated that sales could run fifty times that amount and that annual sales usually ran into the millions.[55]

The tremendous expansion of commercial floriculture depended, of course, upon external factors such as improved transportation, urbanization, and the democratization of consumption, yet internal changes by florists also accounted for much of the growth. More efficient production facilities with superior technology, better marketing, expanded product offerings, and improved accounting methods all assisted the flower industry's expansion. After Henderson's prominent financial success in the 1880s, other florists closely scrutinized his early career in floriculture as well as his investments in new greenhouses. They tracked his production techniques, accounting strategies, and employment practices as well. Although never one of the largest flower growers or wholesalers, Henderson enjoyed a reputation as one of the most canny.

Henderson's market gardening business always included floriculture; thus, in the early years he sold the usual range of potted plants in season. He also placed some of the young boys who worked for him during the warm months on New York City street corners with bouquets to hawk. By the mid-1860s, he regularly cropped winter cut flowers under glass and considered camellias, violets, and carnations the most valuable sorts with roses, tuberoses, and double primroses as secondary staples. The prized white camellia brought in $15 per hundred from December to April, whereas in those months roses would realize only half that amount.[56]

During the 1870s Henderson noticed the profits Boston growers made on roses and gradually expanded his stock, especially as camellias passed out of fashion and roses netted ever higher prices. He mastered the art of "salting down," or holding back, stock for the holiday trade when prices hit their peak and thereby collected handsome profits.[57] By the mid-1880s Henderson's skilled staff forced huge numbers of pot- and bench-grown hybrid perpetuals and hybrid tea roses. The gardeners paid exacting attention to detail in order to produce successful blooms, and Henderson realized $50 per hundred at wholesale between December 15 and January 15.[58] His men marketed a large variety of plants ever more efficiently, and Henderson continued to drop his prices while still making handsome profits. Some smaller growers found the economies of scale and labor rationalization of "Henderson's plant factory" unfathomable and expressed their disbelief to the horticultural press, but his financial achievements remained evident to all.[59] His huge glasshouse complex advertised his success, and further innovations in glass production provided him with an incentive to further update and expand his facilities.

During the 1880s, window glass production expanded and glass prices fell in the United States, even though negotiations between the Knights of Labor and the Glass Manufacturers Association resulted in higher wages for American workers. Much of this expansion may be attributed to glass workers producing finished sheet glass with better furnaces and tools that provided superior support for their molten materials. Also, in 1881–1882, the development of gas fields in Pennsylvania and Ohio caused domestic glass manufacturers to switch from coal to natural gas. This new fuel allowed U.S. adoption of the more efficient tank furnace for the first time in 1886–1887, although Europeans utilized the technology much earlier. Although high tariffs protected the domestic glass industry in the 1880s, purchasers on the Atlantic seaboard still benefited from the quality and price of imported glass.[60]

In 1880 Henderson substantially improved his greenhouse facilities by constructing a huge, new complex based on twenty-by-one-hundred-foot modules with top ventilating sections of sash. The new structures featured larger panes of glass each measuring twelve by sixteen inches. The entire upper ridge of sash moved, providing superior ventilation as well. Also, attaching the multiple glasshouses side by side, in long rows, conserved heat.[61]

About 1887 Henderson purchased an additional new range of pits or low greenhouses from Lord & Burnham, consisting of eleven-foot-wide even span connected pits covering a ninety-by-three-hundred-foot area in glass. The complex featured six-foot-by-four-foot sashes screwed down on the north side, but every alternating panel on the south side attached to a hoisting mechanism, lifting it by as much as four feet. This allowed for full ventilation of plants needing cool temperatures such as violets, pansies, and bulbs. Although the new construction cost $7.50 per running foot, Henderson considered the complex superior to the old glass structures because he no longer needed to place protective woven matting on the exterior surface of panes in bad winter weather. This substantially reduced the cost of broken glass. Although the greenhouses' three boilers burned about $500 in coal annually, their output offset the cost and produced satisfying profits. Their advantages were clear: "Although the area covered by these low houses is considerably less than what we had in [unheated] sunken pits and cold frames, our output of plants has been more than doubled, and at much less expense in labor."[62]

In 1884 Henderson invested $7,000 in a three-hundred-fifty-by-twenty-foot state-of-the-art rose house built to his requirements by Lord & Burnham. Featuring iron framing throughout, and double-thick second-quality French glass in twelve-by-twenty-inch panels, it had as well steam heat running through one-and-a-quarter-inch pipes, supplied by two Lord & Burnham No. 5 boilers. By 1889 experiments with this steam heat convinced Henderson that it provided more efficiently controlled heat for large greenhouse spaces with less fuel and labor than did hot water.[63] The new three-quarter-span structure also boasted top lifting sash (movable panes of glass) along the entire south side of the roof for the best possible ventilation.[64]

Henderson's rose house exemplified a trend in large cities, as florists increasingly specialized in particular flowers they deemed profitable in their area. Their business depended on reaching a substantial population with disposable income interested in having luxury flowers. The 1890 U.S. census showed that 4,659 floral establishments in the country produced over $12 million in plants and $14 million in cut flowers. New York led the country in total value of both plant and cut-flower sales, and its growers easily showed the highest levels of total capital investment, with over $9 million. New Jersey growers, with over $3.5 million invested, primarily split their production between the big urban centers of New York City and Philadelphia.[65]

Henderson's rose growing never matched the business scale of growers devoted primarily to that specialty, but his publications provided information to florists in smaller cities and towns who entered the business. His instructions for propagation and cultivation stressed efficiency and a quick return on investment. For example, he advocated grafting roses on vigorous imported stock rather than rooting sections of material cut from existing plants because the grafting produced a crop more quickly, a particularly important factor when new varieties commanded top prices.[66] Henderson also moved into wholesale floriculture. Like any good commodities trader in the 1880s, he built a reputation for buying low and selling high.[67] His recommendations for rose varieties echoed the desires of wholesale dealers, and, like most veteran growers, he recognized the fiscal wisdom of marketing a limited number of varieties on a consistent basis, so that customers could easily identify favorites.

Henderson's comments on changes in floral fashion in his revised, 1887 edition of *Practical Floriculture* demonstrate how name recognition of rose varieties could raise prices. He noted that in the winter of 1886 the finest buds of "American Beauty," "Paul Neron," "Magna Charta," and "Baroness Rothschild" retailed at $1 each from the beginning of December to the end of February; yet a few years earlier they had sold for even more. Henderson wrote that fashion decreed that flower colors match furniture upholstery. Further, long stems on roses, which constituted the single most popular flower in the trade, became de rigueur. He noted that a flower canopy for a society wedding ran $600, while the floral bill for decorating the entire house ran $5,000. Although this constituted one of the most opulent examples, Henderson assured his readers that $500 to $1,000 made up the usual bill. He concluded, "the flowers for the balls of the American Club of New York in Tweed's palmy days often cost six thousand dollars for a single night."[68] Money easily drove the selection of flowers and their cultivation when such vast sums changed hands. Consumption marked power, and lavish floral displays clearly trumpeted status.

This pattern of increasing consumption fueled the wholesale side of Henderson's floral business, just as the swarms of Irish immigrants filled his laboring ranks. Unlike many employers, Peter Henderson did not despise or denigrate the Irish. He knew they provided some of his best workers and he considered it a sincere compliment when told that he handled a spade like an Irishman. Still, Henderson & Company gave its employees only two holidays annually, New Year's day and St. Patrick's day. The career of one of his most successful workers, Patrick O'Mara, illustrates the interesting correlation that could exist between the lives of impoverished immigrants who filled the great urban areas and the concurrent affluence of the rising middle class brushing up against the extravagance of newly minted millionaires.

Born in County Tipperary in 1858, Patrick O'Mara emigrated from Ireland at the age of eight and began working part time three years later at the Jersey City nursery site, while attending school in the winter. At age fourteen he became a regular employee, with a workday that began at 3:30 to 4:00 in the morning during the growing season. To encourage promptness, the last man at work each day forfeited twenty-five cents, which then went to the man who had showed up first. O'Mara worked in the glasshouses with men who later became successful florists in Ohio, New Jersey, and Chicago, as well as alongside horticulturalists like Sidney Wilkenson, who later headed the large seed and plant firm, Henry A. Dreer. During the busy season Henderson expected sixteen-hour days from his best workmen, for whom he provided steady year-round employment. Still, most gardeners readily sought the work as they made overtime pay for any hours after 6 P.M. From April to June, fifteen to twenty of his best men worked by lamplight until 10 or 11 P.M. to make possible the successive crops filling every square foot of growing space. O'Mara eventually learned the production side of the floral business and in 1882 went on the road as a twenty-four-year-old sales agent for the firm. He called at all the commercial nurseries and large private estates in the Northeast to secure plant and seed orders. Upon Henderson's death in 1890, he became the plant department manager and treasurer of the firm.[69] O'Mara's case remains unusual because he moved from facilitating consumption for the social strata above him to joining its ranks. Meanwhile, the vast majority of Irish immigrants continued to work for pay that left them impoverished.

O'Mara exemplified the most successful of the firm's employees, but the nursery and trial grounds easily employed sixty "hands" during the growing season. In the 1880s Henderson spent several hours a day working there and took particular interest in the plant crosses and trials. His large workforce, with a number of skilled propagators, made possible the development of some of his most successful introductions, including the "Premier" strain of pansies in 1884. The "Giant Butterfly" strain a few years later became one of his favorite flowers.[70] During the last decade of his life he also introduced the "American Wonder" pea (1880), an early forerunner of Iceberg, "Henderson's New York" lettuce (1886), the "Early Hackensack" melon (1889), "Henderson's Succession" cabbage (1888), and "Henderson's Bush" lima bean (1889). In the case of the bush lima bean, he eliminated the need for poles or other supports, making it a profitable crop for large-scale commercial growers.[71]

Henderson's plants accommodated the exigencies of the marketplace and, from a horticultural perspective, became as emblematic of nineteenth-century American industrialization as the smokestacks and massive industrial expansion occurring in the Jersey City area. His "greenhouse factory" grew products almost as standardized as those manufactured by his neighbors. If Colgate & Company produced tons of neatly wrapped Cashmere Bouquet soap bars and the Joseph Dixon Crucible Company spilled four million Ticonderoga lead-pencils from its production lines each year, Henderson matched them with roses boxed by the hundreds and masses of seeds that eventually yielded boxcars filled with produce.[72]

Conclusion: Mass Production in Horticulture

Although flowers often appear in images depicting nineteenth-century domesticity, they usually do not spring to mind when considering the rise of consumer culture. Instead we picture pristine flower beds, bouquets with ruffled edges, and rose-covered bowers. Studies of mass marketing in the nineteenth and early twentieth centuries have primarily focused on machine-made products such as automobiles, sewing machines, and farm machinery. Food products such as Crisco, Jello, and packaged biscuits have drawn attention for their mode of presentation and distribution, but nonbranded comestibles have appeared as incidental items in the rise of department stores and grocery chains.[73] Historian David Hounshell equates the development of mass production with mechanized production of interchangeable components, not with produce and flowers.[74] Yet Peter Henderson grasped the same concepts that informed other business endeavors developing around him. He emulated the example of durable goods manufacturers by featuring views of his own production facilities in his promotional materials.[75] His neighbors fashioned soap cakes and pencils; he produced plants.

The appearance of novel products and major alterations in the presentation of goods makes identification of changes in industrial production and mass marketing innovations more apparent than the application of these approaches to traditional items such as flowers. New technologies in plant growing can take place with little or no consumer awareness. Yet changes in the availability of fruits and vegetables, with concomitant alterations in nutrition, have important consequences for society. Poor nutrition exacerbated by adulterated and contaminated foods resulted in illness and diseases, such as scurvy, among the laboring poor. Better food regulation and inspection eventually improved this situation, but the availability of affordable fresh fruits and vegetables greatly improved the health of urban populations suffering these problems during the nineteenth century. Without greater labor productivity, better technologies, improved seed, and effective fertilizers, adequate and affordable sources of nutrition would have been available to a much smaller segment of urban society. The changes occurring during Henderson's career altered the availability of horticultural products and helped improve the urban diet.

Although more financially successful than many of his peers, Henderson, in his life and activities, echoes international and domestic developments that marked his century. Globalization of the seed trade constituted but one example of this emulation of broader trends. His use of traditional networks to negotiate the pitfalls of a changing and evolving profession coexisted with his ability to rationalize labor and constrain costs. His shrewd understanding of industrial development and expansion of public sector services, such as municipal utilities and the national postal service, provided him with vital resources and opportunities. Henderson's many articles and books, as well as his use of advertising, echoed emerging practices, but he also brilliantly cross-marketed his goods to an evolving American middle class with rapidly expanding consumption patterns. None of these efforts occurred in isolation;

indeed, this biological entrepreneur functioned in an environment that demanded creative and integrated responses.

Overall, Henderson's late-nineteenth-century greenhouses and trial grounds reflected new technologies, burgeoning labor rationalization, and the propensity of the market to become enmeshed in an ever greater number of human activities. The wagons departing Henderson & Sons' nursery grounds carried prolific vegetables, fruits, and flowers prized for their predictability and profitability. Often cultivated under glass, enriched by industrial waste, and protected by chemicals, these products of the industrial garden provided a glimpse of America's future, at least that segment of it linked to and exemplified by our marketing and manipulation of the plant world.

Notes

1. Seedsman, "The Seed Trade: Seeds in Olden Time," *Gardeners' Chronicle* (September 22, 1894): 336; E. Wilson Serpell, "Chemists and Grocers as Seedmen," *Gardeners' Chronicle* (May 2, 1885): 579; E. F. W. H., "Chemists and Grocers as Seedmen," *Gardeners' Chronicle* (May 9, 1885): 610; R. Dean, "Booksellers as Seedmen," *Gardeners' Chronicle* (May 23, 1885): 671; "Changes in the Seed Trade," *Gardeners' Chronicle* (September 29, 1894): 368; Seedsman, "The Seed Trade," *Gardeners' Chronicle* (September 29,1894): 368; Seedsman, "The Seed Trade: Seeds in Olden Time," *Gardeners' Chronicle* (September 22, 1894): 336.
2. P. H., "Obituary, Mr. George Sterling," *Garden* (January 24, 1885): 76.
3. Alfred Henderson, *Peter Henderson, Gardener, Author, Merchant: A Memoir* (New York: Press of McIlroy & Emmet, 1890), 13.
4. Henderson, *Peter Henderson*, 14.
5. James Anderson, "Education of Gardeners," *Gardeners' Chronicle* (April 19, 1879): 496; "Gardening Apprentices," *Gardeners' Chronicle* (September 8, 1894): 288.
6. Peter Henderson, *Essay on Horticultural Progress, Read before the New York Horticultural Society* (New York: Sears & Cole, 1880), 11–12.
7. Wilhelm Miller, "Robert Buist," *The Standard Cyclopedia of Horticulture*, vol. 2, ed. L. H. Bailey (New York: Macmillan Company, 1943), 1567.
8. C. M. Hovey, "Robert Buist's Exotic Nursery," *Magazine of Horticulture* 1 (June 1835): 203–206; C. M. Hovey, "Buist's Exotic Nursery," *Magazine of Horticulture* 5 (February 1839): 62; C. M. Hovey, "Buist's Exotic Nursery," *Magazine of Horticulture* 10 (April 1844): 121–122, 84–85.
9. Henderson, *Practical Floriculture* (1887), 253.
10. Peter Henderson, "Gardeners Emigrating," *Garden* (November 19, 1881): 496; P. Henderson, "Gardeners Emigrating," *Garden* (January 24, 1885): 71; P. Henderson, "Gardeners in America," *Garden* (December 1, 1888): 513.
11. Pearce Davis, *The Development of the American Glass Industry* (Cambridge, Mass.: Harvard University Press, 1949), 95–108.
12. Henderson, *Gardening for Profit* (1887), 31; William H. Shaw. comp., *History of Essex and Hudson Counties, New Jersey*, vol. 2 (Philadelphia: Everts & Peck, 1884), 1158–1159.
13. Shaw, *History of Essex and Hudson Counties*, 1150.
14. Shaw, *History of Essex and Hudson Counties*, 1157–1158; "Orchids in the United States of America," *Gardeners' Chronicle* (April 9, 1892): 460.
15. "Gardeners and the Railway Rates," *Gardeners' Chronicle* (March 4, 1893): 269–272.
16. U.S. Census, 1860, Schedule for the Third Ward of Jersey City in the County of Hudson, New Jersey (June 21, 1860), 60. New Jersey, Hudson County 1860 U.S. Census, population schedule. Micropublication M 653, roll 692. Washington, D.C.: National Archives.
17. Alfred Henderson, *Peter Henderson*, 20.
18. Henderson, *Garden and Farm Topics*, 197, 202.
19. Peter Henderson, *Practical Floriculture: A Guide to the Successful Cultivation of Florists' Plants for the Amateur and Professional Florist* (New York: Orange Judd and Company, 1869), 192.

20. Peter Henderson, *Practical Floriculture: A Guide to the Successful Cultivation of Florists' Plants for the Amateur and Professional Florist* (New York: Orange Judd and Company, 1887, reprint 1907), 15.
21. Henderson, *Practical Floriculture* (1907), 16.
22. Peter Henderson, "Flower Growing in the United States," *Garden* (April 21, 1888): 362.
23. Henderson, *Practical Floriculture* (1887), 70.
24. C. M. Hovey, "Some Remarks on the Size of Flower Pots Usually Employed for Plants, with Hints upon the Importance of Having Some Standard for Classifying the Various Sizes," *Magazine of Horticulture* 5 (November 1839): 46–50.
25. "A few years ago each potter followed his own convenience or ideas in making pots, and the shapes and sizes in the market varied infinitely. This was a source of much annoyance to the commercial florist." From G. Harold Powell, "Flower Pots," *American Gardening* (July 18, 1896): 451.
26. Jim Keeling, *New Terracotta Gardener* (London: Headline, 1993), 29; Primrose Peacock, "Pots with a Past," *Garden* 115 (November 1990): 592; *Descriptive Catalogue of Roses and Stove and Greenhouse Plants* (Boston: Waban Conservatories, 1877); *Illustrated Wholesale and Retail Catalogue* (Boston: Parker & Wood, c. 1880s), 30.
27. "The Standard Pot," *American Florist* (May 15, 1889): 486.
28. "The S. A. F. Standard Pot," Supplement to the *American Florist* (January 1, 1889): n.p., shown opposite 246.
29. "Advertisement," *American Florist* (August 3, 1893): 1360.
30. "The 'Standard' Flower Pot," *Gardeners' Chronicle* (March 23, 1889): 373.
31. "Standard Pots," *American Florist* (February 15, 1889): 297.
32. "Storing Pots," *American Florist* (June 1, 1899): 496.
33. Henderson, *Practical Floriculture* (1869), 61.
34. Henderson, *Gardening for Profit* (1867), 44–56.
35. Henderson, *Gardening for Profit*, 57–65.
36. Henderson, *Practical Floriculture* (1869), 222, 237; Henderson, *Practical Floriculture* (1889), 292.
37. Joan F. Doherty, *Hudson County: The Left Bank* (Northridge, Calif.: Windsor Publications in cooperation with the Hudson County Chamber of Commerce and Industry, 1986), 53.
38. Henderson, *Gardening for Profit* (1888), 20.
39. Henderson, *Gardening for Profit* (1888), 225, 279.
40. Henderson, *Gardening for Profit* (1867), 120–121.
41. Henderson, *Gardening for Profit* (1887), 162–163.
42. Henderson, *Gardening for Profit* (1887), 164.
43. *Everything for the Gardener, Anniversary 1847–1947* (New York: Henderson & Company, 1947), 1.
44. Alfred Henderson, *Peter Henderson*, 21.
45. Last will and testament of Peter Henderson, July 12, 1889, Copy by Alex T. McGill, Surrogate General of the State of New Jersey, Probated April 26, 1890. Property description page 3, seventh item. Notes by J. Owen Grundy, City Historian of Jersey City made in 1974 and located at the Jersey City Free Public Library. My thanks to Joan D. Lovero, Chief Librarian, for her help in obtaining copies of these materials.
46. Charles H. Winfield, *History of the County of Hudson, New Jersey* (New York: Kennard & Hay Stationery Mfg. and Printing Company, 1874), 71.
47. Doherty, *Hudson County*, 66.
48. Winfield, *History of the County of Hudson*, 371.
49. Henderson, *Practical Floriculture* (1869), 188.
50. Henderson, *Practical Floriculture* (1887), 265.
51. Alfred Henderson, *Peter Henderson*, 26–27.
52. Peter Henderson, "Humbugs in Horticulture," an essay read at the annual meeting of the National Association of Nurserymen, Florists, and Seedmen, held in Chicago, June 16, 1880, in *Garden and Farm Topics* (New York: Peter Henderson & Company, 1884), 227–240.
53. Henderson, *Gardening for Pleasure*, 368, 373, 376, 381.
54. Manuscript Population U.S. Census 1880, Schedule for Jersey City, Third Precinct, Sixth District, County of Hudson, New Jersey, June 22, 1880, 62.
55. Peter Henderson, *Essay on Horticultural Progress*, read to the New York Horticultural Society on March 9, 1880 (New York: Sears and Cole, 1880), 11–12.

56. Henderson, *Practical Floriculture* (1869), 130–138.
57. Patrick O'Mara, "Patrick O'Mara's Recollections," *American Florist* (April 22 1899): 1145–1146.
58. Henderson, *Garden and Farm Topics*, 93–94.
59. "Cost of Production," *American Florist* (July 15, 1889): 574; E. Fryer, "Cost of Production," *American Florist* (August 1, 1889): 592; Robert S. Brown, "Cost of Production" *American Florist* (August 1, 1889): 592–593.
60. Davis, *Development of the American Glass Industry*, 120–129.
61. Henderson, *Gardening for Profit* (1886), 73–79.
62. Peter Henderson, "Low Greenhouses or Pits," *American Florist* (August 1, 1889): 593.
63. Henderson, *Practical Floriculture* (1889), 102.
64. Henderson, *Practical Floriculture* (1889), 159–160.
65. "Commercial Floriculture in the United States from the Census of 1891," in *American Florist Company Directory* (Chicago: American Florist Company, 1896), 25–26.
66. Henderson, *Practical Floriculture* (1887), 142–143.
67. Patrick O'Mara, "Patrick O'Mara's Recollections," *American Florist* (April 22, 1899): 1146.
68. Henderson, *Practical Floriculture* (1887), 239.
69. Patrick O'Mara, "Patrick O'Mara's Recollections," *American Florist* (April 2, 1899): 1145–1146; "Officers of the New York Florists' Club," *American Florist* (December 23, 1899): 611.
70. *Everything for the Gardener*, 1.
71. *Everything for the Gardener*, 5.
72. Shaw, *History of Essex and Hudson Counties*, 1159, 1161. Sheilds T. Harwin, *The Colgate Story* (New York: Vantage Press, 1959), 61–64.
73. William Leach, *Land of Desire: Merchants, Power, and the Rise of a New American Culture* (New York: Pantheon Books, 1993), 23; Susan Strasser, *Satisfaction Guaranteed: The Making of the American Mass Market* (New York: Pantheon Books, 1989), 3–28, 32–35, 47–49, 115–119; Richard S. Tedlow, *New and Improved: The Story of Mass Marketing in America* (New York: Basic Books, 1990), 112–258.
74. David Hounshell, *From the American System to Mass Production, 1800–1932* (Baltimore: Johns Hopkins University Press, 1984).
75. I am indebted to Professor Laird for many helpful discussions on nineteenth-century advertising. See particularly, Pamela Walker Laird, *Advertising Progress: American Business and the Rise of Consumer Marketing* (Baltimore: Johns Hopkins University Press, 1998), 96–180.

Biological Innovation in American Wheat Production
Science, Policy, and Environmental Adaptation

ALAN L. OLMSTEAD and PAUL W. RHODE

History celebrates the battlefields whereon we meet our death, but scorns the plowed fields whereby we thrive. It knows the names of the King's bastard children, but cannot tell us the origin of wheat. That is the way of human folly.
—Jean Henri Fabre[1]

American wheat production underwent revolutionary changes during the nineteenth and early twentieth centuries as a progression of mechanical innovations transformed the wheat harvest and made the United States the undisputed world leader in farm mechanization. The stories of the mechanical reaper, the self-raking reaper and the self-binder, the mechanical thresher, and the combine have been told and retold, and the accomplishments of the leading inventors and manufactures are an integral part of American folklore. It was after all Cyrus McCormick who "made bread cheap"! But the machines that he and his fellow inventors produced did much more. By allowing millions of farmers to systematically exchange animal power for human labor, they dramatically lessened the back-breaking toil of farm life and vastly increased labor productivity. According to the conventional, if crude, estimates of the U.S. Department of Agriculture (USDA), it required roughly two hours and twenty minutes of labor to produce a bushel of wheat in 1840, but by the late 1930s only forty minutes of work were needed. While labor productivity was soaring at an unprecedented rate, the output per acre of land harvested was almost constant.[2] These observations—the huge increase of wheat output per unit of labor and the stagnant yields—have led to the erroneous conclusion that, prior to roughly 1940, mechanization was the source of almost all productivity growth and that biological innovations were unimportant.

Willard Cochrane, one of the deans of the agricultural economics profession, offers a succinct statement of this conventional view, noting that mechanization "was the principal, almost the exclusive, form of farm of technological advance" between 1820 and 1920.[3] Yujiro Hayami and Vernon Ruttan repeatedly echo this theme in their analysis of technological change in the American wheat industry: "the advances in mechanical technology were not accompanied by parallel advances in biological technology. Nor were the advances in labor productivity accompanied by comparable advances in land productivity."[4] This view is also a part of the mantra of most economic historians as reflected in the recent survey of American agricultural development before 1900 by Jeremy Atack, Fred Bateman, and William Parker in the authoritative *Cambridge Economic History of the United States*: "Mechanization generated the most dramatic changes in nineteenth-century agricultural productivity" and "Land abundance . . . contributed to the general lack of interest in land productivity by nineteenth-century American farmers."[5] It is also forcefully stated in Atack and Bateman's influential book on antebellum Northern farming:

> The great improvement in acreage yields lay almost a century into the future when chemical fertilizers, hybrid seeds, irrigation, and various scientific developments became available to farm operators. Some technological devices designed to raise labor productivity were, however, becoming available during the nineteenth century. Mechanical rather than chemical or biological, these improvements operated primarily through their effect on the usage of labor.[6]

This argument is now enshrined in most economic history textbooks.[7]

The existing literature would have us believe that before the development of a sophisticated understanding of genetics, biological knowledge in agriculture essentially stood still, generating little or no boost to productivity. This leads to the popular picture of nineteenth-century agriculture as a world of unchanging cropping patterns and cultural practices, a world where each farmer sowed grain that he himself grew and that his father grew before him, a world of a happy, organic balance between cultivators and their natural environment.[8]

This essay argues that much of the conventional wisdom is bunk and must be scrapped. Contrary to the consensus opinion, the nineteenth and early twentieth centuries witnessed a stream of "biological" innovations that rivaled the importance of mechanical changes to agricultural productivity growth.[9] These new biological technologies addressed two distinct classes of problems. First, researchers and wheat farmers made great strides in combating the growing threat of yield-sapping insects and diseases, many of which were the unintended consequences of biological globalization. With the large-scale importation of Eurasian crops to North America came hitchhikers who fed on and destroyed those crops. In the absence of vigorous efforts to maintain wheat yields in the face of these evolving threats, land and labor productivity would have been significantly lower.[10] In effect farmers practiced what today would be termed integrated pest management (IPM) with the sensitive details of the farming systems evolving in response to new threats and changing knowledge. In addition there was a relentless campaign to discover and develop new wheat varieties and cultural methods to allow the wheat frontier to expand into the Northern Prairies, the Great Plains, and the Pacific Coast states.[11] Without these land-augmenting technologies, western yields would have been significantly lower, and vast areas of the Great Plains would not have been able to sustain commercial wheat production. These insights suggest a revised set of stylized facts that is more consistent with the historical record. Consistent with the old view we agree that (1) impressive changes in mechanical technologies dramatically increased labor productivity and that (2) yields per acre were roughly constant. But to these cornerstones one must add (3) farmers demonstrated at least as much interest in biological innovations as they did in machinery, (4) the impact of biological innovations rivaled the importance of mechanical changes, and (5) in the absence of these biological innovations land and labor productivity would have been much lower that what was actually obtained. The remainder of the essay will demonstrate our case.

Investments in Maintaining Yields

> *Now, HERE, you see, it takes all the running YOU can do, to keep in the same place. If you want to get somewhere else, you must run at least twice as fast as that!*
> —The Red Queen in *Through the Looking Glass*[12]

D. Gale Johnson and Robert L. Gustafson's imaginative and pathbreaking analysis of the sources of productivity growth in U.S. grain production in the post–World War I era still sets the standard for modern scholarship. In an effort to determine the importance of varietal changes in the period between 1928 and 1954, they regressed average yield per seeded acre and average yield per harvested acre on nine independent variables, including an index of newness of varieties seeded. They decomposed the United States into eastern and western states because the United States as a whole is too heterogeneous an area with respect to wheat production to be treated in a single analysis. Using the regression estimates, they constructed estimates of the net effect of varietal newness on the regional average yields. For the western region they found that between 1928 and 1954 wheat yields increased by 2.45 bushels from 11.7 to 14.15. Of this increase approximately 60 percent was due to the introduction of new varieties.

Although informative for the period they were studying, Johnson and Gustafson's methodology would seriously understate the longer-run impact of new varieties for two important reasons. First, decomposing the country into two regions essentially assumes away one of the most significant innovations in the entire history of American agriculture—the introduction, selection, and development of new varieties suitable for the western states.[13] Second, Johnson and Gustafson's formal analysis does not take into account the substantial decreases in yields that would have occurred due to the onset of diseases and pests if varieties had remained static. As the following statement makes clear, they were keenly aware of this latter problem. "Our analysis of varietal changes in wheat seems to imply that much of the research on new varieties constitutes to a considerable extent a maintenance operation.... In the absence of the research and the adoption of new varieties it is quite clear that the yields of the small grains would have declined over time."[14] Johnson and Gustafson proceed to offer an example as to how important such "maintenance operations" might be:

> The heavy attacks of black stem rust on durum wheat in 1952, 1953, and 1954 indicate the very considerable necessity of continually developing new varieties. The average yield of durum wheat per seeded acre for the decade 1942–51 was 14.5 bushels; in 1952 the yield was 9.7; in 1953, 6.2; and in 1954, 3.0. During the same three years other spring wheat yields were roughly comparable to the long-time average. By 1956 rust resistant varieties were available on a significant scale and yield had returned to normal levels. In 1958 a record yield of 23.8 bushels per acre harvested was achieved despite the fact that climatic conditions were such that rust losses would have been heavy had it not been for the rust-resistant varieties.[15]

An important method of combating diseases has been to shift to more resistant varieties. Wheat varieties are prone to "wearing out," because overtime they become more susceptible to diseases and pests (more accurately, diseases and pests evolve so as to become more destructive of existing varieties). Today, besides developing more resistant varieties, a sophisticated network of research and extension services regularly sample the rusts present in the wheat fields in order to be able to recommend relatively resistant varieties. For the post–World War II period, Johnson and Gustafson note that "for the United States as a whole, the USDA quinquennial wheat variety surveys indicate that, on average during the survey years 1944, 1949, and 1954, around 40% or more of U.S. wheat acreage was seeded with varieties not grown, or grown in only limited amounts, five years previously."[16]

Building on Edmund Russell's introductory remarks, not only are biological innovations alive, they necessarily generate destructive coevolutionary processes. Wheat farmers were cursed by the Red Queen's dictum: they had to run hard just to stay in place. As pioneering plant pathologist E. C. Stakman memorably put it in his 1947 *American Scientist* article, "Plant Diseases Are Shifty Enemies."[17] The same could be said of insects and other pests. Without significant investments in maintenance operations, grain yields would have plummeted as the plant's enemies evolved.[18] Continued cultivation of wheat in a region invited infestations of the Hessian fly, the midge, the chinch bug, and a score of other destructive insects. A number of diseases, but especially rusts and smuts, also could destroy a crop, and in many areas, winterkill and drought were serious threats. Before modern science transformed agricultural production functions, were farmers simply at the mercy of whatever plague happened to arrive at their door? In fact, well before the U.S. Civil War, scientists and farmers, through a process of careful observation, developed an understanding of the life cycle of the major pests. In addition, there is clear evidence that farmers repeatedly changed varieties and cultural practices in an attempt to ward off some of the harmful effects of insects and plant diseases. Tracing the responses to just a few of the important threats will demonstrate that these nonmechanical innovations had a significant impact on productivity in the nineteenth century.

In most areas, the most destructive diseases affecting wheat were rusts, wind-blown fungi that attacked the plant's stems and leaves. Rust attacks could cause the plants to lodge and produce shriveled grain.[19] In the span of a couple weeks stem rust could destroy what had promised to be a healthy crop. There were two fundamental ways that a wheat variety might avoid rust damage. First, it might have genetic resistance to the rust races currently in the area. Finding such varieties was a top priority. Before the modern age, this was a haphazard process, but breeders made significant progress. Second, a variety might mature before the rust did much damage (although under more ideal conditions, early maturation often compromised quality and yield). Since winter wheats ripened much earlier than spring wheats, the former were generally less vulnerable to damage.[20] One of the great achievements of wheat breeders before 1940 was the development of hardier winter wheats, allowing many parts of Kansas, Nebraska, Iowa, Wisconsin, and Illinois to shift out of spring varieties around 1900.

Problems with rust were not new. As early as the 1660s, the New England settlers were enacting a scenario that would be repeated thousands of times as farmers sought to match crops to their local conditions. Early introductions of English winter wheat failed in the harsh New England winters. After some trial and error, farmers succeeded in growing spring varieties. But in 1664, black stem rust appeared in Massachusetts, badly blasting the wheat crop by 1665. Farmers attempted to substitute earlier maturing winter wheats without much success. The inability to find winter hardy, rust-resistant varieties largely explains why New England never emerged as a serious wheat-producing region.[21] The high incidence of leaf rust in the southeastern United States is a major reason why little wheat was grown in that region despite generations of attempts. In addition, stem rust attacks forced large sections of Iowa and Texas to at least temporarily abandon wheat production in the late nineteenth century.[22]

In the late-nineteenth and early-twentieth centuries normal stem rust losses are estimated at 5–10 percent of the wheat crop.[23] Regional epidemics in 1878, 1892, 1894, 1904, 1914, 1916, 1923, 1925, 1935, and 1937 pushed losses much higher. The 1916 stem rust epidemic is estimated to have destroyed about 200 million bushels in the United States (over 30 percent of the harvested crop) and 100 million bushels in Canada.[24] In many locales the entire crop was lost. The emergence of vast concentrations of wheat in the Great Plains increased the breeding ground for rusts (and other enemies) and thus the frequency and severity of rust epidemics.[25] The spread of grain cultivation to the Southern Plains (Oklahoma and Texas) provided warm overwintering grounds for the fungi that would blow north in the spring and summer, further increasing rust problems in the early twentieth century. The added incidence of rust is just one reason why agronomists maintain that the wheat-growing environment had seriously deteriorated by the early twentieth century.[26]

Given the advances after World War II, the early efforts to control rusts seem primitive. But that was not the perspective as of 1940, when E. C. Large proclaimed that the "greatest single undertaking in the history of applied Plant Pathology was to be the attack on the Rust diseases of cereals."[27] What accomplishments so excited Large? A systemic analysis of rusts in the United States dates back to the contributions of Mark Alfred Carleton in the 1890s. Carleton tested over one thousand wheat varieties for yield, winter hardiness, rust and insect resistance, and for other qualities. The work of numerous other American scientists, along with research in Australia, Canada, and Europe, unlocked many of the mysteries of rust diseases. Aided by the rediscovery of Mendel's laws around 1900 and the publication of Johannsen's pure-line theory in 1901, this research accelerated the development of rust-resistant hybrids.[28]

There is clear evidence that farmers and wheat breeders were systematically developing and adopting more rust-resistant and earlier maturing varieties. For its day, Red Fife, which gained such favor in the northern Great Plains, had excellent rust-resistant qualities and was early ripening. Early Manitoba wheat farmers noted that Fife matured ten days earlier than the Prairie Du Chien variety that it replaced.[29] Marquis, which followed Red Fife, further cut the ripening period by seven to ten days, thereby providing significant rust protection. Kubanka proved remarkably resistant to the epidemic of 1904 that hammered the Bluestem and Fife crops.[30] When rusts evolved to attack Kubanka, it was replaced by Mindan (1918), which in turn was replaced in 1943 by Carleton and Stewart. At the time of their release these two varieties were highly resistant to the prevailing stem rust races. They maintained their resistance until race 15B suddenly made them obsolete.[31] A similar progression took place in the hard winter wheat belt because the new Turkey wheats that became the dominant variety by 1900 also had excellent rust resistant qualities when first introduced. Subsequent releases, including Kharkof (1900), Kanred (1917), and Blackhull (1917), were chosen in part for their rust resistance.[32]

A better understanding of the stem-rust lifecycle allowed farmers and scientists to attack its breeding ground in barberry bushes.[33] In 1660 farmers in Rouen, France, observed that wheat growing near barberry bushes was more apt to be damaged by stem rust and took steps to tear out the bushes. In 1726 Connecticut passed a law empowering town meetings to eradicate the bushes and Massachusetts and Rhode Island enacted similar legislation against the barberry in the mid-eighteenth century. (In an important sense, the colonists's problem was of their own making. The barberry was purportedly "one of the few ornamental plants that could be introduced from Europe as a plant that would stand transportation a long distance on shipboard.")[34]

In 1865 Anton De Bary scientifically demonstrated the role of barberry bushes as a host. Yet this knowledge was slow to diffuse. Hamilton suggests that the widespread presence of barberry bushes may have contributed to the stem rush epidemic that devastated the Minnesota wheat crop in 1878. It was not until 1918 that Minnesota outlawed the barberry bush. This was part of a larger cooperative federal-state campaign initiated in the aftermath of the 1916 rust epidemic to eradicate barberry bushes across the thirteen north central states. "Between 1918 and 1939 over twenty-two million bushes had been destroyed."[35] It is unlikely that the early scientists promoting the eradication program fully understood the barberry's role in the propagation of rusts. Not only did the barberry provide a home for the rust to carry over and multiply, the bush was the breeding ground where rusts mutated and developed new races.[36] Alan Roelfs estimates that the eradication program delayed the disease's onset by about ten days and, by removing the site of the rust's sexual reproduction, significantly slowed the evolution of new destructive races.[37]

Taken as a whole the century of biological changes prior to the modern biological revolution had an enormous impact on limiting the damage that rust otherwise would cause. The introduction of new varieties of spring wheat along with better cultural methods probably cut the ripening time by about twenty days; the destruction of barberry bushes effectively gave farmers up to another ten days of protection and reduced the rate of rust mutation. In addition, the new wheat varieties almost always had better resistance to rust relative to the varieties they replaced. Parallel changes took place in the winter wheat areas and the introduction of hard winter wheats allowed vast areas to convert from growing spring to winter varieties that could be harvested much earlier in the summer (see below).

In addition to rusts, various smut fungi did great damage to wheat throughout North America. Stinking smut (or bunt) was the most destructive. "In a ripe but bunted ear of wheat the grains were swollen and black, still whole, but with all their inner substance transformed into a pulverulent mass."[38] Milder cases damaged the grain and lowered its value. In 1908, Dondlinger noted that "formerly at least one-fifth of the cereal crops was annually destroyed by smut."[39] In addition, Gussow and Conners observed that "'previous to 1900 bunt was alarmingly serious and threatened to be a limiting factor in wheat production'" in southern Canada.[40] Even if Dondlinger's figure is an exaggeration, both of these accounts suggest that the damage from smut was declining by the turn of the century.[41] This was a direct result of scientific advances and farmer education. In an exhaustive series of experiments in the mid-1700s, Mathieu Tillet of France proved smut was a seed-borne disease and developed a number of treatments. Other researchers built on this discovery, leading to increasingly effective chemicals. In the nineteenth century, American farmers were known to soak seeds in hot water to control loose smut and employed lime and copper sulfate solutions to fight stinking smut. By 1900 cheaper formaldehyde solutions became available, and by the early 1920s mercury solutions and carbon carbonate dusts came on the market. There were still losses to smut, but they were far lower than before.[42]

Yet another disease that was particularly damaging in the soft wheat belt was wheat scab (or head blight), which in some years destroyed about 5 percent of the crop nationally. The disease both harms wheat seedlings and attacks the heads, causing shrunken and aborted kernels.[43] As with many other diseases, scab increased in importance over time as its domain grew. As an example, it probably did not enter Minnesota until corn became an important crop around 1900. A 1919 epidemic destroyed eighty million bushels of wheat and reduced the quality of many more in Kansas, Nebraska, and the Dakotas. Among the recommended control measures was to keep the fields clean and to plant when the soil was cool, that is to plant winter wheat as late as possible and spring wheat as early as possible.[44] By the early twentieth century, scientists had determined that growing wheat in rotation with corn intensified scab outbreaks in wheat. The spores carried over and multiplied in cornfields, causing considerable damage and subsequently migrated back to the wheat fields. Thus the fight against scab yielded the interesting realization that some crop rotations increased rather than reduced diseases, as was commonly believed.[45]

Insects represented another arrow in the Red Queen's quiver. The Hessian fly was the most destructive of the scores of insects that attack wheat. Its spread reduced yields and led to wholesale changes in the varieties planted and in cultural practices. The conventional wisdom asserts that the Hessian fly entered the United States at Long Island in 1776 in the straw of Hessian mercenaries. From there, it spread into Pennsylvania in 1786, began to cause damage in Saratoga, New York (about two hundred miles from its origin) in 1789, entered Virginia by 1794, swept across the Alleghenies by 1797, hit Ohio by the mid-1820s, Michigan in 1843, Illinois by 1844, Kansas by 1871, and reached the Pacific Coast in 1884. The new scourge, appropriately named *Cecidomyia destructor*, shifted American wheat farmers onto a significantly lower production possibility frontier.[46] In many areas, the appearance of the fly had such devastating effects that it initially induced total abandonment of the crop.

By carefully studying the fly's behavior, farmers learned that the pest might have several broods, and that the most damaging (for winter wheat) was the fall brood whose maggots sucked sap from the young plants. The infested plants were always stunted and often killed or susceptible to lodging.[47] Gradually, farmers also learned that they could reduce the damage by sowing winter wheat late (or for spring wheat, early) and by better cleaning their fields to reduce the carryover of the fly population. Planting late delayed the harvest, increasing the danger from rust, but most farmers were willing to take this risk. As a dramatic example of the extent of the cultural changes, one local account from Connecticut indicates that by 1811 the date of planting had shifted from the third week in August to the end of September or early October. Further south in New Jersey, the sowing date moved from late August to early October.[48] The fly also induced a search for new varieties that had stronger stocks to resist the maggots or that could be sown late. Shortly after the first serious attack in 1779, Long Island farmers adopted "a yellow-bearded, Southern variety of wheat, which seemed to be less affected by the attacks of the fly."[49] Among the notable proponents of the yellow-bearded wheat was George Washington, who along with other luminaries such as James Madison, weighed in with suggestions for responding to the new insect threat.[50] By far the most important biological innovation was the introduction of Mediterranean wheat from Europe in 1819.[51] This variety proved suitable for late planting and gained wide favor by the 1840s and 1850s.

Just when American farmers were learning to live with the Hessian fly, a new scourge appeared. An insect that contemporary observers called the grain midge first entered Vermont from Canada in the 1820s. This one insect had such a profound effect that the 1860 Census of Agriculture devoted more attention to it than to the mechanical reaper. The census traced the midge's gradual spread, describing the horrible damage it wrought. It appeared in Washington County, New York, in 1830 and by 1832 "had so multiplied as to completely destroy the crop in many fields." In 1834 and 1835 the midge moved south into Rensselaer and Saratoga Counties, "devastating the wheatfields." In 1835 and 1836, "over all the territory to which it had extended . . . , it was so extremely destructive that further attempts to cultivate grain were abandoned."[52] The New York State Agricultural Society estimated that in 1854 the midge caused at least $15 million in damage to the state's wheat crop. Still in 1854 the problem had not yet reached its zenith because it was in that year the pest was first reported in the fertile Genesee valley. "In 1856 it destroyed from one-half to two-thirds of the crop on the uplands, and nearly all on the flats. In 1857 it was still worse, taking over two-thirds of the crop." The census reported that the midge also caused problems in Pennsylvania, but that further to the south the insect did little damage "owing, it is thought, to the warmer climate."[53] Between 1849 and 1859, wheat production in New York fell by 44 percent. The census directly attributed most of this decline to the effects of the midge, as "spring crops and winter barley took the place of wheat."[54]

Initially, farmers "knew little of the habits of this minute insect, and were unable to offer it any resistance."[55] Once again they adjusted their cultural practices to survive the midge. At first, there was a widespread shift from winter to spring wheats, which even if successful in avoiding the insect, offered significantly lower yields than the premidge winter varieties. Farmers faced a dilemma because the key to fighting the Hessian fly was to delay planting winter wheat, but the trick with the midge was to harvest as early as possible. All else equal, this required planting earlier. Thus the arrival of the midge further constricted the available options by creating smaller windows in which planting and harvesting had to take place. In New York the sowing date, which had been pushed from August to late September or early October because of the Hessian fly, now had to be recalibrated to the first three weeks of September because of the midge.[56]

Experience with midge infestations showed that "the injury has been almost entirely confined to the high quality 'white' varieties, the Mediterranean escaping altogether."[57] By the 1850s, Mediterranean had become the dominant variety in the United States even though its flour quality and yield (in the absence of insects) were inferior to many abandoned varieties.[58] Although the 1860 census called the midge the "greatest of all pests which has infested the wheat-crop," adjustments in cultural practices, including plowing deep, burning the chaff from infected fields, and rotating crops, soon demoted it to a lesser status.[59]

Meanwhile, the battle against the Hessian fly intensified, as countless farmers and researchers investigated the fly's behavior and tested cultural practices and wheat varieties to limit its damage. Out of necessity farmers adopted so-called fly-safe varieties that allowed for late planting and, gradually, researchers publicized "fly-safe" dates for every nook and cranny that grew wheat. The recommended dates varied by about two months with latitude, longitude, elevation, soil conditions, rainfall, and wheat varieties. As noted above, the planting decision involved a delicate balancing of several threats, but as wheat culture moved onto the Great Plains the problem got even dicier. Planting late to avoid the fly made the crop more susceptible to winterkill and reduced yield potential because the root system had less time to develop. Delaying the harvest exposed the crop to heat, drought, grasshoppers, and other enemies.[60] As a 1923 Kansas report noted, "the proper time of seeding must be determined for each locality by experimental sowings extending over a period of years."[61] Planting decisions had to be fine-tuned depending on the seasonal conditions and the assessment of the fly population in each field.[62] Preventive measures had a collective dimension because the benefits of destroying volunteer wheat and cleaning infected fields of stubble were spread throughout the area.

Despite considerable precautions, there were local fly outbreaks every year and serious regional infestations roughly every five to six years. As examples, in 1900 over one half of wheat acreage in Ohio and Indiana was abandoned due to fly damage and yields on the harvested land fell by about 60 percent. The following year the fly destroyed over half of New York's wheat crop. Kansas experienced six serious outbreaks between 1884 and 1913 with losses peaking at about 27 percent of the crop.[63] Damage tended to be more serious with unseasonably warm falls, in wet years, and in years with large volunteer crops. Nationally, estimates of annual Hessian fly losses around 1900 hover at 10 percent of the wheat crop.[64] In 1938, USDA entomologist J. A. Hyslop noted the "general adoption, throughout the greater part of the regions infested by the hessian fly, of the practice of planting wheat after the fly-free date has materially reduced" the losses from 6.0 percent of the crop over the 1923–27 period to about 2.2 percent over the 1928–35 period.[65]

What if the conventional wisdom proclaiming a dearth of biological innovations is correct, and farmers in fact made no changes to combat the fly? Numerous accounts from the late eighteenth and early nineteenth centuries tell us that farmers who did not adjust simply lost their crops.[66] For later years, experiment station investigations repeatedly show that moving the planting date a week or two earlier typically led to heavy losses. One Kansas study is particularly noteworthy because it was based on the experiences of a large number of real farms. It showed a close correlation between regional fly losses and the proportion of the wheat sown before the fly-free date.[67] Another Kansas study reported what happened in the absence of normal precautions such as planting early and destroying volunteer wheat. In a controlled test, the wheat on the improperly managed field was nearly destroyed and only produced about one-fifth the yield of the field following standard guidelines.[68] Studies conducted in numerous other states also found that in most seasons early-sown wheat suffered moderate to heavy damage, while wheat sown later escaped fly infestation. As an example, a study conducted at eight locations over eight years in Illinois showed that on average wheat sown after the fly-safe date yielded 29 percent more than wheat sown before the date.[69]

More recent studies by modern agronomists show similar results. As an example, in 1981 when researchers took no precautions on test plots near Colfax, Washington, the entire crop was destroyed.[70] To gain perspective, we asked three senior agronomists who specialized in wheat culture what would have happened, given the conditions prevailing in the early twentieth century, if farmers had not followed the normal precautions. Their collective response was "those farmers would not have had a wheat crop worth harvesting."[71] These findings lend credence to Dondlinger's 1908 assessment that "preventive measures reduce the annual loss from the Hessian fly by an amount estimated from $100,000,000 to $200,000,000 for the wheat crop alone."[72] In 1909 the value of the U.S. wheat crop was approximately $674 million, so by this crude accounting, biological investments to control just one pest increased national wheat yields relative to what they would have been by between 18 and 42 percent.[73]

As wheat culture expanded several other pests, including chinch bugs, grasshoppers, and greenbugs, became growing concerns. For the most part these insects posed only minor problems as of 1839. Chinch bugs were first noticed sucking the sap of wheat plants in Orange County, North Carolina, around 1783. By 1790, it had spread through North Carolina and Virginia, causing substantial damage to wheat and corn. Over the next half century, the chinch bug spread from the south Atlantic states into the vast grain fields of the Midwest. There were notable outbreaks in the Carolinas and Virginia in 1839, in Illinois in 1844–45 and 1854–58, Iowa in 1847, and Indiana in 1848 and 1854. During the major infestations of 1863–65, the bug wiped out three-fourths of the wheat crop and one-half of the corn crop in Illinois and the surrounding states. Serious regional outbreaks occurred in 1868, 1871, 1874, 1887, and 1892–97. Taking 1887 as an example, the insect reduced the wheat crop by an estimated 50 percent in Iowa, 20 percent in Minnesota, 19 percent in Kansas, and 13 percent in Illinois. According to Horton and Satterthwait, over the entire 1850–1915 period, the "average annual losses sustained by the most heavily infested States" were about 5 percent of the wheat crop. Nationally, the reduction in yield averaged about 2.5 percent. But these losses were in spite of control measures that included destroying the bug's hibernating places in bunch grasses, leaves and litter, protecting fields with barrier strips, and spraying infected areas in the late spring.[74]

Grasshoppers represented another growing problem. Recurring plagues were all too familiar during the formative stage of agriculture on the Great Plains, contributing to widespread farm abandonment in the Southern Plains in the 1880s and 1890s. Because grasshoppers thrived in drier climates, their damage significantly increased in severity as the wheat frontier moved west. Severe grasshopper outbreaks occurred at least every decade, with each lasting from one to six years. Annual losses to all crops in the decade 1925–34 averaged about $25 million, but crude estimates for the nineteenth century suggest even higher losses. Given the location of the attacks the destruction would have been concentrated in wheat. Grasshoppers were not effectively controlled until the 1940s, but starting in 1885 a series of increasingly effective chemical baits became available that at least dampened grasshopper damage. In addition, settlement began to encroach on their breeding grounds.[75]

Greenbugs, or spring grain aphids, which first gained notice in Virginia in 1882, were still another member of the parade of new pests. These European immigrants were considered unimportant before the 1890s, but their proclivity for sucking the sap of wheat plants and for spreading Barley Yellow Dwarf virus had increasingly devastating effects after the turn of the twentieth century. An outbreak in 1907 destroyed fifty million bushels of wheat and oats in the Southern Plains and led Texas wheat-growers to abandon as much as 70 percent of their acreage. Smaller outbreaks occurred in 1904, 1911, and 1916. Over this period, insecticides proved ineffectual in controlling the aphid, although natural enemies such as a wasp-like parasite did provide a check under the right climatic conditions. The recommended control measure was cultural and collective in nature: destroying volunteer grain crops during the summer and early fall to prevent survival between the harvest and next year's planting. By the mid-twentieth century, this relative newcomer had replaced the Hessian fly as the number one insect pest affecting wheat.[76]

The Red Queen had yet another arrow in her quiver, because during the period under investigation there was a serious deterioration in the weed environment in part due to new introductions from other parts of the world. Referring to the Northern Great Plains, Salmon and his colleagues assert, "weeds were not an important factor on the new lands until near the end of the century," and for California they note that "previous to 1900 any improvements in per acre yield resulting from a choice of better varieties and from the increasing use of fallow probably were more than offset by the increase in weeds."[77] Along with bindweed and wild oats, among the most damaging was Russian thistle, a tumble weed that entered the United States in the mid-1870s. The "best authorities" place and date the thistle's introduction to Scotland, South Dakota, around 1873. The weed spread to Iowa, Nebraska, and North Dakota by 1888, to Minnesota, Wisconsin, Illinois, and Indiana by 1890–91, and Kansas, Montana, and Idaho by 1894. Adapting to the times, the thistle hitchhiked rides on the railroad, reaching as far east as New York and as far west as California by the mid-1890s. Where it became established, the weed caused crop losses estimated between 15 and 20 percent. An Illinois observer noted: "No other weed has caused such widespread discussion, or been the subject of such great fear." In the 1890s numerous states and the USDA initiated programs to destroy the weeds. We have a natural experiment that suggests what might have happened without control measures. In Russia, with no similar collective efforts, "the cultivation of crops has been abandoned over large areas."[78] By the early twentieth century, USDA experts estimated that weeds reduced the yield of spring wheat by 12–15 percent and of winter wheat by 5–8 percent.[79]

This discussion has touched on only a handful, albeit some of the most important, of the hundreds of insects, diseases, and weeds in the Red Queen's arsenal in her war on wheat.[80] To fully comprehend this issue historically, it is worth noting that the great advances in modern science since the 1940s have not entirely freed the grain farmers of today from such threats. In 1986 a new insect threat, the Russian wheat aphis, entered the United States from Mexico and by 1988 had spread north through the Western Grain Belt to the Canadian border. Its appetite for wheat caused losses totaling $891 million over the 1987–93 period.[81] In recent years, fusarium head blight (or scab) has reemerged as the major yield-sapping disease in the Northern Plains. Over the 1990s, scab has destroyed over $2.5 billion in wheat (over 500 million bushels) in the United States and at least $520 million in Canada. What had been a "minor problem" has placed North America's "breadbasket under siege."[82]

Our cursory discussion of wheat pests sheds light on issues that have long interested historians. The boll weevil's march across the South dramatically altered the fortunes of American cotton farmers, leading to widespread failure and forcing wholesale changes in cultural practices and cotton varieties. But for all the attention given to the boll weevil, only in its worst years did it depress cotton yields by as much as the Hessian fly affected wheat yields in a normal year. The comparison here is with the weevil in the first decades of the twentieth century and the fly before 1910. According to Osband, yield losses to the weevil reached a maximum of 11 percent in 1932.[83]

A better understanding of the pest environment in northern agriculture also bears on interpreting the causes on international yield differences. Numerous observers have noted that wheat farmers in many parts of Europe achieved significantly higher yields than their American counterparts, and attributed the difference to the more labor-intensive methods and the greater attention given to maintaining land fertility in labor-abundant Europe. This explanation probably has considerable merit, but it is incomplete because it fails to take into account the dramatically different threats that American farmers faced. C. L. Marlatt was one of the USDA's foremost entomologists at the turn of the century. He noted that "our system of growing the same grain crops over vast areas year after year furnishes at once the very best conditions for the multiplication of the insect enemies of such crops. In addition to this is the fact that America, with its long, hot summers, presents the most favorable conditions for the multiplication of most insects. These two reasons undoubtedly account for the far greater losses experienced in this country as compared with Europe, the summers of which are very cool and short."[84] Marlatt's observation about insects also applied to many of the diseases that were most destructive to wheat.

Turn-of-the-century observers clearly identified the collective action problem facing individual farmers. An individual who decided to adopt rotation schemes along the lines practiced in parts of Europe might reap little benefit because the insects and diseases would simply migrate from nearby fields. There were also important dynamic implications to the differing pest and disease environments. All else equal, wheat grown on land with a high nitrogen content—prime bottomland, land that had been fertilized, or land that had been left fallow—took considerably longer to mature than wheat grown on lower quality land or on land that had been cropped in wheat the previous year. This is why the census of 1860 (as cited above) noted that the midge caused more destruction on the "flats" than on the "uplands" in the Genesee valley. The problem was not limited to New York. On the Canadian Prairie Red Fife ripened about eight to ten days earlier on stubble land than on fallowed land.[85] Thus wheat farmers everywhere faced a conundrum. If they took obvious steps to increase yields, they greatly increased the risk that rust and insects might wipe out the crop. In the spring wheat belt farmers also increased the risk of losing their crop to early frosts. On the Great Plains farmers discovered that if land was plowed and then harrowed to keep down the weeds and left fallow it would produce an excellent wheat crop even in dry years when land that had been planted continuously yielded almost nothing. But to use this technique delayed the harvest about ten days. One of the contributions of earlier-ripening wheats like Marquis was that it allowed farmers in the Northern Plains to employ this dry farming technique with far greater assurance that they would beat the late summer or fall frosts. Marquis was one of a long list of improved wheat varieties that facilitated western settlement.

The Introduction of New Wheat Varieties

> *The greatest service which can be rendered any country is to add a useful plant to its culture; especially, a bread grain.* —Thomas Jefferson, 1821[86]

In addition to helping hold wheat pests at bay, biological innovations were crucial for the expansion of the wheat frontier. An understanding of the westward and northward movement of the wheat belt strikes at the logic underlying the received wisdom that biological changes were of little consequence. As wheat farmers moved onto new lands they invariably confronted different and often harsher environments. The stylized facts that rest on the observation that yields per acre changed little over a century simply ignore the changing composition and quality of the geo-climatic conditions that framers confronted. Leading agronomists have long understood this problem. As an example, Salmon noted that "yields per acre are often used to measure or indicate technological improvements. They are reasonably good indices in countries in which acreage remains fairly constant or where the productivity of the new acreage does not materially differ from the old. They may be misleading, however, in a country such as the United States, where the acreage has greatly increased in areas where the conditions for growth are quite different. If an improvement reduces cost per acre, thereby permitting a larger expansion on less production land, average over-all acre yields may actually be reduced."[87]

Over the 1839–1909 period, U.S. wheat production increased almost eightfold, rising from roughly 85 million to 640 million bushels. The rapid growth in output was crucially dependent on the western expansion of cultivation. These geographic shifts are illustrated in Figure 1, which maps the distribution of U.S. wheat output in 1839 and 1909, and in Table 1, which shows the changing geographic center (mean and median) of production over the same period.[88] In 1839, the center was located east of Wheeling, (West) Virginia. Cultivation was concentrated in Ohio and upstate New York; relatively little was grown as far west as Illinois. By 1909, the center of production had moved roughly nine hundred miles west to the Iowa/Nebraska borderlands. The core areas of the modern wheat belt had emerged in an area stretching from Oklahoma and Kansas in the south to the Dakotas in the north (as well as the Canadian Prairies). Another important concentration appeared in the Inland Empire of the Pacific Northwest. The western shift was so overwhelming that "new areas" account for 64 percent of 1909 output and 74 percent of the growth since 1839. More generally, the area west of the Appalachian Mountains, which had comprised less than one-half of output in 1839, comprised 92 percent of output by 1909.

Table 2.1 The Changing Geographic Center of Wheat Production, 1839–1919

	Latitude			Longitude			Approximate	Miles of Movement		
	deg	min	sec	deg	min	sec				
Mean Location										
1839	39	43	43	80	56	0	27 miles SE of Wheeling, WV	1839 – 1849	65
1849	40	14	18	81	58	49	56 miles NE of Columbus, OH	1849 – 1859	214
1859	39	59	59	86	1	38	17 miles NE of Indianapolis, IN	1859 – 1869	153
1869	40	39	17	88	48	40	75 miles NE of Springfield, IL	1869 – 1879	89
1879	40	36	14	90	30	46	75 miles NW of Springfield, IL	1879 – 1889	157
1889	39	33	53	93	9	18	141 miles SE of Des Moines, IA	1889 – 1899	173
1899	41	39	19	94	59	23	55 miles NE of Omaha, NB	1899 – 1909	76
1909	42	24	26	96	4	49	77 miles NW of Omaha, NB	1909 – 1919	126
1919	40	36	20	95	42	39	50 miles SE of Omaha, NB	Total 1839–1909	921

Median Location									
1839	40	0	0	80	40	0	38 miles E of Wheeling, WV	1839 – 1849	34
1849	40	23	59	81	2	22	107 miles NE of Columbus, OH	1849 – 1859	238
1859	40	3	19	85	32	15	38 miles NE of Indianapolis, IN	1859 – 1869	134
1869	40	36	56	87	58	36	105 miles NE of Springfield, IL	1869 – 1879	42
1879	40	20	15	88	41	11	63 miles NE of Springfield, IL	1879 – 1889	160
1889	40	18	15	91	43	9	132 miles SW of Des Moines, IA	1889 – 1899	212
1899	41	10	51	95	36	36	19 miles SE of Omaha, NB	1899 – 1909	78
1909	41	16	0	97	7	0	182 miles SW of Omaha, NB	Total 1839–1909	988

Note: The Total Miles of Movement between 1839 and 1909 is based on the starting and ending locations and, due to variations in the direction of movement, is less than the sum of decadal Miles of Movement

Source: The 1839 data are from Lee A. Craig, Michael R. Haines, and Thomas Weiss. "U.S. and Censuses of Agriculture, by County, 1840–1880." [Computer file] Unpublished files graciously provided by the authors, Dept. of Economics, North 2000. The 1909 data are Thirteenth Census of the United States Taken in the Year 1910 Vol. 6–7. Agriculture: 1909 and 1910, Reports by States, With Washington, D.C.: GPO, 1913. The county-seat location data are from Robert P. Sechrist, Basic geographic and historic data for interfacing IC PSR data sets, 1620–1983) [United States] [Computer file]. ICPSR version. Baton Rouge, L.A.: Robert P. Sechrist, Louisiana State University [producer], 1984. Ann Arbor, MI: Inter-university Consortium for Political and Social Research [distributor], 2000. All other years are reported in U.S. Bureau of the Census. Statistical Atlas of the United States. Washington, D.C.: GPO, 1925, p. 22.

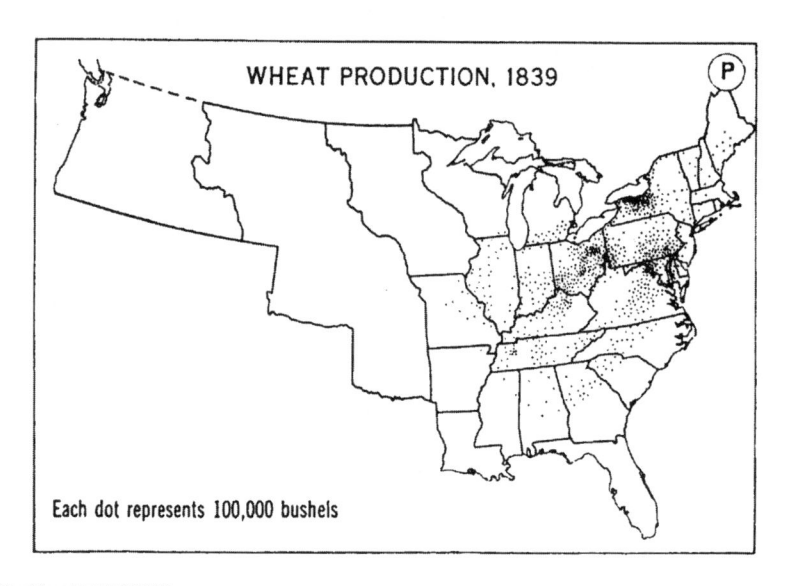

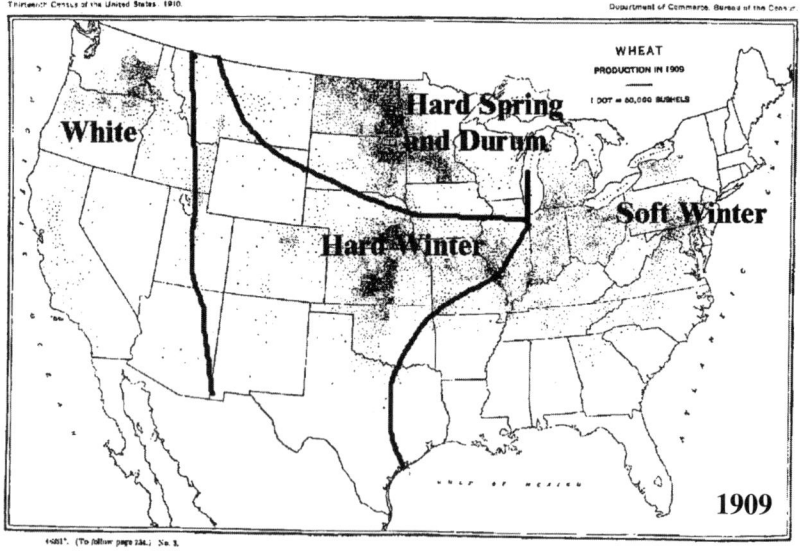

Fig. 2.1 Wheat Production in the United States, 1839 and 1909.
Sources: Compiled from Charles O. Paullin, *Atlas of the Historical Geography of the United States* (Washington, D.C.: Carnegie Institution 1932), Plate 143P; U.S. Bureau of the Census. Thirteen Census of the United States Taken in the Year 1910, vol. 5. Agriculture (Washington, D.C.: GPO, 1913), Plate no. 3.

Table 2.2 Weather Indicators in Old and New Regions

	Precipitation (inches)	Mean low temperature (degrees F)	Mean high temperature (degrees F)	Frost free days
Wooster, OH	36.2	39.1	58.5	155
Dickinson, ND	16.2	27.7	53.9	120
Ft. Hays, KS	22.8	40.1	67.3	170

Sources: Charles W. Collins, *Ohio, an Atlas* (Madison, WI: American Printing & Publishing, Inc., 1975) pp. 26, 34; Lowell R. Goodman and R. Jerry Eidem, *The Atlas of North Dakota*. (Fargo: North Dakota Studies, Inc., 1976) p. 18; Homer E. Socolofsky and Huber Self, *Historical Atlas of Kansas* (Norman: University of Oklahoma Press, 1972) p. 4; Midwestern Regional Climate Center, "Historical Climate Summaries for Wooster_Exp_Stn, OH." [Online], Midwest Climate Watch home page, 26 January 2000. <http://mcc.sws.uiuc.edu/Summary/Html/339312.html> (21 February 2001); "Dickinson Exp Stn, ND (322188) Period of Record Monthly Climate Summary." [Online], High Plains Regional Climate Center home page, 5 January 2001. <http://www.hprcc.unl.edu/cgi-bin/cli_perl_lib/cliRECtM.pl?nd2188> (21 February 2001); "Hays 1 S, KS (143527) Period of Record Monthly Climate Summary." [Online], High Plains Regional Climate Center home page, 5 January 2001. <http://www.hprcc.unl.edu/cgi-bin/cli_perl_lib/cliRECtM.pl?nd2188> (21 February 2001). no. 4 (1994): 864–883; "Weather Extremes—Hays, K.S." [Online], Agricultural Research Center-Hays home page, 9 February 2001. <http://www.oznet.ksu.edu/wkarc/Arch/W_Extremes.htm> (21 February 2001). "Climatological Summary for Hays, KS." [Online], Agricultural Research Center-Hays home page, 9 February 2001. <http://www.oznet.ksu.edu/wkarc/Arch/W_History.htm> (21 February 2001).

Figure 2.1, which also shows different types of wheat grown in the four major wheat regions of the United States, illustrates the significance of this shift in the locus of production. As wheat culture moved onto the Northern Prairies, Great Plains, and Pacific Coast, it began to confront climatic conditions far different from those prevailing in the East.[89] These new regions possessed such different geo-climatic conditions that according to the USDA's leading wheat specialist "they are as different from each other as though they lay in different continents."[90]

Climate data support this assertion. Table 2.2 shows the average precipitation, the mean average high and low temperatures, and the length of the frost-free growing season at three agricultural experiment stations. These are relatively coarse indicators of the climatic conditions relevant for wheat production, but they serve to emphasize the substantial regional differences.[91] Annual data indicate that the driest year in the past one hundred years at the Wooster experiment station in central Ohio was wetter than the average years at the stations in Ft. Hays, Kansas, and Dickinson, North Dakota. Furthermore, the coldest year on record in Ohio was warmer than the average year in North Dakota. As a result, the pioneers suffered repeated crop failures when they attempted to grow the standard eastern varieties under the normal conditions of the plains, except in protected river valleys.[92]

The successful spread of the crop across the vast tracts extending from the Texas panhandle through Kansas to the Dakotas and Canadian Prairies was dependent on the introduction of hard red winter and hard red spring wheats, which were entirely new to North America. Over the late-nineteenth century, the premier hard spring wheat cultivated in North America was Red Fife (which appears identical to a variety known as Galician in Europe). According to the most widely accepted account, David Fife of Otonabee, Ontario, selected the grain-stock from a single wheat plant grown on his farm in 1842. The original seed was included in a sample that Fife received from a Scottish source from a cargo of winter wheat shipped from Danzig to Glasgow. It was not introduced into the United States until the mid-1850s.[93] Red Fife is recognized as the first true hard spring wheat in North America and became the basis for the spread of the wheat frontier into Wisconsin, Minnesota, the Dakotas, and Canada. It also provided much of the germ stock for later wheat innovations, including Marquis. At the time of the first reliable survey of wheat varieties in 1919, North Dakota, South Dakota, and Minnesota grew hard red spring and durum wheats to the virtual exclusion of all other variety classes.

Another notable breakthrough was the introduction of "Turkey" wheat, a hard red winter variety suited to Kansas, Nebraska, Oklahoma, and the surrounding region. The standard account credits German Mennonites, migrating to the region from southern Russia, with the introduction of this strain in 1873.[94] Malin's careful treatment describes the long process of adaptation and experimentation, with the new varieties gaining widespread acceptance only in the 1890s. In 1919, Turkey type wheat made up about "83 percent of the wheat acreage in Nebraska, 82 percent in Kansas, 67 percent in Colorado, 69 percent in Oklahoma, and 34 percent in Texas. It . . . made up 30 percent of total wheat acreage and 99 percent of the hard winter wheat acreage in the U.S."[95] A similar story holds for the Pacific Coast: the main varieties grown in California and the Pacific Northwest differed in nature and origin (Chile and Australia) from those cultivated in the humid East in 1839.

Wheat cultivation in the East was also in a constant state of flux, with many varieties being tried and abandoned, and others taking root where they proved better suited to evolving local conditions. The most notable change in the East in the mid-nineteenth century was the replacement of soft white varieties by soft reds. Leading this transition was Mediterranean, a late-sown variety introduced from Europe in 1819, which gained wide favor (for reasons described above) during in the 1840s and 1850s. The field of competing varieties was large and ever changing. Danhof notes that around 1840 a survey listed forty-one varieties being grown in New York State, "of which, nine winter wheats and nine spring wheats were most important."[96] In 1857, the Ohio State Board of Agriculture cataloged 111 varieties (96 winter, 15 spring) grown locally in recent years, detailing the time of ripening, performance in different soils and climates, flour quality, and resistance to enemies. Of the eighty-six varieties that we could date, 28 percent had been introduced into Ohio within the previous five years.[97]

This evidence suggests that current rapid turnover in wheat varieties, which many contemporaries view as a product of modern science, has nineteenth-century antecedents.[98] In the past as today, new wheat varieties could be secured by (1) introduction from other regions; (2) selection of naturally occurring mutations and crosses; and (3) deliberate hybridization. The balance across methods has shifted in modern times, but it is important to recall that the commercial spread of hybrid wheat began before 1870.[99]

Since the days of Washington and Jefferson, the U.S. government has been active in the search for new wheat varieties. The 1854 Commissioner of Patents report noted that "a considerable share of the money appropriated by Congress for Agricultural purposes has been devoted to the procurement and distribution of seeds, roots, and cuttings."[100] The report described fourteen varieties of wheat recently imported from nine different countries. In 1866 the newly formed Department of Agriculture tested 122 varieties (55 winter and 67 spring) including "nine from Glasgow, eight from the Royal Agricultural Exhibition at Vienna . . . several varieties from Germany," and a number from the Mediterranean and Black Seas.[101]

Private breeders were also at work. In 1862, Abraham Fultz of Mifflin County, Pennsylvania, found three spikes of bald wheat in a field of Lancaster wheat, a variant of Mediterranean. The selected seed proved hardy, ripened relatively early, and produced semi-hard, red grains of good quality. "Fultz" was so advantageous that it spread quickly, with the USDA distributing the seed by 1871. Even before this date, Garrett Clawson of Seneca, New York, had selected several superior heads from a field of Fultz that yielded a good white wheat—White Clawson or Goldcoin (1865). By 1886 S. M. Schindel, a seedman in Hagerstown, Maryland, hybridized Fultz and Lancaster to produce Fulcaster, which was "considerably resistant to rust and drought." It soon competed with Fultz as the most popular soft red winter wheat. Other hybrids, such as Diehl-Mediterranean (1884) and Fultzo-Mediterranean (1898), also gained favor in the East.[102] In the Northern Plains, breeders such as J. B. Power, L. H. Haynes, and W. M. Hays were active, producing improved varieties during this period.

As a rule breeders and farmers were looking for varieties that increased yields, were more resistant to lodging and plant enemies, and as the wheat belt pushed westward and northward, varieties that were more tolerant of heat and drought and less subject to winterkill.[103] The general progression in varieties allowed the North American wheat belt to push hundreds of miles northward and westward, and significantly reduced the risks everywhere. One of the most important of the early-twentieth-century innovations was Marquis, which was bred in Canada by Charles Saunders who crossed Red Fife with Red Calcutta. According to Tony Ward's analysis of Canadian experiment station data, changes in cultural methods and varieties shortened the ripening period by twelve days between 1885 and 1910. Given the region's harsh and variable climate, this was often the difference between success and failure. Kenneth Norrie's work also emphasizes the key contribution of these biological developments to the settlement of the Canadian prairies between 1870 and 1911.[104]

The introduction of Marquis and various durum varieties to the United States illustrates the rapid spread of new varieties in the early twentieth century. The USDA introduced and tested Marquis seed in 1912–13. By 1916, Marquis was the leading variety in the Northern Grain Belt.[105] This was not an isolated case. As a result of extensive exploratory campaigns on the Russian Plains, USDA cerealist Mark Alfred Carleton introduced Kubanka and several other durum varieties in 1900.[106] These varieties proved to be hardy spring wheats and, at the time, relatively rust resistant. By 1903 durum production, which was concentrated in Minnesota and the Dakotas, approached seven million bushels. In 1904, the region's Fife and Bluestem crops succumbed to a rust epidemic with an estimated loss of 25–40 million bushels, but the durum crop was unaffected. By 1906, durum production soared to fifty million bushels.[107] The Northern Great Plains witnessed a wholesale transformation of its wheat stocks in the late 1910s. Overall, the production share of the traditional varieties such as Velvet Chaff, Bluestem, and Fife fell from 84 percent in 1914 to under 13 percent by 1921 as the new Marquis and Durum varieties took hold. These rates of diffusion are comparable with those publicized by Griliches for the spread of hybrid corn in the Midwest during the 1930s.

The national turnover of varieties is evident in USDA surveys of wheat distribution, first systematically collected in 1919 and reported thereafter roughly every five years until 1984. Using the 1919 survey together with information on the date of introduction/release of specific varieties, we can gain a clearer picture of the changing U.S. wheat germstock.[108] In that year, roughly 24.2 percent of U.S. wheat acreage was in hard red spring wheat, 6.4 percent in durum, 32.0 percent in hard red winter, 30.1 percent in soft red winter, and 7.1 percent in white. It is important to recall that in 1839 there was essentially no commercial production of durum or the hard reds, which comprised 62.8 percent of the 1919 total. Table 2.3 provides further evidence of the age distribution of wheat in 1919. Of the 133 varieties that could be dated, the acreage-weighted mean "vintage" was 1881, or less than forty years old. The median was 1873, which corresponded to the introduction of Turkey. This is not surprising given that Turkey was the largest single type, making up almost 30 percent of total acreage. Note that even the soft red winter varieties experienced significant turnover. Their mean "vintage" was 1868. And of the top four soft red winter

Table 2.3 Vintage of U.S. Wheat Varieties in 1919

Decade of introduction	Percent of acreage	Decade	Percent
Before 1800	0.2	1860–69	6.7
1800–09	0.2	1870–79	31.6
1810–19	3.6	1880–89	9.7
1820–29	0.7	1890–99	8.7
1830–39	1.7	1900–09	9.8
1840–49	1.2	1910–19	17.0
1850–59	2.0	Unknown	6.9

Sources: Clark et al., "Classification"; "GrainGenes," [Online database].

wheats in 1919—Fultz, Fulcaster, Mediterranean, and Poole—only Mediterranean was introduced before 1839.[109] The key results are that in 1919, well before the usual dating of the onset of the biological revolution, roughly 80 percent of U.S. wheat acreage consisted of varieties that did not exist in North America before 1873, and less than 8 percent was planted in varieties dating earlier than 1840.

Farmers in the Great Plains, mountain states, and Pacific Coast regions showed a strong revealed preference for varieties different from those grown in the wheat belt of 1839. But were the advantages of the new wheats large or small? On this issue we have some evidence, albeit fragmentary. The controlled settings of the experiment station variety trials provide perhaps the best information. For example, from the late 1880s on, the stations in Minnesota and North Dakota cooperated to test hundreds of spring wheat varieties in the Northern Plains. Because the agronomists rapidly dropped unsuccessful varieties after one to three years, the eastern stocks rarely even appeared in these trials. During the 1892–94 period, they did include China Tea, an early maturing soft spring wheat, in their Red River valley test plots. China Tea's average yields were about 88 percent of the leading Fife and Bluestem varieties. But this result is incomplete because of China Tea's extremely low quality. It was consistently classed a "reject," suitable only for animal feed and subject to almost 50 percent price discounts. The 1892–93 Fargo trials also included Lost Nation, a soft spring wheat popular in the 1870s and 1880s. Its yields were only 80 percent of Red Fife, and it was considered less reliable.[110] In addition, Lost Nation's quality was well below the Fife's, resulting in a roughly 10 percent price discount. The disadvantages of the soft wheats would have been much greater in the drier and colder wheat-growing areas west of the Red River valley.

These experimental results left the Minnesota officials a "little disappointed" because they would "heartily welcome" a soft spring variety that generated sufficiently high yields. To provide perspective, these officials estimated that soft wheats of standard grade would have to outyield their "famous" hard wheats by five bushels per acre to overcome the quality differential.[111] The North Dakota officials left no doubt where matters stood in their state:

> North Dakota is beyond the northern limit of winter wheat. . . . Not only has the growing of the wheat been limited to spring varieties but almost exclusively to those kinds hardest in berry and strongest in gluten. Even among these the limitations has included little else than Fife. . . .
> Fife . . . is the great wheat of this region. . . . Averaging all conditions, this wheat is a superior yielder (and) . . . is as hardy as any variety of spring wheat . . . the value of this wheat can hardy be overstated.[112]

These results help explain why by the early twentieth century effectively all of the wheat grown in Minnesota and the Dakotas consisted of durum or hard spring wheat varieties. Moreover, the contrasts between China Tea or Lost Nation with Red Fife, as large as they are, significantly understate the extent of technological change because by 1909 Red Fife had been largely replaced by yet superior varieties, including various durum wheats, Bluestem, and Preston. Production data for the 1914–21 period indicate that (consistent with earlier experiment station results) the durum yields were roughly one-third (32 percent) higher than Fife and the newer hard spring wheats outyielded Fife by about 16 percent.[113]

Early settlers in Kansas experimented with scores of soft winter varieties common to the eastern states.[114] According to the Kansas State Board of Agriculture, "as long as farming was confined to eastern Kansas these [soft] varieties did fairly well, but when settlement moved westward it was found they would not survive the cold winters and hot, dry summers of the plains."[115] The evidence on winterkill, that is, wheat losses due to cold, lends credence to this view. Data for four east-central counties for 1885–90 show that over 42 percent of the planted acres were abandoned. For the decade 1911–20, after the adoption of hard winter wheat, the winterkill rate in these counties averaged about 20 percent.[116]

Drawing on decades of research, Salmon and his colleagues noted that for Kansas "the soft winter varieties then grown yielded no more than two-thirds as much, and the spring wheat no more than one-third or one-half as much, as the TURKEY wheat grown somewhat later."[117] In 1920, Salmon concluded that without these new varieties, "the wheat crop of Kansas today would be no more than half what it is, and the farmers of Nebraska, Montana and Iowa would have no choice but to grow spring wheat," which offered much lower yields.[118]

By the eve of World War I, Nebraska had emerged as the nation's fourth-leading wheat producer. Its farmers experienced many of the same challenges as growers in Kansas.

In Nebraska spring wheat predominated until after 1900, and winterkilling of the soft winter wheat was even more severe than in Kansas. Some measure of the benefit derived from the general culture of TURKEY wheat in Nebraska after 1900 is afforded by comparing its average yield with that of spring wheat at the North Platte Station in western Nebraska. During the twenty-eight-year period ending in 1939, as reported by Quisenberry et al. (1940), winter wheat yielded on the average 20.6 bushels as compared with 14.3 for spring wheat, a gain of more than 44 per cent. At Lincoln, in eastern Nebraska, the corresponding gain for a this 31-year period is 14.2 bushels, or 96 per cent.[119]

The movement in statewide yields bolsters this evidence. Yields had averaged about 12.5 bushels per acre for 1870–1900, but jumped by about 40 percent to 17.5 bushels in 1900–1909. At the time scientists attributed the vast majority of this increase to the substitution of Turkey Red for spring wheats.[120]

Clark and Martin's analysis of field tests conducted across the Great Plains and in the Pacific Northwest between 1906 and the early 1920s offers further evidence that hard winter wheat outperformed soft winter varieties in yield, days to maturity, and survival rates.[121] Their summary finding was that "hard red winter wheat is now the principal crop in many sections of limited rainfall, including much of Kansas and Nebraska, Western Oklahoma, Northeastern Colorado, Central Montana, and the drier portions of the Columbia Basin of Oregon and Washington. In these areas farming was not practiced or was exceedingly hazardous before this class of wheat was grown."[122]

Wherever it is feasible, it is preferable for farmers to grow winter wheat instead of spring wheat. Winter wheat generally offers higher yields and is much less subject to damage from insects and diseases. The problem is that in colder climates winter wheat suffers high losses to winterkill. The agronomy literature commonly recognizes that the development of more hearty winter varieties that could be grown in harsher climates was a great achievement, but just how much land was affected by this fundamental change in farming practices? County-level data from the Census of Agriculture helps to map the northern shift in the "spring wheat"–"winter wheat" frontier in the Plains and Prairie states. Data breaking down wheat production by spring and winter varieties are available in the published census for 1869 and 1929.[123] Table 2.4 reports estimates (derived from regression analysis) for each degree of longitude between 87° and 105° of the latitude where spring wheat output equaled winter wheat output in 1869 and 1929.[124] The results are striking; in both years, except in isolated pockets, spring wheat output exceeded winter wheat output north of the estimated frontier, and spring wheat output fell below winter wheat output south of the frontier. In most places the break was sharp with a narrow transition zone; farmers grew little winter wheat just thirty miles above the demarcation line, and they grew little spring wheat thirty miles below the line. In 1869, the frontier generally followed the fortieth parallel for longitudes between 87° and 94°. Then it swept down to the southwest across eastern Kansas. (Given the prevailing limits of wheat culti-

Table 2.4 The Northern Shift of the Spring Wheat-Winter Wheat Frontier, 1869–1929

	Latitude of equal output (measured in decimals)	
Longitude	1869	1929
87	42.63	43.39
88	40.98	43.33
89	40.55	43.38
90	40.86	43.63
91	40.86	43.62
92	40.53	43.57
93	40.53	43.74
94	39.71	44.25
95	38.27	42.96
96	33.57	43.03
97	33.64	43.07
98		42.69
99		43.38
100		43.05
101		42.64
102		41.37
103		41.72
104		40.81
105		36.19

Note: North of the frontier for any given longitude, spring wheat output exceeds winter wheat output except in isolated pockets. Sources: The 1869 data were provided by Craig, Haines, and Weiss. "Development, Health, Nutrition." The 1929 data are from U.S. Bureau of the Census. Fifteenth Census of the United States: 1930. Agriculture. Vol. II, Reports by States, With Statistics for Counties and a Summary for the United States. Pt. 1, The Northern States, Pt. 2, The Southern States, Pt. 3, The Western States (Washington, D.C.: GPO, 1932).

vation, the frontier cannot be mapped for higher longitudes in 1869.) By 1929, the spring wheat frontier had shifted dramatically to the north and west. In that year, the frontier followed roughly the forty-third parallel between 87° and 100° and then took a southwest course. Thus, over this sixty-year period, the frontier crept northward across most of Kansas and Iowa, as well as southern Nebraska. Collectively the region between the 1869 and 1929 "spring wheat"–"winter wheat" frontiers accounted for almost 30 percent of U.S. wheat output in 1929!

An examination of the spread of wheat culture in the Pacific states supports this general view of the crucial importance of varietal adaptation. By the end of the nineteenth century the Inland Empire, comprising parts of Idaho, eastern Washington, and Oregon, had emerged as a major wheat producer. In 1909, combined production in these regions rivaled that of Minnesota. Although in national accounts this region is often treated as a distinct wheat growing area, it was far from a homogeneous entity. To the contrary, large differences in rainfall, soils, and the relatively rapid changes in elevation (and thus temperature and growing season) contributed to intraregional differences that parallel the contrasts separating the drier Great Plains from the more humid East. Early settlers experimented with wheat culture in the coastal regions, with the first recorded wheat harvest in 1825 at Fort Vancouver. Circa 1839, Club wheats were the dominant varieties, but these were generally not suited for the harsher climates in the semiarid zones to the east where, in the view of one of the region's leading agronomists, "wheat growing was formerly considered impossible."[125] Gradually, farmers accomplished the impossible by identifying a number of soft spring wheat varieties appropriate for local conditions. By the mid-1890s, a survey of the types of wheat grown in Washington showed a varied subregional pattern with the variety of choice closely matched to a given area's expected rainfall. In areas with more than twenty inches, Little Club dominated; in areas with about eighteen inches of rainfall Red Chaff was the clear favorite; and areas with less than seventeen inches specialized in growing Pacific Bluestem (this was an entirely different variety from the Bluestem later grown in the Great Plains). This pattern reflected the proven superiority of each variety to the local microclimates.

Cultural systems also evolved to match specific local conditions. Farmers in selected areas developed a cultural system similar to that found in California by planting spring wheats in the fall when the weather permitted. This increased yields in normal years, but often had disastrous results when cold weather hit, creating an urgent demand for more hardy winter varieties. Turkey strains made significant inroads around 1900 because they offered enormous yields "when the season was favorable."[126] In addition, a number of the newer spring wheats such as Marquis became popular for short periods until replaced by yet better varieties—many developed locally by hybridization and others, such as the famous Baart and Federation wheats, imported from Australia in 1900 and 1914 respectively. The addition of new wheats increased the ability of farmers to match varieties to local conditions and push production into even more arid zones. As an example Baart replaced Bluestem in the driest areas, which allowed production in regions with as little as ten inches of rain.[127] The varieties of choice continued to change, and for 1918–19 there were ten varieties, most concentrated in clearly defined zones, contributing more than one million bushels each to Washington's two-year output.[128] In total, the survey listed twelve varieties, which together accounted for 93.5 percent of the state's entire output. None of these varieties existed in the United States in 1839. Roughly 73 percent of the acreage surveyed was planted in varieties that probably had not entered Washington until after 1890 and about 47 percent would not have been available to Washington farmers until after 1900.

In part because of the need for regionally adapted varieties, Washington developed one of the most impressive wheat research programs in the world. Around 1890 the Washington State Experiment Station began the process of collecting and testing hard winter wheats. All the winter wheats had serious problems because the straw was too weak to consistently withstand the high winds, and they were prone to high shattering losses. W. J. Spillman began work to cross the best spring and winter varieties in 1899, and in 1907 the experiment station began the release of his new hybrids.[129] "During the season of 1908 there were almost one thousand new or selected varieties growing on the Experiment Station farm."[130] Between 1911 and 1926, "647 varieties and selections have been introduced from outside sources and 1240 produced by hybridization, all of which have been included in varietal tests." Out of these nearly two thousand varieties, the experiment station saw fit to continue testing only forty-seven in 1926.[131] Similar, albeit less extensive, activities were under way in Oregon and Idaho.

The California experience over the late nineteenth century perhaps best exemplifies what happened in the absence of biological innovation. After learning to cultivate Sonora and Club wheats in the 1850s, 1860s, and 1870s, California grain growers appear to have focused their innovative efforts almost entirely on mechanization. They pioneered the adoption of labor-saving gang plows and combined harvesters, but purportedly did little to improve cultural practices, introduce new varieties, or even maintain the quality of their seedstock. The result was such sharply declining yields in many areas that wheat, formerly the state's leading staple, ceased to be a paying crop and was virtually abandoned. Acreage had ranged between two and three million during the 1880s and 1890s but dropped to roughly one-half million by around 1910. Only after this collapse did the state's agricultural research establishment, which had focused primarily on horticultural and viticultural activities, begin to devote serious attention to biological innovations in grain culture.[132]

Conclusion

> *The word "productivity" pushes a historian toward economics, but the phrase "productivity growth" pushes an economist toward history.*
> —William N. Parker[133]

The nineteenth-century revolution that mechanized northern agriculture has been well documented, but the complementary revolution in biological and cultural technologies has been largely ignored. Yet it was biological innovations that made possible the maintenance of yields in the face of an unstable and deteriorating pest environment, and in spite of the wholesale shift in the locus of production onto inferior lands—lands which in many cases could not have sustained commercial wheat production given the biological technologies available in the middle of the nineteenth century. If, as the literature assumes, generations of wheat farmers had simply followed in their fathers' footsteps (apart from adopting labor saving machinery), their crops would have been ravished. The types of problems that earlier generations of farmers faced continue into the present because modern agricultural scientists estimate that even with vastly improved scientific methods over 41 percent of all wheat research was needed to simply maintain yields in recent decades.[134]

Contemporaries seldom anticipated the biological innovations that transformed North American wheat culture. As an example, John Klippart, who was arguably the foremost authority on wheat culture of his day, argued in 1860 that the United States would soon have to import wheat. His pessimism was based on the view (largely correct at the time) that the nation had reached the limits of its wheat producing lands given the climate farther west.[135] Klippart obviously was familiar with the mechanical reaper and thresher, and he would not have been surprised by the next generation of harvesting equipment—the self-binder. These are the machines that the standard accounts assert made the settlement of the West possible. What so colored Klippart's vision was his inability to foretell the wholesale changes in the genetic makeup of the wheat varieties that would become available to North American farmers. Mechanical inventions certainly lowered the cost of growing wheat in the West, but the binding constraint was biological. Without a biological revolution (assisted by the transportation revolution), the centers of wheat production in the United States and Canada could not have assumed their late-nineteenth-century dimensions. Drawing on sophisticated models of global climate change helps explain Klippart's myopia and provides a sense of the magnitude of the nineteenth-century achievement, because the cross-sectional climatic changes that farmers encountered in moving wheat production from New York and Ohio to Kansas and North Dakota exceeded the changes projected for the Great Plains over the next one hundred years.[136]

The longer run impact of the nineteenth century carried over into the Green Revolution era, because much of the genetic stock that modern wheat breeders put into the blender to produce the first generations of post–World War II hybrids came from Turkey wheat and other late-nineteenth- and early-twentieth-century introductions from around the world. In 1969 eleven varieties of hard red winter wheat were grown on one million or more acres. Turkey was important in the pedigree for all of these varieties. The semidwarf characteristics that are the hallmark of the Green Revolution in the United States derive from a Japanese variety called Norin 10. Yet one of the parents of Norin 10 was Turkey, which the Japanese had imported from the United States around 1890.[137] More generally, our findings suggest that the high rate of return to agricultural research is not just a modern phenomenon beginning with the spread of hybrid corn.[138] Mark Carleton's introductions of foreign wheat varieties and Charles Saunders' creation of Marquis are beacons of wise government investments. As noted in the introduction, Cyrus McCormick has long been eulogized as the man who "made bread cheap." But he needed considerable help. It is time that we add the names of Mark Carleton, Charles Saunders, David Fife, Cyrus Pringle, and the other researchers who revolutionized American wheat production to the high pantheon of nineteenth-century inventors.

The new wheat varieties that these individuals gave to North American farmers represented radically new forms of capital equipment that revolutionized the location and efficiency of wheat production, just as the steam engine, the Bessemer process, and electricity revolutionized the structure and location of industry. By allowing wheat production to move into more hostile climates, the new wheat technologies significantly contributed to the pressure on eastern farmers to abandon wheat and seek other crops and production systems. The ripple was also felt in Europe, because without the widespread adoption of Red Fife, Turkey, and other new varieties, the grain invasion described by Kevin O'Rourke and others would not have been possible.[139] But for the new agricultural technologies to be effective, millions of small farms had to experiment and fine-tune their production processes both to ward off pests and diseases and to adapt the new and improved varieties to myriad geo-climatic niches that define American agriculture.[140] Wheat was not an exception. A cursory look at other crops and livestock shows similar patterns of biological innovation during the pre-1940 era. This result should not be a surprise because similar forces were at work. Major innovations were necessary for most crops and livestock to facilitate western settlement and to maintain yields in the face of new pests and diseases.

To this point we have argued that describing the pre-1940 era as a period of labor-saving mechanization while dubbing the post-1940 era as a period of biological change is misleading because of the tremendous biological achievements in the earlier period. However, the contrast is also misleading because of what happened after 1940. It is true that the post–World War II period has been a revolutionary period of land-augmenting productivity changes, as reflected in the increase in the rate of growth in yields per acre. But contrasting the period 1910–40 with 1940–80 shows that the increase in the rate of growth of output per unit of agricultural labor after 1940 was roughly three times as great as the increase in the rate of growth in yields. Thus if one were forced to make a distinction, and one thought it appropriate to actually pay attention to the data, it would make much more sense to call the earlier period the era of biological change and the later period the era of mechanical change.

Notes

We have received valuable comments from Greg Clark, Jack Goldstone, D. Gale Johnson, Bruce Johnston, Frank Lewis, Joel Mokyr, Jose Morilla, Philip Pardey, Vicente Pinilla, James Simpson, Vernon Ruttan, Philip Scranton, and from the seminar participants at UC Davis,

Stanford University, Northwestern University, the University of Minnesota, the All-Chicago Economic History Group, Triangle Economic History Workshop, the University of Alcala, the University of Zaragoza, the Victoria Department of Natural Resources and Environment and the Victorian Branch of the Australian Agricultural and Resource Economics Society, Melbourne, Australia, the University of Agricultural Sciences, Vienna, the University of Ljubljana, and the conference participants at the Australian Agricultural and Resource Economics Society Conference, Christchurch, New Zealand, the Economic History Association at Philadelphia, Pennsylvania, and the Rutgers Center for Historical Analysis Conference on "Industrializing Organisms." Lisa Cappellari, Susana Iranzo, and Shelagh Mackay provided assistance on this project. Lee Craig generously shared county-level data from the 1839 Census. Several plant scientists, including Calvin Qualset, Charles Schaller, and Robert Webster, provided valuable perspectives. As has become custom, we owe special thanks to Julian Alston and Peter Lindert for their insights, advice, and encouragement. Work on this essay was facilitated by a fellowship granted by the International Centre for Economic Research (ICER) in Turin, Italy.

1. Fabre (1823–1915) was a French entomologist and philosopher. See Kenneth D. Kephart, "Commercial Wheat Cultivars of the United States" [Online], Ver. 1.1 June 1993. Retrieved February 20, 2001, from gopher://greengenes.cit.cornell.edu/11/.cwc. Introduction.
2. Annual yield data only go back to 1866.
3. Willard W. Cochrane, *The Development of American Agriculture: A Historical Analysis* (Minneapolis: University of Minnesota Press, 1979), 200, also see 107; Griliches' treatment is less emphatic, but appears to lead to the same general conclusion. See Zvi Griliches, "Agriculture: Productivity and Technology," in *International Encyclopedia of the Social Sciences*, Vol. 1 (New York: Macmillan and Free Press, 1968), 241–245.
4. Yujiro Hayami and Vernon Ruttan, *Agricultural Development: An International Perspective*, rev. and exp. ed. (Baltimore: Johns Hopkins University Press, 1985), 209.
5. Jeremy Atack, Fred Bateman, and William N. Parker, "The Farm, the Farmer, and the Market," in *The Cambridge Economic History of the United States*, Vol. II, *The Long Nineteenth Century* eds. Stanley L. Engerman and Robert E. Gallman (New York: Cambridge University Press, 2000), 245–284. They note: "Land productivity, measured by the yield per acre, changed relatively little for most crops during the nineteenth century" (259).
6. Jeremy Atack and Fred Bateman, *To Their Own Soil: Agriculture in the Antebellum North* (Ames: Iowa State University Press, 1987), 186; also see Wayne D. Rasmussen, "The Impact of Technological Change on American Agriculture, 1862–1962," *Journal of Economic History* 22, no. 4 (1962): 578–591.
7. For example, see Gary M. Walton and Hugh Rockoff, *History of the American Economy* (Orlando: Dryden Press, 1998), 334; Sidney Ratner, James H. Soltow, and Richard Sylla, *The Evolution of the American Economy* (New York: Macmillan Publishing, 1993), 264–265; Jeremy Atack and Peter Passell, *A New Economic View of American History From Colonial Times to 1940*, 2nd ed. (New York: W. W. Norton, 1994), 280–282; Jonathan R. T. Hughes, *American Economic History* (Glenview, Ill.: Scott, Foresman, 1987), 275–276: The theme is also standard fare in the USDA's treatment of productivity growth. See Ralph A. Loomis and Glen T. Barton, "Productivity of Agriculture: United States," *U.S. Department of Agriculture, Technical Bulletin*, no. 1238 (1961): 6–8.
8. See Jim Stanelle, "Certified: This Seed Will Inspire Trust and Confidence," *Seed World* (June 1999): 18 for a statement of this view.
9. In the context of the literature the term "biological change" encompasses nonmechanical activities that modify the growing environment. In addition to strictly biological innovations such as improved plant varieties, "biological changes" include changes in cultural practices, irrigation systems, fertilizers, and chemicals. For a more formal estimate of the effects of biological innovations on labor and land productivity, see Alan L. Olmstead and Paul W. Rhode, "The Red Queen and the Hard Reds: Productivity Growth in American Wheat: 1800–1940," *Journal of Economic History* 62 (2002): 929–966.
10. Several USDA economists have promoted the general view that mechanical technologies dominated biological innovations in the pre-1940 era. For example see Loomis and Barton, "Productivity," 6–8. In an excellent article on biological innovation in wheat, another USDA economist, Dana Dalrymple, hits on this issue noting the "effect of some yield-increasing technologies may have been masked" by disease or other problems, but he fails to develop the implications of this insight. Instead he repeats the standard mantra that

"mechanical technologies were of major importance well before biological technologies." The key point is that just because yields were relatively constant does not necessarily imply that biological innovation was of minor importance. See Dana G. Dalrymple, "Changes in Wheat Varieties and Yields in the United States, 1919–1984," *Agricultural History* 62, no. 4 (1988): 20–21.

11. When discussing wheat, modern agronomists have abandoned the term "variety" and adopted the term "cultivars" in its place because of the subtle distinctions as to what properly constitutes a distinct variety. Because the historical literature we cite consistently refers to "varieties," we have chosen to use the dated terminology.
12. Lewis Carroll, *Through the Looking Glass* (New York: McGraw-Hill, 1923), 37.
13. These terms have precise meanings. "Introduction" refers to the importation of a new variety from another country. "Selection" refers to the varieties, which resulted from natural mutations and were identified and carefully increased by breeders. "Developed" refers to varieties created by hybridization. A new variety was likely the result of more than one process. Introduced varieties may have been recently selected or developed in another country. Developed varieties had to be selected after the initial process of hybridization.
14. D. Gale Johnson and Robert L. Gustafson, *Grain Yields and the American Food Supply: An Analysis of Yield Changes and Possibilities* (Chicago: University of Chicago Press, 1962), 120. The term "maintenance yields" probably dates to the early work of Vernon Ruttan.
15. Johnson and Gustafson, *Grain Yields and the American Food Supply*.
16. Johnson and Gustafson, *Grain Yields and the American Food Supply*, 119.
17. E. C. Stakman, "Plant Diseases Are Shifty Enemies," *American Scientist* 35 (1947): 321–350.
18. Carroll, *Through the Looking Glass*, 37; L. Van Valen, "A New Evolutionary Law," *Evolutionary Theory* 1 (1973): 1–30.
19. W. Q. Loegering, C. O. Johnston, and J. W. Hendrix, "Wheat Rusts," in K. S. Quisenberry and L. P. Reitz, eds., *Wheat Improvement* (Madison, Wisc.: American Society of Agronomy, 1967), 307–335. Stem and leaf rusts thrive in the hot, humid climates and attack wheat in most grain-growing regions of North America. Stripe rust thrives in cooler climates and in most years is limited to the Mountain and Pacific regions.
20. It is useful to clarify the basic nomenclature of wheat. The primary distinction is between winter (-habit) and spring (-habit) wheats. ("Habit" is added because the distinction does not depend strictly on the growing season.) Winter-habit wheat requires a period of vernalization, that is, prolonged exposure to cold temperatures, to shift into its reproductive stage. This typically involves sowing in fall and allowing the seedlings to emerge before winter. During the cold period, the winter-habit wheat goes dormant but remains exposed to risks of winterkill. The grain is harvested in the late spring or early summer. Spring-habit wheat grows continuously without a period of vernalization. In Europe and North America, farmers in cold regions often sow spring-habit wheat shortly before the last freeze, harvesting the crop in mid- to late summer. But it is interesting to note that varieties with spring-habits were also used in areas with mild winters, such as the Mediterranean and California. There, the wheat was planted in the fall and grew without interruption. (There is a third, less important category of facultative wheat that is intermediate in cold tolerance but does not require vernalization to flower and develop grain.) Note that a longer growing season is generally associated with greater yield potential, but also involves greater exposure to weather risks, diseases, and insects. Other important distinctions refer to the kernel's texture (soft, semi-hard, and hard) and color (white versus red). Hard wheats, which were relatively drought-resistant, outperform soft wheats in the more arid areas. The rough-and-ready dividing line was between the 30 and 35 inches of precipitation. See S. C. Salmon, "Climate and Small Grain," in U.S. Department of Agriculture *Yearbook, 1941* (Washington, D.C.: GPO, 1942), 334–335. East of the Mississippi, soft white and red wheats were prevalent, whereas in the Great Plains hard reds traditionally dominated. Durum wheat, which became popular in selected regions of the northern Great Plains after 1900, is a distinct species from common wheat, with distinct flour quality and uses.
21. Lyman Carrier, *The Beginnings of Agriculture in America* (New York: Clarkson N. Potter, 1973), 147. Clay dates the arrival of the blast in New England in 1660. See Clarence Albert Clay, *A History of Maine Agriculture, 1604–1860*, University of Maine Studies, 2nd series, no. 68 (Onono, Maine: University Press, 1954), 38; Percy Wells Bidwell and John I. Falconer, *History of Agriculture in the Northern United States, 1620–1860* (Washington, D.C.: Carnegie Institute of Washington, 1925), 13–14; Charles L. Flint, "Progress in Agriculture,"

in L. Stebbins, ed., *Eighty Years' Progress of the United States* (New York: New National Publishing House, 1864), 72–73.

22. Mark Alfred Carleton, "Cereal Rusts of the United States: A Physiological Investigation," *U.S. Department of Agriculture, Division of Vegetable Physiology and Pathology Bulletin*, no. 16 (1899): 13–19; Mark Alfred Carleton, "The Basis for the Improvement of American Wheats," *U.S. Department of Agriculture, Division of Vegetable Physiology and Pathology Bulletin*, no. 24 (1900): 11–22.

23. Such estimates are fraught with uncertainty. In his prize-winning 1857 essay, H. Y. Hind attributed greater losses to rust than to any other enemy of wheat. And in 1899, Carleton echoed this conclusion, arguing "the average annual loss from rust throughout the United States far exceeds that due to any other enemy, insect or fungous, and often equals those from all others combined." See H. Y. Hind, *Essay on the Insects, Mites, and Ticks in the United States* (Toronto: Lovell and Gibson, 1857), 110; Carleton, "Cereal Rusts," 19. But, alas neither Carleton nor anybody else had any hard data, and there was probably a tendency to exaggerate losses by extrapolating from examining heavily damaged fields. Beginning in 1918 the USDA's *Plant Disease Reporter* began collecting estimates by polling plant specialists about the damage in each state or region. These estimates show national stem rust damage averaged around 3 percent over the 1919–39 period, with peak losses of 23 percent in 1935. National leaf rust damage averaged around 2 percent, with a 9.6 percent peak in 1938. Alan P. Roelfs, "Estimated Losses Caused by Rust in Small Grain Cereals in the United States, 1918–76," *U.S. Department of Agriculture Miscellaneous Publication*, no. 1363 (1978) summarizes these results for the period 1918–76. Whereas Carleton and others may have overestimated the losses to disease, there is good reason to think that the formal estimates seriously understate the losses. Subsequent studies suggest that it is likely that the scientists reporting the incidence of disease did not fully recognize the damage caused and tended to report only damage in excess of normal damage. As an example, Chester argues that the estimates of the losses to leaf rust for the years 1900–1935 "must be regarded as gross under estimates." Instead of averaging about 1.5 percent he claims annual losses were at least 5 percent and maybe much higher. See K. Starr Chester, "Plant Disease Losses: Their Appraisal and Interpretation," *Plant Disease Reporter Supplement*, no. 193 (1950): 189–362, especially pp. 210–212. For our needs actual losses are less important than understanding what would have happened without changing varieties and cultural methods.

24. Alan P. Roelfs, "Effects of Barberry Eradication on Stem Rust in the United States," *Plant Disease* 66, no. 2 (1982): 177–181; J. D. Miller, R. M. Hosford, R. W. Stack, and G. D. Statler, eds., "Diseases of Durum Wheat," in Giuseppe Fabriani and Claudia Lintas, eds., *Durum Wheat: Chemistry and Technology* (St. Paul, Minn.: American Association of Cereal Chemists, 1988), 75–83; Mark Alfred Carleton, "Hard Wheats Winning Their Way," in U.S. Department of Agriculture, *Yearbook 1914* (Washington, D.C.: GPO, 1915), 407–408; Peter Tracy Dondlinger, *The Book of Wheat: An Economic History and Practical Manual of the Wheat Industry* (New York: Orange Judd, 1908), 167–168.

25. R. F. Peterson, *Wheat* (London: Leonard Hill Books, 1965), 201–204. Systematic efforts to estimate and record losses to rust only began after the epidemic of 1916. See Laura M. Hamilton, "Stem Rust in the Spring Wheat Area in 1878," *Minnesota History* 20, no. 2 (1939): 157. USDA reports in the period reveal the average size of individual wheat fields in the Dakotas were over one hundred acres, four to five times larger than those common in the east.

26. "Stem and leaf rust, foot rots, scab, and most other diseases appear to have been relatively unimportant in comparison with later periods." S. C. Salmon, O. R. Mathews, and R. W. Luekel, "A Half Century of Wheat Improvement in the United States," in *Advances in Agronomy*, Vol. 5, ed. A. G. Norman (New York: Academic Press, 1953), 16.

27. E. C. Large, *The Advance of Fungi* (New York: Henry Holt, 1940), 292.

28. Large, *Advance of Fungi*, 292–312; Salmon, Mathews, and Luekel, "Half Century of Wheat," 113–114. For a treatment of the early history of rust research see William R. Bushnell and Alan P. Roelfs, eds., *The Cereal Rusts* (Orlando, Fla.: Academic Press, 1984), 3–38.

29. It probably had direct resistance also because when it was first selected "it proved at harvest to be entirely free from rust, when all wheat in the neighborhood was badly rusted." See Carleton, "Hard Wheats," 393.

30. Carleton, "Hard Wheats," 407–408.

31. Miller et al., "Diseases of Durum Wheat," 69–92.

32. T. S. Cox, J. P. Shroyer, Liu Ben-Hui, R. G. Sears, and T. J. Martin, "Genetic Improvement in Agronomic Traits of Hard Red Winter Wheat Cultivars From 1919–1987," *Crop Science* 28 (1988): 756–760. Kanred ranked first out of 150 varieties tested for stem and leaf rust. See S. C. Salmon, "Developing Better Varieties of Wheat for Kansas," in *Wheat in Kansas* (Topeka: Kansas State Board of Agriculture,1920), 214, 228.

    According to Norman E. Borlaug, who received the 1970 Nobel Peace Prize for his achievements as a wheat breeder, it had been his dream as a University of Minnesota graduate student in the late 1930s and early 1940s to transfer the genes that made rice uniquely immune to rust into wheat, oats, barley, rye, maize, and the other susceptible cereal crops. See Norman E. Borlaug Center for Southern Crop Improvement Retrieved May 23, 2002, from http://ipgb.tamu.edu/borlaug.html.

33. Not only did the barberry provide a home for the stem rust to carry over and multiply, the rust passed through its sexual recombination stage on the barberry. Thus, the bush was the breeding ground where rusts mutated and hybridized to developed new races.

34. Testimony of Dr. K. F. Kellerman, U.S. House of Representatives, Agricultural Department Appropriations Bill, (Washington, D.C.: G.P.O, 1935), 380. Kellerman adds: "when the New England wave of settlement started west the same sentimental reason that led them to bring the barberry bush from Europe caused them to carry the barberry bushes with them as they went into the Western States. In the spring-wheat States, it has proven to be very destructive to the wheat farmer."

35. Carleton R. Ball, "The History of American Wheat Improvement," *Agricultural History* 4, no. 2 (1930): 48–71; Hamilton, "Stem Rust," 156–164; Large, *Advance of Fungi*, 121–146; Robert B. Elwood, Lloyd E. Arnold, D. Clarence Schmutz, et al, *Changes in Technology and Labor Requirements in Crop Production: Wheat and Oats*, Studies of Changing Techniques and Employment in Agriculture, Report no. A-10 (Philadelphia: Works Progress Administration, 1939), 80; Salmon, Mathews, and Luekel, "Half Century of Wheat," 123; C. Lee Campbell and David L. Long, "The Campaign to Eradicate the Common Barberry in the United States," in Paul D. Peterson, ed., *Stem Rust of Wheat: From Ancient Enemy to Modern Foe* (St. Paul, Minn.: APS Press, 2001), 16–50.

36. Robert Webster, interview by authors, Davis, California, May 27, 2000. There is some controversy as to the effectiveness of this campaign. Rust spores are wind-blown and will migrate over wide areas, thus it was important to destroy most of the barberries to be effective. In addition, in bad years rusts absolutely devastated the wheat crops of Australia where the barberry did not exist.

37. The rust spores can migrate over vast areas. But now new races have to migrate from Texas and Mexico, or from barberry bushes still remaining in mountainous regions. Roelfs, "Barberry Eradication," 177–181; Webster, interview by authors.

38. Large, *Advance of Fungi*, 70.

39. Dondlinger, *Book of Wheat*, 162.

40. As reported in Salmon, Mathews, and Luekel, "Half Century of Wheat," 16. They also discuss the rise of the bunt problem in the Pacific Northwest after 1900. "Nowhere in the United States and probably nowhere in the world has bunt been so serious or so difficult to control." Salmon, Mathews, and Luekel, "Half Century of Wheat," 84.

41. Dondlinger's estimate may be credible because bunt losses in modern tests with untreated seeds often exceeded 20 percent. Charles Schaller, interview by authors, Davis, California, April 25, 2000.

42. Large, *Advance of Fungi*, 70–82; Salmon, Mathews, and Luekel, "Half Century of Wheat," 125–126; E. M. Freeman and E. C. Stakman, "Smuts of Grain Crops," *University of Minnesota Agricultural Experiment Station Bulletin*, no. 122 (1911): 33–64; Andrew Boss, L. B. Bassett, C. P. Bull, et al., "Seed Grain," in Minnesota Agricultural Experiment Station, *Seventeenth Annual Report* (Delano, Minn.: The Station, 1909), 370–379.

43. Salmon, Mathews, and Luekel, "Half Century of Wheat," 127–129.

44. Aaron G. Johnson and James G. Dickson, "Wheat Scab and Its Control," *Farmers' Bulletin*, no. 1217 (1921): 3.

45. Jean McInnes and Raymond Fogelman, "Wheat Scab in Minnesota," *University of Minnesota Agricultural Experiment Station Technical Bulletin*, no. 18 (1923): 1–43. Salmon, Mathews, and Luekel, "Half Century of Wheat," 16. It was also well known that corn provided a green bridge allowing chinch bugs to carry over. Later discoveries demonstrated other harmful effects of wheat-corn rotations. As an example, planting winter wheat down

wind from mature corn significantly increased the migration of aphids carrying barley yellow dwarf virus. R. James Cook and Roger J. Veseth, *Wheat Health Management* (St. Paul, Minn.: APS Press, 1991), 57, 76–77.

46. Although 1776 is the widely accepted date of entry, Fletcher asserts that the Hessian fly was in New York and New Jersey before the revolution. Stevenson Whitcomb Fletcher, *Pennsylvania Agriculture and Country Life, 1640–1840* (Harrisburg: Pennsylvania Historical and Museum Commission, 1950), 147. T. J. Headlee and J. B. Parker, "The Hessian Fly," *Kansas Agricultural Experiment Station Bulletin*, no. 188 (1913): 88; C. L. Marlatt, "The Principal Insect Enemies of Growing Wheat," *Farmers' Bulletin*, no. 132 (1901): 14; F. M. Webster, "The Hessian Fly," *Ohio Agricultural Experiment Station Bulletin*, no. 107 (1899): 259–260. Other sources offer slightly different chronologies, but the above account gives a general picture of its spread. There are several strains of the Hessian fly and later treatments often focus on *Mayetiola destructor*. R. G. Dahms, "Insects Attacking Wheat," in *Wheat Improvement*, eds. K. S. Quisenberry and L. P. Reitz (Madison, Wisc.: American Society of Agronomy, 1967), 428–431.
47. Headlee and Parker, "Hessian Fly," 113.
48. Bidwell and Falconer, *History of Agriculture*, 96; Hubert G. Schmidt, *Agriculture in New Jersey: A Three-Hundred-Year History* (New Brunswick, N.J.: Rutgers University Press, 1973), 92.
49. Webster, "Hessian Fly," 259.
50. See "The Hessian Fly," c. 1997. From *George Washington: Pioneer Farmer*, retrieved May 23, 2001, from http://www.mountvernon.org/pioneer/lite/farms/hessian.html; Elizabeth A. Mosimann, "The Library of the Philadelphia Society for Promoting Agriculture: Its History and Conservation," retrieved May 23, 2002, from Philadelphia Society for Promoting Agriculture http://www.library.upenn.edu/special/pspa/collections/collection2.html.
51. As is often the case there are conflicting stories as to this wheat's origin. Klose claims that it came from the Mediterranean islands, but the *Ohio Agricultural Report for 1857* asserts that the variety came from Hesse via Holland. Fletcher has it entering Pennsylvania in about 1835 via a farmer in New Jersey, who in turn supposedly had purchased it from a U.S. navy officer who had imported it from the Mediterranean region. See Nelson Klose, *America's Crop Heritage: The History of Foreign Plant Introduction by the Federal Government* (Ames: Iowa State College Press, 1950), 66; Ohio State Board of Agriculture, *Annual Report for the Year 1857* (Columbus: State Printer, 1858), 700–701; Fletcher, *Pennsylvania Agriculture*, 148.
52. U.S. Bureau of the Census, *Agriculture of the United States in 1860: compiled from the Original Returns of the Eighth Census* (Washington, D.C.: GPO, 1864), xxxi–xlv, quote from xxxiii; Ulysses Prentiss Hedrick, *A History of Agriculture in the State of New York* (New York: Agricultural Society, 1933), 332–335.
53. Census, *Agriculture 1860*, xxxiv.
54. Census, *Agriculture 1860*, xxxv.
55. Census, *Agriculture 1860*.
56. Census, *Agriculture 1860*, xl, xxxv; Bidwell and Falconer, *History of Agriculture*, 239. In fact, moving to an earlier planting date might allow for a slightly earlier harvest, but it was more important to find earlier ripening varieties.
57. Census, *Agriculture 1860*, xxxiv.
58. The often-repeated quote of "an excellent farmer" found in the *Ohio State Board of Agriculture Report* for 1857 that "three-fourths, if not nine-tenths of the wheat raised in this country is the Mediterranean variety," appears to be an exaggeration. This report does note several other varieties that were abandoned in New York because of their susceptibility to the midge. A review of the *Patent Commissioner Reports* for the late 1840s and the early 1850s suggests that Mediterranean was the most common variety. Ohio State Board of Agriculture, *Annual Report*, 685; U.S. Patent Office, *Annual Report of the Commissioner Patents. Agriculture*, Vols. 1847–1855 (Washington, D.C.: GPO, 1848–1856).
59. Census, *Agriculture*, xxxiii. Much later tests would confirm that Mediterranean in fact had fly-resistant qualities. Salmon, Mathews, and Luekel, "Half Century of Wheat," 98.
60. Delaying the planting date for winter wheat could markedly limit the chances for a buildup of other pests, including the greenbug, the Russian wheat aphid, aphid vectors of barley yellow dwarf virus, and the curl mite vector of wheat streak mosaic virus. Cook and Veseth, *Wheat Health*, 84.
61. James W. McColloch, "The Hessian Fly in Kansas," *Kansas Agricultural Experiment Station Technical Bulletin*, no. 11 (1923): 80.
62. Headlee and Parker, *Hessian Fly*, 135.

63. C. L. Marlatt, "The Annual Losses Occasioned by Destructive Insects in the United States," in U.S. Department of Agriculture, *Yearbook for 1904* (Washington, D.C.: GPO, 1905), 461–474. For a small sample of the studies conducted and for estimates of state losses in bad years see I. P. Roberts, V. Slingerland, and J. L. Stone, "The Hessian Fly: Its Ravages in New York in 1901," *Cornell University Agricultural Experiment Station Bulletin*, no. 194 (1901): 239–260; F. M. Webster, "The Hessian Fly in Ohio in 1899 and 1900," *Ohio Agricultural Experiment Station Bulletin*, no. 119 (1899): 239–247; Headlee and Parker, *Hessian Fly*, 113; McColloch, "Hessian Fly," 7–9.
64. Marlatt, "Principal Insect Enemies," 13. Numerous other sources place the actual losses in this general magnitude ranging between $50 and $100 million. Dondlinger asserts that 10 percent is a lower bound estimate. Dondlinger, *Book of Wheat*, 172–173. The direct estimates of losses to the fly are almost surely lower bound, because when farmers adjusted varieties and cultural practices to avoid the fly there were real costs in terms of yield losses and increased exposure to winterkill and rust. The USDA estimated in 1904 that annual wheat losses to all insects were "at least 20 percent of the crop." Marlatt, "Annual Losses," 468.
65. J. A. Hyslop, *Losses Occasioned by Insects, Mites, and Ticks in the United States* (Washington, D.C.: U.S. Bureau of Entomology and Plant Quarantine, Division of Insect Pest Survey and Information, 1938), 9.
66. As an example, see Fletcher, *Pennsylvania Agriculture*, 147–148.
67. Headlee and Parker, "Hessian Fly," 115.
68. McColloch, "Hessian Fly," 91–94.
69. Clell Lee Metcalf and Wesley Pillsbury Flint, *Destructive and Useful Insects: Their Habits and Control*, 4th ed. (New York: McGraw Hill, 1962), 410–411.
70. Cook and Veseth, *Wheat Health*, 56.
71. Schaller, interview by authors, Davis, California, February 27, 2001; Robert Webster, interview by authors, Davis, California, April 25, 2000; and Calvin Qualset, interview by authors, Davis, California, April 25, 2000. The fly would not have been a serious problem in every year, but the level of destruction in most years probably would have been higher, and severe outbreaks would have been more frequent and more widespread.
72. Dondlinger, *Book of Wheat*, 174. Our discussion concentrates on wheat, but the Hessian fly and many other insects and diseases attacked other crops. The lessons learned for one crop often applied to others.
73. No effort is made to account for the price response due to difference in output.
74. L. O. Howard, "The Chinch Bug," *U.S. Department of Agriculture Division of Entomology Bulletin*, no. 17 (1888): 7–9; J. R. Horton and A. F. Satterthwait, "The Chinch Bug and Its Control," *Farmers' Bulletin*, no. 1223 (1922): 4; Benjamin Horace Hibbard, "The History of Agriculture in Dane County Wisconsin: A Thesis Submitted for the Degree of Doctor of Philosophy, University of Wisconsin, 1902," *Bulletin of the University of Wisconsin* 101, *Economics and Political Science Series* 1, no. 2 (1904): 131. Hibbard noted: "About 1860 the chinch bug on his northern march had reached Wisconsin; this was certainly ominous, but no considerable share of the crop was destroyed until 1864, in which year and for two succeeding years, the insect made a clean sweep of the wheat fields."
75. Charles V. Riley, *The Locust Plague in the United States* (Chicago: Rand McNally, 1877), 29–54; G. J. Haeussler, "Insects as Destroyers," in *Yearbook of Agriculture, 1952* (Washington, D.C.: GPO, 1953), 141–146; Salmon, Mathews, and Luekel, "Half Century of Wheat," 136–139.
76. W. R. Walton, "The Green-Bug or Spring Grain-Aphis: How to Prevent Its Periodical Outbreaks," *Farmers' Bulletin* no. 1217 (1921): 7–8. As with rust, the spread of grain cultivation to the Southern Plains worsened insect problems by creating a homebase to overwinter. These enemies of wheat could then fly or blow north in the spring and summer.
77. Salmon, Mathews, and Luekel, "Half Century of Wheat," 16, 19.
78. George Perkins Clinton, "Russian Thistle and Some Plants that Are Mistaken for It," *Illinois Agricultural Experiment Station Bulletin*, no. 39 (1895): 87–97; Dondlinger, *Book of Wheat*, 151–152.
79. H. R. Cates, "The Weed Problem in American Agriculture," *Yearbook 1917, U.S. Dept. of Agriculture* (Washington, D.C.: GPO, 1918), 205.
80. An overall assessment of wheat yield reductions is available in the USDA *Losses in Agriculture* series. The report for the 1942–51 period estimates that plant diseases reduced wheat yields by 6.6 percent and values by 7.5 percent. Insects caused 2 percent losses and weeds

10.2 percent. See U.S. Agricultural Research Service, "Losses in Agriculture: A Preliminary Appraisal for Review," *ARS*, no. 20–21 (1954): 14–15, 64–65, 88. An update for the 1951–60 period placed the losses from plant diseases at 14 percent, from insects at 6 percent, and from weeds at 12 percent. U.S. Agricultural Research Service. "Losses in Agriculture," *Agriculture Handbook*, no. 291 (1965): 4, 14, 41, 56.

81. William P. Morrison and Frank B. Pearis, "Response Model Concept and Economic Impact," in *Response Model for an Introduced Pest—The Russian Wheat Aphid*, eds. Shannon S. Quisenberry and Frank B. Pearis (Lanham, Md.: Entomological Society of America, 1998), 1–11.

82. M. McMullen, R. Jones, and D. Gallenberg, "Scab of Wheat and Barley: A Re-Emerging Disease of Devastating Impact," *Plant Disease* 81 (1997): 1340; C. E. Windels, "Economic and Social Impacts of Fusarium Head Blight: Changing Farms and Rural Communities in the Northern Great Plains," *Phytopathology* 90 (2000): 17–21.

83. Kent Osband, "The Boll Weevil Versus 'King Cotton,'" *Journal of Economic History* 45, no. 3 (1985): 627–643. Marlatt, "Annual Losses," 468.

84. C. L. Marlatt, "Principal Insect Enemies," 5. Marlatt further noted that the smaller holdings and more intensive cultivation in Europe lead to more careful inspection and more aggressive responses to insect outbreaks.

85. Paul de Kruif, *Hunger Fighters* (New York: Harcourt, Brace & World, 1928), 41–42.

86. John P. Foley, ed., *The Jefferson Cyclopedia*, Item 6677 (New York: Funk & Wagnalls, 1900), 697.

87. Salmon, Mathews, and Luekel, "Half Century of Wheat," 5. Salmon and his co-authors hint at a larger problem because the use of average national yields to measure land productivity is subject to obvious conceptual difficulties. The following reasoning, for which we thank Frank Lewis, helps illustrate some of the sample selection problems involved. Suppose potential wheat land may be ranked along a scale according to its yield capacity. Given prevailing farm prices and costs, there will be a minimum yield for which it is profitable to devote the land to wheat cultivation. Land ranked below this threshold will go uncultivated and the average measured yield is based only on land above the profitable-cultivation threshold.

   Now consider the effect of a yield-increasing biological innovation, which like many of those considered in this essay, disproportionately increases yields on low yielding lands. This will raise more land above the threshold, pushing out the frontier of wheat cultivation, and increase total production. Although the innovation will raise productivity on low-yielding land, it need not have a positive effect of measured yields. Indeed, if the effects of the biological innovation are limited to low-yielding lands close to the threshold, average measured yields can actually fall. Also note the other cost-reducing innovations, such as mechanization, can lower the threshold yield necessary for profitable cultivation. The frontier of cultivation will expand and measured yields will fall, even in the absence of changes in the productivity of a specific acre of land.

88. We calculated the 1839 and 1909 center from census county-level production data and the location of the county's seat. The 1839 data are from Lee A. Craig, Michael R. Haines, and Thomas Weiss, *U.S. Censuses of Agriculture, by County, 1840–1880*, unpublished computer files graciously provided by the authors, Department of Economics, North Carolina State University 2000; and Lee A. Craig, Michael R. Haines, and Thomas Weiss, "Development, Health, Nutrition, and Mortality: The Case of the 'Antebellum Puzzle' in the United States," *National Bureau of Economic Research Working Paper Series on Historical Factors in Long Run Growth*, Historical Paper #130 (2000). Those for 1909 data are from U.S. Bureau of the Census, *Thirteenth Census of the United States Taken in the Year 1910*, Vols. 6–7. The information for 1849–1899 and 1919 (mean only) are from U.S. Bureau of the Census, *Statistical Atlas of the United States* (Washington, D.C.: GPO, 1913), 22. The county seat location data are from Robert P. Sechrist, *Basic Geographic and Historic Data for Interfacing ICPSR Data Sets, 1620–1983 [United States]* [computer file]. ICPSR version (Baton Rouge, La.: Robert P. Sechrist, Louisiana State Univ. [producer], 1984; Ann Arbor, Mich.: Inter-university Consortium for Political and Social Research [distributor], 2000). The data include only U.S. production. As a result, the changes do not capture the spread of grain cultivation onto the Canadian Prairies.

89. The classic story is of early members in Selkirk colony who settled on the Red and Assiniboine Rivers near Lake Winnipeg. Winter wheat, first tried in 1811–12, proved a failure. The fields were resown with spring wheat, which, due to drought and cultural problems,

also failed. In 1813–14, they resupplied with a small amount of spring wheat seed from Fort Alexander, which yielded sufficient grain for the colony to continue cultivation. But in 1819, a locust plague completely devastated the colony's wheat crop, leaving it without seed. During the dead of winter, a band of the settlers traveled to Prairie du Chien on the upper Mississippi River to secure a replacement stock of 250 bushels. This spring wheat performed well in 1820 and provided the basis for future cultivation and variety selection. Over the next several decades, the region's farmers experimented with varieties imported from England, Ireland, and the Ukraine. Around 1857, immigrants for Ontario introduced Red Fife, which had the decided advantage of maturing about ten days earlier than the variety from Prairie du Chien. Stanley N. Murray, *The Valley Comes of Age: A History of Agriculture in the Valley of the Red River of the North, 1812–1920* (Fargo: North Dakota Institute for Regional Studies, 1967), 37; John Perry Pritchett, *The Red River Valley, 1811–1849: A Regional Study* (New Haven, Conn.: Yale University Press, 1942), 113, 228. But the Selkirk colonists were re-enacting the experience of many pioneers in earlier settled areas.

90. Carleton, *Improvement of American Wheats*, 9. The four general wheat regions shown in the lower panel of Figure 2.1 represent gross demarcations because each of these areas contained important subregions. As a sign of the evolving state of knowledge on matching varieties with regions, in 1898 Carleton thought that durum varieties would be best suited for Texas and Oklahoma (because of durum's drought tolerance), but by 1909 the durum domain had been firmly established in the Dakotas and Minnesota.

91. To take a few examples, the season distribution as well as total inches of precipitation matter for the plant's moisture intake. Variations of temperature with the season also mattered significantly. High summer temperatures of the Great Plains states rushed the grain ripening process, causing the growing season to end well before the first frost. Furthermore, the potential for freezing ambient air temperatures to cause winterkill depends on whether the soil is covered by an insulating blanket of snow. Generally, the crowns of even the most-hardy wheats will be winter-killed if exposed to temperatures of −9 to −11° F. And high summer temperatures have adverse effects on yields as well. Photosynthesis stops around 82–85° F and above 90–95° F the plants may experience negative growth. See Cook and Veseth, *Wheat Health*, 21–24.

92. J. Allen Clark and John H. Martin, "Varietal Experiments with Hard Red Winter Wheats in the Dry Areas of the Western United States," *U.S. Department of Agriculture Bulletin*, no. 1276 (1925): 1.

93. The date of its entry into the United States is often given as 1860, but Wisconsin farmer J. W. Clark, writing in that year noted "Fife" had been grown extensively in the Northwest "for the last three seasons." J. W. Clark, "Characteristics of 'Fife' Spring Wheat," *Country Gentleman* 16, no. 18 (1860): 282–283; and Ohio State Board of Agriculture, *Annual Report for the Year 1857*, 759, notes that "Fife" had been introduced into Ohio by the mid-1850s.

94. Ball, "American Wheat Improvements," 63. The Mennonites had introduced Turkey into southern Russia only in 1860. Bernhard Warkentin, one of the early Mennonite settlers in Kansas, reportedly imported 25,000 bushels of seed from Russia and had as many as 300 test plots near his home in Kansas. In 1904 black rust destroyed a large part of the soft wheat, but the new Russian wheat was hardly affected. Harley J. Stucky, *A Century of Russian Mennonite History in America* (North Newton, Kans.: Mennonite Press, 1973), 27–30.

95. Karl S. Quisenberry and L. P. Reitz, "Turkey Wheat: The Cornerstone of an Empire," *Agricultural History* 48, no. 10 (1974): 98–114. Improvements in flour milling technologies contributed to the spread of hard red wheat, thereby creating an example of the synergism of biological and mechanical innovations. Using the traditional stone-grinding methods, millers found hard red wheat yielded darker, less valuable flour than the softer white wheat varieties. The introduction of the middling purifier (to separate the bran from the flour) in 1870 and the new roller grinding process in 1878 allowed millers to make high-quality flour from the new varieties. Over this period, flour from hard red wheat, which had formerly sold at a substantial discount relative to that ground from white winter wheat, began to sell at a premium. Henry A. Knopf, "Changes in Wheat Production in the United States," Ph.D. thesis, Cornell University (1967), 233; James C. Malin, *Winter Wheat in the Golden Belt of Kansas* (Lawrence, Kans.: University of Kansas Press, 1944), 188–189.

96. R. G. Danhof, *Changes in Agriculture: The Northern United States, 1820–1870* (Cambridge, Mass.: Harvard University Press, 1969), 157.

97. Ohio State Board of Agriculture, *Annual Report*, 737–761. Given that there was often much confusion regarding wheat names, it is likely that some varieties were listed under different names.
98. Johnson and Gustafson, *Grain Yields*, 119; Philip G. Pardey, Julian M. Alston, Jason E. Christian, and S. Fan, *Hidden Harvest: U.S. Benefits from International Research Aid*. Food Policy Report (Washington, D.C.: International Food Policy Research Institute, 1996), 8–12; Dalrymple, "Changes in Wheat," 23–27.
99. Large, *Advance of Fungi*, 302–304. In the United States, the first wheat hybrid is usually traced back to 1870 when Cyrus G. Pringle marketed Champaign, but Todd dates the first hybrid wheats to the 1840s. Sereno Edwards Todd, *The American Wheat Culturist: A Practical Treatise on the Culture of Wheat, Embracing a Brief History and Botanical Description of Wheat, with Full Practical Details for Selecting Seed, Producing New Varieties, and Cultivating on Different Kinds of Soil* (New York: Taintor Brothers, 1868), 40–46; Ball, "American Wheat Improvement," 48–71.
100. Patent Office, *Annual Report, 1854*, v, x–xiii.
101. U.S. Department of Agriculture, *Report of the Commissioner of Agriculture for the Year 1866* (Washington, D.C.: GPO, 1867), 8.
102. Carleton, "Improvement of American Wheats," 65, 70; J. Allen Clark, John H. Martin, and Carleton R. Ball, "Classification of American Wheat Varieties," *U.S. Department of Agriculture Bulletin*, no. 1074 (1922): 83–85, 135, 160; F. L. Patterson and R. E. Allan, "Soft Wheat Breeding in the United States," in *Soft Wheat: Production, Breeding, Milling, and Uses*, ed. W. T. Yamazaki and C. T. Greenwood (St. Paul, Minn.: American Association of Cereal Chemists, 1981), 36–41.
103. The economics literature focuses on yields as a summary measure of biological improvement in wheat. But breeders and farmers were also keenly interested in a number of other economically significant characteristics unrelated to yield including milling quality, protein and gluten content, color, baking quality, and the percentage of the kernel weight that was converted to flour.
104. K. H. Norrie, "The Rate of Settlement of the Canadian Prairies, 1870–1911," *Journal of Economic History* 45, no. 3 (1985): 410–427; Tony Ward, "The Origins of the Canadian Wheat Boom, 1880–1910," *Canadian Journal of Economics* 27, no. 4 (1994): 864–883. Ward's regression estimates capture other effects besides the switch to Marquis. He notes for example that the time of ripening of Red Fife declined over the period also and that changes in cultural techniques such as employing grain drills also reduced the time of ripening. A. H. Buller, *Essays on Wheat* (New York: Macmillan, 1919), 175–176, credits Marquis with giving adopters about one extra week between harvest and freezeup (which put an end to fall plowing).
105. Clark, Martin, and Ball, "Classification of American Wheat," 90–91.
106. This was not the first introduction of durum wheat. In 1853 and 1854 durum varieties were imported from what is now Algeria, Turkey, and Palestine. In 1864, the USDA introduced, tested, and distributed Arnautka wheat, but it evidently never gained commercial favor. Russian immigrants are credited for introducing the commercial stock of this variety in 1898. The USDA obtained the seed in 1900 and increased it for commercial distribution. Ball and Clark, "Experiments with Durum Wheat," 3–7; J. Allen Clark, John H. Martin, and Ralph W. Smith, "Varietal Experiments with Spring Wheat on the Northern Great Plains," *U.S. Department of Agriculture Bulletin*, no. 878 (1920): 8–9.
107. As another example, in 1900 Carleton also returned from Russia with Kharkof, a hard winter wheat adapted to the cold, dry climate in western and northern Kansas. By 1914 it accounted for about one-half of the entire Kansas crop. Carleton, "Hard Wheats," 404–408.
108. Clark, Martin, and Ball, "Classification of American Wheat," 1–238. A wheat's "vintage" is measured since first introduction. It often took a decade for new varieties to be tested on farms and begin to gain acceptance (in the case of Turkey general acceptance took over 20 years), so the mean number of years since general availability would have been much less.
109. J. Allen Clark and K. S. Quisenberry, "Distribution of the Varieties and Classes of Wheat in the United States in 1929," *U.S. Department of Agriculture Circular*, no. 283 (1933): 37.
110. China Tea, also known as Black Tea, Siberian, Java, and Early Java was imported to New York from Switzerland around 1837. Clark, Martin, and Ball, "Classification of American Wheat," 140–141. Given that it takes several years to increase the seed, the variety could not have been widely available in 1839. Thus using China Tea as our 1839 reference variety biases the case against biological innovation.

111. "Grain and Forage Crops," *North Dakota Agricultural Experiment Station (Fargo) Bulletin*, no. 10 (May 1893): 5–10; "Grain and Forage Crops," *North Dakota Agricultural Experiment Station (Fargo) Bulletin*, no. 11 (November 1893): 1–17; Minnesota Agricultural Experiment Station, *1894 Annual Report* (St. Paul, Minn.: Pioneer Press, 1895), 253–261.
112. "Grain," May 1893, 1, 12–13.
113. By 1940 several more generations of new varieties became available on the northern Great Plains.
114. Malin, *Winter Wheat*, 96–101.
115. Salmon, "Better Varieties," 210.
116. Clearly, many factors could account for the decline, but both Malin and the Kansas State Board of Agriculture credit the new hard winter wheat varieties for improving the survival rate. Malin, *Winter Wheat*, 156–159; winter kill rates for 1911–20 are calculated from Salmon, "Better Varieties," 78–79; for national winterkill data see Salmon, Mathews, and Luekel, "Half Century of Wheat," 6. The approximately twenty-year effort of farmers in Kansas to discover which varieties of wheat were best suited for a given region was simply a reenactment of what settlers in other regions of the country had experienced. As an example, in the 1840s pioneer farmers attempted to grow winter wheat on the Wisconsin prairie. Repeated failures due to winterkill eventually forced the adoption of spring varieties. Hibbard, *History of Agriculture*, 125–126.
117. Salmon, Mathews, and Leukel, "Half Century of Wheat," 14.
118. Salmon, "Better Varieties," 211–212.
119. Salmon, Mathews, and Leukel, "Half Century of Wheat," 16.
120. E. G. Montgomery, "Wheat Breeding Experiments," *Nebraska Agricultural Experiment Station Bulletin*, no. 125 (1912): 4–7. Also see T. A. Kiesselback, "Winter Wheat Investigations," *Nebraska Agricultural Experiment Station Research Bulletin*, no. 31 (1925): 6–7, 103, 107. The definition of Turkey Red lacks precision. There were several strains of Turkey Red, including Malakoff, Kharkov, Crimean, and Beloglina. All were Turkey-type wheats that had been adapted for Nebraska conditions.
121. Clark and Martin, "Varietal Experiments," 1–47. Caution is in order in interpreting their results because many of the soft varieties tested were themselves developed as hybrids between Turkey and other varieties in order to be suitable for more arid conditions.
122. Clark and Martin, "Varietal Experiments," 1. Besides Fife, another important variety grown in the Northern Great Plains was Haynes Bluestem. This was a hard red spring wheat derived from an eastern semihard, red, winter wheat. L. H. Haynes of Fargo, North Dakota, developed the variety through selection by 1885. The Minnesota experiment station further improved the variety, creating a pure-line variety, Minn. no. 169, by the late 1890s. Clark, Martin, and Ball, "Classification," 124–125.
123. The 1869 data were provided by Lee Craig, Michael Haines, and Thomas Weiss, "Development, Health, Nutrition and Mortality: The Case of the 'Antebellum Puzzle' in the United States," NBER Working Paper No. 0130 (October 2000). The 1929 data are from U.S. Bureau of the Census, *Fifteenth Census of the United States: 1930. Agriculture*. Vol. II, *Reports by States, With Statistics for Counties and a Summary for the United States*. Pt. 1, *The Northern States*, Pt. 2, *The Southern States*, Pt. 3, *The Western States* (Washington, D.C.: GPO, 1932). State-level data were published for 1909, 1919, and 1924, but they do not provide the detail we seek. The USDA did produce its own "dot" maps for the latter years and these maps largely support our findings.
124. To derive the estimates, we performed logit regressions for the winter wheat share of wheat output in each county in a given longitude grouping. We again used the county seat as a measure of the county's location. For each degree of longitude, we used the latitude where the winter wheat share equaled one-half.
125. E. E. Elliott and C. W. Lawrence, "Some New Hybrid Wheats," *Washington Agricultural Experiment Station Popular Bulletin*, no. 9 (1908): 1.
126. W. J. Spillman, "The Hybrid Wheats," *Washington Agricultural Experiment Station Bulletin*, no. 89 (1909): 7.
127. Elliott and Lawrence, "New Hybrid Wheats," 1–8.
128. E. G. Schaeffer, E. F. Gaines, and O. E. Barbee, "Two Important Varieties of Winter Wheat: A Comparison of Red Russian and Hybrid 128," *Washington Agricultural Experiment Station Bulletin*, no. 159 (1921): 7–9.
129. Spillman, "Hybrid Wheats," 7–8, 22–27; Salmon, Mathews, and Luekel, "Half Century of Wheat," 16–17, 84–86. Spillman noted that "the advantages in yield of the hybrids is even

greater in the tests made by farmers, after the seeds were distributed to them, than on these experimental plats. It appears to be safe to say that the hybrids will yield an average of five bushels more per acre, under general field conditions, than Red Russian, and in many tests, under field conditions, the difference has been as much as ten bushels, and in some cases even more." Spillman, "Hybrid Wheats," 25.

130. Elliott and Lawrence, "New Hybrid Wheats," 4.
131. E. G. Schafer, E. F. Gaines, and O. E. Barbee, "Wheat Varieties in Washington," *Washington Agricultural Experiment Station Bulletin*, no. 207 (1926): 5.
132. See George Wright Shaw, "How to Increase the Yield of Wheat in California," *California Agricultural Experiment Station Bulletin*, no. 211 (1911): 255–257; Henry F. Blanchard, "Improvement of the Wheat Crop in California," *U.S. Department of Agriculture, Bureau of Plant Industry Bulletin*, no. 178 (1910): 1–5, and Paul W. Rhode, "Learning, Capital Accumulation, and the Transformation of California Agriculture," *Journal of Economic History* 55, no. 4 (1995): 775–777, 786–787.
133. William N. Parker, "Productivity Growth in American Grain Framing: An Analysis of Its 19th-Century Sources," in *The Reinterpretation of American Economic History*, ed. Robert W. Fogel and Stanley L. Engerman (New York: Harper & Row, 1971), 176.
134. Edward O. Adusei and George W. Norton, "The Magnitude of Agricultural Maintenance Research in the USA," *Journal of Production Agriculture* 3, no. 1 (1990): 1–6.
135. John H. Klippart, *The Wheat Plant: Its Origin, Culture, Growth, Development, Composition, Varieties, Diseases, Etc.* (New York: A. O. Moore, 1860), 297–327.
136. Richard M. Adams, Brian H. Hurd, and John Reilly, "A Review of Impacts to U.S. Agricultural Resources," February 1999, retrieved May 28, 2002, from Pew Center for Climate Change at http://www.pewclimate.org/projects/env_argiculture.pdf, 1–13.
137. Quisenberry and Reitz, "Turkey Wheat," 110.
138. In their meta-analysis of the literature on rates of return to agricultural R&D, Alston, Marra, Pardey, and Wyatt reported an overall mean across 1,128 observations of 65 percent per annum. As well as the average, they discussed the large range of reported rates of return, and a general tendency for the rates of return to be biased up as a result of commonly used estimation methods. Alston and Pardey suggest that these biases notwithstanding, agricultural research has nevertheless been a highly profitable investment. J. M. Alston, M. C. Marra, P. G. Pardey, and T. J. Wyatt, "Research Returns Redux: A Meta-Analysis of Agricultural R&D Evaluations," *Australian Journal of Agriculture and Resource Economics* 44, no. 2 (2000): 185–216; Philip G. Pardey and Julian M. Alston, "Attribution and Related Problems in Assessing the Returns to Agriculture R & D," *Agricultural Economics* 25(2001): 141–152.
139. Kevin H. O'Rourke, "The European Grain Invasion, 1870–1913," *Journal of Economic History* 57, no. 4 (1997): 775–801.
140. This emphasis on small-scale, farm-specific adaptations generating an importance source of productivity growth is consistent with Engerman and Sokoloff's findings that a wide range of early-nineteenth-century "manufacturing industries were able to raise productivity at nearly modern rates" without significant capital deepening. The importance of learning by doing as a source of productivity growth in industry has long been appreciated. Given the need for farmers to match cultural methods to specific soil and climatic conditions one would suspect that learning by doing would be even more important in the agricultural sector. Stanley L. Engerman and Kenneth L. Sokoloff, "Factor Endowments, Institutions, and Differential Paths of Growth Among New World Economies," in *How Latin America Fell Behind: Essays on the Economic Histories of Brazil and Mexico, 1800–1914*, ed. Stephen Haber (Stanford, Calif.: Stanford University Press, 1997), 283.

Creating an Industrial Plant
The Biotechnology of Sugar Production in Cuba

MARK J. SMITH

In a 1947 pamphlet the Sugar Research Foundation promoted sugar as "a natural food and useful chemical," extracted from sugar-bearing plants, "exactly as made by nature, without changing in any way its structure or composition."[1] Indeed, sucrose as produced by plants and the granulated sugar that humans consume are both $C_{12}H_{22}O_{11}$, despite differences in form. However, making the plant extract "useful" meant more than turning it into a food product, something that had been done by humans for millennia. In modern times sugar came to be a commodity of mass consumption and a focus of industrial capitalism. In the words of one sugar executive, it was about "the bountiness of nature and the profitable business."[2] By the twentieth century, making sugar both "useful" and "profitable" required integrating the biological processes of the sugarcane with increasingly sophisticated mechanical processes and scientific knowledge under modern corporate control.

Sugar requires neither science nor technology to be enjoyed. It is common practice in sugar-growing regions to chew pieces of the cane, thus accessing the sweet, liquid sucrose contained in the plant's pulpy inner fibers. Humans have long consumed sucrose directly in this fashion. Over time, societies from Indonesia to the Americas learned to crush the canes and boil the juice to obtain crystalline sugar that could be transported and sold. For centuries, sugar makers using rudimentary production methods manufactured relatively small quantities of the product, marketing it as a drug or luxury sweetener for the very rich. During the nineteenth century, however, the white sweetener became a food staple of near-universal consumption, as sugar makers industrialized the production of what had become a popular commodity. While making sugar has always been about biology, in the industrializing era, sugar producers sought to integrate these natural processes with the new technologies and scientific knowledge available to those with the capital to put them into play.

Yet is sugarcane an agricultural crop or an industrial raw material? Certainly, *saccharum officinarum* is a highly domesticated plant in which biological processes convert sunlight, nutrients, and water into sucrose. On the other hand, the finished sugar product comes from a modern factory where sophisticated technology and chemical analysis combine to turn the cane juice into the far more valuable, and palatable, crystalline sugar. Further, unlike most other food products, the synthesis of the biological and technological elements takes place in a field-factory complex, a mechanism for sugar making that emerged along with the rise of the corporation. Known as *centrals*, these large-scale enterprises were industrial centers in the midst of vast agricultural operations where plants, science, and technology collectively resulted in a product destined for world markets. Humans have long used mechanical power in agriculture, but industrial sugar producers raised the interaction between organisms and technology to a new level, such that the working landscape encompassed both field and factory as a single, highly efficient unit.

This essay traces the development of sugarcane as an "industrial plant," that is, an organism linked inextricably to mechanical processes, modern science, and the capitalist goals of industrial society. Sugarcane's development accompanied changes in the "industrial plant" as used in its other sense, as a factory that converted the cane into the finished product and stood, of necessity, at the center of the entire agroindustrial enterprise. Even as sugar corporations sought to integrate the agricultural landscape with modern technology, biological factors established the conditions that influenced the productive approaches they employed. This essay shows that the development of modern industry was not only about mechanical innovation, but also depended on the basic natural processes of species that were in many ways "actors" in the drama of industrialization.

While there has been much debate in recent years about genetic interventions in plant cells, raising a host of questions about the implications for agriculture and industry, the changes discussed here took place at a different level. Technology, science, and modern corporate capitalism combined to alter a whole landscape and the relationships among its occupants. Sugar companies and their surrogates rationalized vast areas and placed a modern factory at their center. Although some efforts at breeding more productive varieties of cane took place, the plant itself remained largely unchanged and, in fact, continued to set the parameters in which industrial production of sugar took place.

To a considerable extent, the island of Cuba served as a proving ground for the development of sugar biotechnology. From its origins in the late eighteenth century, the Cuban sugar industry adopted, adapted, and invented processes of producing cane sugar from sugarcane, which represented the revolution in technology and scientific analysis that was taking place throughout the sugar sector. By the early twentieth century Cuba was the world's leading producer of sugar and the main supplier to one of the world's leading sugar consumers, the United States. Among the largest operations in Cuba was the Central Manatí, a state-of-the-art facility in eastern Cuba established in 1912 by Manuel Rionda, an entrepreneur based in New York whose family had long ties to the Cuban sugar industry. In just its first eighteen months the

Manatí Sugar Company created a model sugar operation: highly mechanized, efficient, and very profitable.

When Rionda and his investors formed their company, the time clearly was ripe for a large-scale sugar operation based in Cuba. World sugar prices and consumption were climbing in 1912. The island had, in Rionda's words, "real good sugar lands"[3] and a close geographic and economic relationship with the United States that made for a favorable business climate. Equally important was the availability of proven technology and corporate management that would allow the company to build an "ideal sugar factory" where its managers could apply "scientific methods to such an extent to get the very highest results."[4] While this discussion explores the development of sugar biotechnology generally, it also draws on the specific case of the Central Manatí to illustrate the process by which biology and modern industry combined to turn nature into profit.

Making Sugar

Making sugar begins with the biological processes of the plant. *Saccharum officinarum,* the most commonly cultivated sugarcane species, is a giant perennial grass consisting mostly of water, fiber, and from 7 to 20 percent sugars. The leaves of the plant turn sunlight, water, and nutrients into glucose and fructose, used for the plant's own metabolism. The plant also combines these simple sugars into sucrose, which is stored in the stem until needed for growth, at which time the cane turns the sucrose back into simple sugars. Environmental and cultural factors strongly influence the plant's ability to produce and store sucrose. Sugarcane grows best in warm tropical regions in soils rich in organic matter. Appearances would suggest that treeless savannahs are best for cane agriculture as they seem to require the least manpower in terms of clearing and planting. However, such lands are often deficient in nutrients, containing minimal humus, and require intensive plowing to kill the existing grasses that would compete with sugarcane. In his 1928 manual of sugarcane cultivation, Franklin Earle identified "dense virgin forest" as especially desirable for productive cane culture, an assessment that has implications for both a region's ecology as well as for the agricultural practices employed by cane farmers.[5]

Of supreme importance for sugar production is the plant's requirement for climates with ample rainfall and pronounced wet and dry seasons. Rainfall is essential when the cane is growing, but cloudy skies over long periods inhibit the mature plant's ability to photosynthesize the simple sugar needed for its own growth. In such conditions, the plant halts primary production of simple sugar and inverts the stored sucrose into fructose and glucose as it attempts to continue its life cycle. The goal of sugar production, of course, is to obtain as much of the sweet sucrose as possible. Sugarcane agriculturalists exploit the biological processes of the plant by harvesting when the cane is ripe, that is,

Fig. 3.1 The operators of the Central Manatí conducted ongoing research into sugar cane growth and agricultural techniques. Here cane growth is documented with the man provided to give scale. On the left is first-year cane, while on the right is the slightly smaller second-year cane. Braga Brothers Collection, Department of Special Collections, George A. Smathers Libraries, University of Florida.

when the sucrose stored in the stalks of the plant is at its peak volume. These biological processes determine in large measure the need for a limited, intense harvest period during dry, sunny months when sucrose content is highest and before the onset of rainy weather when sucrose inversion accelerates. Further, the sucrose in the cut canes rapidly evaporates and degrades, making it essential that extraction of the juice closely follows the harvest, preferably within a day. Unlike most other agricultural crops, sugarcane cannot be stored prior to processing into its finished product. Thus, biological processes that are largely beyond human control determine to a great extent the sucrose content of the plants. Given this, the history of sugar production has been marked by attempts to tap into these biological processes efficiently and effectively, thus yielding the valuable commodity that sugar can become.

Humans have cultivated sugarcane for millennia, yet the approach to growing the crop changed little until the expansion of the industry in the late nineteenth century. Although the plants flower and produce seed, farmers historically have propagated the crop vegetatively. This means that, after the soil is prepared, farmers plant short sections of cane in furrows or holes and cover them with soil. Sprouts appear a few weeks after planting, and the cane is ready for harvest fifteen to eighteen months later, depending for the most part on weather conditions. When the cane is deemed to have reached its maximum sucrose content, farmers give the order to harvest, and workers cut the stalks just above ground level. Harvesting has relied entirely on human power for most of its history, with mechanical harvesters only making their appearance in the middle decades of the twentieth century, although some regions have continued to use machetes up to the present. Workers cut the plants at the base of the stalk and as close to the ground as possible. It is important to avoid leaving any of the plant's leaf nodes above ground as they are likely to sprout too soon, throwing off the plantation's balanced production schedule. As a perennial, sugar plants sprout again after their first cutting, producing new growth, or ratoons, from the same root stock. The first growth produces the most sugar, with sucrose content declining as the plants age. Eventually, harvesting plants becomes uneconomical and farmers replant the fields with new root stock.

While commercial sugar production begins with agriculture, it nevertheless requires mechanical processes to yield a commodity. Extracting the liquid sucrose involves crushing or grinding the cane stalks to release the juice trapped inside the cells of the plant. Crude milling of the canes began centuries ago, using screw presses or millstones of a kind similar to those used for grain. Such mills typically lacked the power and sophistication to extract more than a small portion of the juice from the tough, woody stalks. Later sugar producers fed the canes between wood, stone, or metal rollers driven by human, animal, wind, or water power; such mechanical devices were the norm by about the early sixteenth century. Although innovations in mill technology improved the yield of sugar from the plants through the eighteenth century, the processing of cane was slow, laborious, and relatively inefficient at extracting the juice.

After obtaining the juice, producers take the liquid through various steps to remove water and impurities, aiming to reach the stage at which the liquid sucrose crystallizes into sugar that can be stored and transported. Historically, boiling reduced the juice to liquid sucrose until it began to form crystals within a slurry of molasses. The molasses was allowed to drain away to be used for alcoholic beverages or livestock feed, which left behind raw sugar of varying degrees of purity. Refinements in crystallizing sugar occurred through the late eighteenth and early nineteenth centuries; but the process of boiling and draining the juice provided, as with milling technology, little opportunity for individual operations to produce sugar on what we would think of today as an industrial scale.

Nevertheless, sugar production expanded as more tropical and subtropical regions began to cultivate cane. Major producers such as Cuba, Brazil, and Java exported large amounts of the product. Consumption rose as the price fell, especially in industrializing cities where sugar came to be a staple in urban workers' diets. Per capita consumption in Great Britain, for example, tripled over the course of the eighteenth century.[6] Producers took advantage of the growing market potential for the commodity by increasing output. While sugar makers adopted some technological innovations in the factory, they generally responded as agriculturalists: they planted more cane. Technology followed agriculture by providing a way to process the growing sugarcane tonnage through the construction of additional mills at or near the newly planted fields. As agriculturalists, sugar entrepreneurs directed most of their efforts to the crop, generally making only minor modifications to factories to improve their efficiency. Horizontal iron rollers became the standard milling apparatus in the eighteenth century, but the size and capacity of such mills was constrained by the power that was available to drive them.[7] Since the botanical processes of the cane plant required a short, intense harvest period, and since the technology available to sugar makers through the early nineteenth century limited the processing capacity of an individual sugar factory, the focus of improvement consisted largely of expanding the fields of cane and maintaining their productivity. Sugar producers brought new lands into cultivation and saw irrigation, shifting cultivation, manuring, and other agricultural techniques as the primary methods for increasing output.

By the mid-nineteenth century the impending abolition of slavery, increased global competition, and the innovations of industrialization combined to bring about revolutionary changes in sugar production. Cuban sugar planters, who had relied on slaves almost exclusively in both field and mill, saw their primary labor source threatened by the decline of slavery in the first half of the century. In addition, new competition from other cane sugar producers and, more importantly, from the rapid development of the beet sugar industry in Europe and the United States, brought serious threats to the cane sugar industry's primary markets. Along with these changes in the economic climate for sugar production, new technologies and scientific approaches made it possible for the sugar industry to expand dramatically while meeting the challenges of the era.

The Revolution in Sugar Biotechnology

Of first importance in revolutionizing sugar production was the introduction and widespread adoption of steam power. In Cuba, destined to become the world's premier sugar producer, steam made its appearance as early as the 1790s, but sugar enterprises adopted the new machines slowly. No doubt sugar planters considered the new technology unreliable at first, with expert mechanics hard to come by. Still, by the mid-nineteenth century a British visitor to Havana reported that "steam engines and engineers were coming over daily from America,"[8] and 70 percent of the island's sugar mills used steam by 1860.[9] Most other sugar zones in the Americas followed Cuba's lead with Brazil, Louisiana, and the English Caribbean abandoning water, wind, and animal power by the end of the century.

Although the earliest steam engines were notoriously unreliable, refinements and increasing availability of parts and technical skill gave the machines immense advantage over the old mills. Among steam power's chief benefits was the ability to turn the milling apparatus with much greater crushing and grinding force than could animal or water power. Mills exert as much as five hundred tons of pressure on the cane as it passes between the rollers, and the higher pressures meant a greater yield of juice from the same amount of cane, especially important as planters adopted cane varieties with tougher, woodier stems. Along with the introduction of steam came the abandonment of older varieties with thinner, easier-to-crush stems, necessary for primitive mills with relatively low crushing power. Cuban growers shifted from the older Creole cane to a variety known as *Otahití*, said to have the advantage of "a stem which is more woody, thicker, and consequently richer in combustible matter."[10] One of the producer's chief difficulties was declining forest resources that accompanied the expansion of the cane fields. Without forests and firewood, fuel was a critical problem in a process that required continuous boiling of both the juice and, later, the water for the steam engines. Sugar planters had long used cane refuse (bagasse) to fuel the boiling house, but they considered the plant material from old varieties of cane as a less-effective fuel than wood or woodier varieties of cane adopted later. Steam power made the burning of the bagasse more efficient in that the greater crushing power of the mills meant drier, more combustible refuse and made it possible to adopt varieties of cane that produced an alternative source of fuel in the factory.

The new technology also dramatically increased the speed at which a mill could extract juice from cut cane. Since cane requires processing soon after harvest, the addition of high-speed mills meant that more cane could be processed per day during the harvest period. Cane passing through the steam-powered mills, therefore, was fresher and had a higher sucrose content. In addition, large steam mills could process cane from more extensive fields, contributing to the enlargement of the plantations.

Refinements to the milling apparatus complemented the introduction of steam power. In the nineteenth century the grinding process involved three iron rollers, laid horizontally and forming a triangle when viewed from the end. Some mills included a system of knives or drums rotating at different speeds to shred the cane before it was fed between the rollers. Multiple milling allowed the cane to be passed through rollers more than once, providing for further extraction of juice. All these innovations involved heavy iron machinery, for a single roller might weigh a half ton and required the substantial power made available by steam or electricity. Although expensive, the technology allowed a much higher percentage of juice to be extracted from the cane. Although the earliest mills of the nineteenth century extracted 40 to 70 percent of the juice, later factories reached extraction rates over 90 percent.[11] By the early twentieth century factories were shifting from steam to electricity to power the milling machinery, with steam continuing to provide the heat for boiling the juice.

The processing of the extracted liquid also underwent substantial change in the nineteenth century. Modern sugar companies began to apply the scientific principle by which the boiling point of a liquid declines under lower atmospheric pressure. Therefore, open kettles gave way to a system of processing the cane juice in closed containers from which the air could be extracted. These vacuum pans dramatically lowered fuel costs in several of the processing steps, from evaporation to crystallization. The cane juice passed through the closed containers called effects, which first reduced the water content from about 85 to 40 percent. Subsequent effects reduced the sugar syrup until there was insufficient water to hold the dissolved sugar. At that point the sugar began to crystallize into a heavy liquid known as massecuite, which is piped to the next stage of the operation. Cuban producers adopted the vacuum pans as early as 1835, and by the end of the century they were standard equipment in the factories.

Cane sugar producers also adopted centrifugal processing of the liquid, an innovation introduced by their competitors in the beet sugar industry, replacing the slow, inefficient method of draining the molasses from the crystallized sugar. The centrifuge consisted of a perforated metal drum or wire mesh basket that spun the reduced sugar syrup at high speed, throwing off the liquid molasses while the crystal sugar remained behind. Patented in 1837 and adapted for sugar processing by, among others, Sir Henry Bessemer, centrifuges not only yielded a drier, more pure sugar, but greatly accelerated the process of removing molasses and impurities, requiring only a few minutes to purge the liquid from the crystals, compared to two to three weeks for the old method of drainage.

As technological processes became highly refined and efficient, producers sought to apply the principles of modern chemical analysis in order to reduce losses of sucrose in processing the product. The goal was to reach the industry standard: sugar that was 96 percent pure. For most of sugar's history there was little or no effort to analyze and control the chemical processes of its production. In the 1780s some effort was made to assess the water content of the cane juice as it passed through the various steps of processing. By doing so, sugar analysts could gauge more accurately the "strike point," the instant at which the sugar began to crystallize. As sugar producers sought greater control and rationalization of sugar production, they adopted scientific methods of increasing sophistication. Chemical analysis of the juice at various stages from field to finished product allowed companies to determine the water content of the liquid, the amount of impurities, and the vital percentage of sucrose. Mill operators introduced a device known as a polariscope to assess the sucrose content of the cane juice, seen as essential in controlling the modern factory. Adopted by the sugar industry in the 1840s, the polariscope was based on the principle that light is bent when passed through liquid or crystals. By measuring the degree of deflection, or polarization, as light passed through liquid at various stages in factory production, sugar analysts could determine not only how the process was proceeding, but would also have an overall measure of factory efficiency. Armed with such information, producers sought to "balance" their processes and "tighten up the mill," to produce the highest yields.[12]

These changes meant that sugar factories became larger, more productive, and more efficient at extracting liquid from the cane plants and at making sugar. Output from individual sugar operations grew enormously as the number of mills declined. In Cuba, production climbed from an annual average of about six hundred tons per factory in 1870 to about two thousand tons in the 1890s.[13] By the 1920s, some large Cuban centrals could boast of producing more than sixty thousand tons of sugar in a single harvest season.

The technological and scientific revolution in the factory allowed these larger mills to consume huge amounts of cane, and of course they required ever-larger amounts of their raw material. The biggest mills of 1860 could grind the cane from no more than about twelve hundred acres. By 1890 a modern factory could consume the cane from as many as 6,500 acres, and a few decades later the fields for a large central might encompass more than fifty thousand acres. The cultural practices on these fields, however large, had changed little over the centuries, with improvements consisting of the application of long-known agricultural techniques and irrigation. Yet as factories grew in size the imperative for a reliable source of raw material became ever more acute. Generally, the initial response to the need for more cane was simply to expand the fields. Plantation owners abandoned their small mills and directed their crop to the large, centrally located mills with their sophisticated technology and voracious appetite for cane. For cane planters in Cuba the late nineteenth century was complicated by a long civil war in the 1870s and a period of low prices brought on by increasing competition, especially from beet sugar. As the technology advanced, many of the plantation owners were unable to acquire the expensive modern machinery that was becoming the standard in the industry. Without modern mills many producers became specialists in cane growing, known in Cuba as *colonos*, those who sold their crop to the new factories for a percentage of the mill output. While some of the cane specialists were independent landowners, with widely varying cane acreages, others rented land from large plantations or directly from the corporate-owned sugar operations. The colonos sought to maximize the amount of cane delivered to the mill by growing cane exclusively, concentrating on the agricultural side of the industry and leaving the manufacturing to the large centrals. Near the end of the century, sugar makers began to bind the colonos by contract and debt to sell their cane exclusively to a particular mill at a specified price. Both those who owned their land and those who rented received a fixed amount of sugar, or its value in cash, for a specified weight of cane delivered to the mill. Not only did the colono system help to ensure a continuous supply of raw material to the mill, it fostered further milling improvements and mechanization by providing an incentive to increase the amount of sugar extracted from the cane.

Since the contract between the company and the colono specified the price paid by cane weight, the factory could increase profits by extracting more sugar from a given dollar value of cane.[14]

After the Cuban war of independence ended in 1898, U.S. funding flooded into the Cuban sugar industry such that by 1902 North Americans controlled roughly 40 percent of the island's sugar production. While U.S. capital fostered the expansion of Cuban sugar production, the actual number of mills continued to decline. By 1914 there were 173 centrals, perhaps 10 percent of the number in the mid-nineteenth century. These were clearly industrial operations, the biggest of which could grind some two thousand tons of cane each day. The operations had high fixed costs tied up in their machinery and manpower, and needed to be ensured of a continuous raw material supply in order to remain productive and profitable. The corporate owners saw controlling cane production as essential, and land concentration was the key to achieving such control. Many colonos who had previously owned their lands now became tenants, farming the sugar company's land and providing the cane to the mill at specified rates, often tied to the market price of sugar. This system gave the manufacturers the means to keep the mills supplied with cane while passing some of the burden of sugar price declines to the cane farmer.

As the cultivated land surrounding the central factories expanded, the problem became how to move the cut cane from the often-distant fields to the mill before the sucrose content declined. Compared with other agricultural crops, sugarcane is bulky, the valuable part is held in its heavy, woody stem, and it cannot be stored prior to processing. Railroads were the obvious solution for the conditions created by the botanical properties of the cane plant, and railroad expansion accompanied the emergence of the central mills throughout the sugar sector. Cuba had been at the forefront of railroad technology since the early nineteenth century, and with the expansion of the fields, rails and trains came to characterize the landscape in the sugar growing regions. Centrals and large plantations built their own private railroad lines not only to transport their crop from field to factory, but also to tie factories to the island-wide rail network and beyond, to world markets. In addition, after the war of 1898 the U.S. military government reorganized rail transportation and repaired miles of track that had been damaged or destroyed in the fighting or by neglect. The extension of rail lines eastward opened up areas to cane production that previously had not been cultivated for any export crops. While the sugar zone in Cuba had largely been confined to the western end of the island, around Havana and Matanzas, growers expanded eastward rapidly in the early twentieth century, clearing large areas of the forested land seen as most suitable for cane cultivation.

As expansion of the fields and improvements in transporting the crop helped to enhance the productivity of the factory by ensuring access to its raw material, growers also began to give greater attention to field productivity in the late nineteenth century. For decades virtually all cane grown in most regions was a single variety known by various names, most often in the West as Creole. This cane was notable for its soft, easy-to-mill tissue but, compared to other varieties, produced low tonnage per acre. In his important text on sugarcane culture, Earle notes that the Creole variety "has no possible commercial value" in modern sugar production due to its moderate sucrose content and susceptibility to disease.[15] Yet, as noted above, the more powerful mills of the nineteenth century fostered a shift to varieties with woodier stems, which both provided an alternative fuel and a higher juice content. Varietal adoptions proceeded throughout the century, as different regions shifted to types of naturally occurring cane deemed better for their particular environmental circumstances. For example, Louisiana turned to a striped "ribbon" cane more suitable to the cooler climate, while Cuba shifted to Otaheití, considered superior for millability. The Otaheití variety, also known as "Bourbon," "Caña Blanca," or "Lahaina," remained the primary choice in Cuba for several decades, but a long series of experiments fostered a shift to Crystalina. Well adapted to a wide range of soils and having a good combination of millability, sucrose content, and large yields per acre, Crystalina soon came to be one of the most widely planted commercial varieties of cane. By the 1920s the entire Cuban sugar industry came to be based on this one variety.

As single varieties came to dominate vast areas of cultivated fields the potential for "sudden failure" arose.[16] A leading sugarcane scientist pointed out "the unwisdom of basing the entire cane industry of a country on any one variety."[17] Such threats came to reality as the decades passed and the fields aged. In various sugar zones soil problems and disease led growers to search for ways to avoid the catastrophe of the collapse of their entire crop. Failure threatened the Cuban crop by the 1870s as the long-cultivated soils in the West became compacted and depleted of nutrients, so cane growers adopted new varieties with root systems better adapted to the declining soil conditions.

At the forefront of what was becoming a "varietal revolution" were agricultural experiment stations and industry-sponsored research programs. While cane growers generally propagated the plant from cuttings, sugarcane scientists in the 1880s recognized that some varieties set seeds that could be used to develop new breeds of cane. Still, new varieties could be hybridized with specific characteristics to address particular problems and to improve yields in both field and factory. Spurred by research in Java and Barbados, sugar regions around the world developed modern breeding programs, and growers soon had access to new, hybrid varieties, as well as other cultural information that emerged from scientific research. The programs seemed to evolve in response to problems with disease and other sources of crop failures. Java's program developed following a major outbreak of the serah, a hereditary infection of the cane plant, and research in Puerto Rico may have been stimulated by the sudden and serious threat of the mosaic virus. Louisiana growers developed their research agenda through the sponsorship of the Louisiana Sugar Planters' Association, which convinced the U.S. Department of Agriculture to begin an experimentation program in the 1880s.

Cuba, although one of the major players in technological innovation, lagged behind other regions in sugarcane research, possibly because new lands for cane cultivation, where disease problems appeared less often, were available well into the twentieth century. In the 1920s a major outbreak of a cane disease known as mosaic, however, encouraged the country to expand their research efforts. Cuban sugar producers sought government support to develop "the completest and most modern facilities for conducting and applying scientific researches in the improvement of agriculture and industrial methods, with special attention to sugar cane and its products." The new station and related "technological, agricultural, and industrial institute" would be supported with a tax on sugar consumed in Cuba, providing funding of $728,000 annually.[18] Among the research efforts undertaken in Cuba were attempts to determine the susceptibility of the island's dominant Crystalina variety to the mosaic disease and to investigate new varieties. Wrote one sugar grower: "Crystalina cane . . . seems to have practically played out in the last few years. I don't believe that we should rely entirely on one variety."[19] The finding that field productivity of Crystalina declined by 84 percent after infection by mosaic led to the development and adoption of new hybrid varieties with greater disease resistance and higher productivity on the aging fields. By 1928 F. S. Earle's textbook on cane culture contained a list of several hundred natural and hybrid varieties, a list the author describes as of then "incomplete."[20] While the threat of disease outbreaks fostered the first interest in breeding programs, the research addressed the development of varieties for specific climate conditions, growth characteristics, yields per acre, and improved millability. Within a few decades the vast majority of cultivated cane in the world came from varieties developed through cane research programs.[21]

Although most of the research conducted in the fields, factories, and experiment stations concentrated on producing sugar for human consumption, efforts to find other, nonfood uses for both cane and its extracts emerged. By the 1940s the Sugar Research Foundation could boast of sugar's use in "scores of curious and unsuspected ways." The foundation sponsored or undertook research "directed to finding greater usefulness for sugar as a chemical." Sugar found its way into "hair tonics and shoe polishes, in adhesives, photographic materials and explosives, in tanning leather and silvering mirrors."[22] Molasses, one of the sugar factory's chief byproducts, had long been used for the production of alcohol, for rum and other beverages as well as an important industrial solvent and basic raw material for synthetic rubber. Bagasse had been a primary fuel source in the boiling house since the industry began, but in the twentieth century sugar companies experimented with using the mill byproduct for wallboard and cellulose paper. Several Cuban firms launched projects to use bagasse, including a paper-making operation called Celulosa Cubana, S.A., developed by the family of sugar entrepreneur Manuel Rionda. Also, sugarcane stalks have a waxy coating that sugar producers heretofore considered an impurity to be removed during processing. But researchers determined that the waxy residue could serve as a substitute for other commercially valuable wax products.[23] The list of sugar and sugarcane-related nonfood products was quite long by the 1940s; the Sugar Research Foundation listed dozens of industrial and household substances resulting from sugar research programs and corporate development.[24]

Sugar Biotechnology at the Central Manatí

To a considerable extent, the modern corporation presided over the changes in both factory and field. Some scholars have seen the emergence of the *central*, the large-scale factory drawing cane from surrounding, separately owned cane farms, as the end of "the self-contained, or vertically integrated, structure of the sugar mill/plantation complex."[25] Indeed, from a business standpoint the processes of sugar production reflected a separation between the agricultural and industrial operations, with increasing specialization among the farming, manufacturing, marketing, and management activities. On the other hand, with the advances in both technology and agriculture, the modern central represented an integration of field and factory on a scale that went far beyond the mill/plantation of earlier times. The central developed by Manuel Rionda at Manatí serves as an example of such integration in the modern sugar industry.

Prior to the formation of the Manatí Sugar Company, Rionda and his family had developed and operated several other large centrals in Cuba. But, according to Rionda, the company's success was due largely to the vast acreage of cane and highly profitable market for Cuban sugar, the latter thanks primarily to the privileged place of Cuban exports in the U.S. tariff structure. By 1912 he was expressing grave concerns about how his company was "exhausting our lands," "wasting cane," and failing to apply "scientific methods," all of which should not be tolerated in "any large industrial corporation." He complained to his lawyer that "our abilities, economic methods, and modern machinery have no reasons to be credited with any of the financial success of the company in the past." Therefore, a new sugar company should be created to address both the manufacturing and agricultural aspects of sugar production, as Rionda said, to get "more sugar out of a ton of cane and more cane by the acre of land."[26]

The scale of the Manatí project is clear from the company's overall development costs. Rionda capitalized the company with nearly $2 million from various investors, and set about buying land, ordering equipment, constructing the buildings, and creating the infrastructure for the estate. By 1917, appraisers valued the company's properties at $13,172,746, including over $5.3 million in land and $3.6 million in the sugar factory. The railroad lines and rolling stock

Fig. 3.2 The sugar factory at Manatí. The main building covered over 100,000 square feet and rose to a height of 106 feet, not including the 200-foot-high brick and steel chimney. The facility included state-of-the-art sugar technology designed by the Honolulu Iron Works. Braga Brothers Collection, Department of Special Collections, George A. Smathers Libraries, University of Florida.

came in at about $2.3 million, with the remainder made up by accessory property including fencing, hoists, telephone lines, and pumping equipment for the wells. Although a modern, state-of-the-art industrial operation, the company's property included 983 oxen, as well as mules, horses, and carts.[27]

The tract assembled by Rionda was six to seven miles wide and about twenty-five miles long, encompassing just over 95,000 acres in eastern Cuba on the northern coast of the island bordering Manatí Bay. Described as generally flat, or "slightly rolling," the territory included some grasslands of little agricultural value, but thousands of acres were suitable for cane cultivation. Inspectors hired by the company found large areas of woodlands with high quality soils containing substantial humus and deemed "most fertile."[28] In the course of preparing the land for cultivation, the company would have thousands of acres destined for cane fields slashed and burned, replacing a diverse vegetation with the single species on which they would base their entire corporate enterprise.

The pattern of planting and harvesting cane illustrates the relationship between the botanical processes of sugarcane and the industrial-scale mechanization of the factory. Rionda intended to have about 3,300 acres sown in cane by the end of 1912, with the first harvest coming a little more than a year later. The colonos planted additional fields by June 1913, giving the central just over ten thousand acres of cane for the first harvest season, or *zafra*. Sugar growers refer to the cane planted in the fall as *frío*, for the cold-weather period of its harvest. Manatí's management expected their *frío* cane, planted in the fall of 1912, to produce about 112,000 tons of cane for grinding between December 1913 and March 1914. The cane sown in the first half of 1913, called *primavera* cane, would produce about 87,000 tons for grinding in the spring of 1914. By sowing their cane in this *frío-primavera* pattern, the company took advantage of sugarcane's process of producing sucrose and ensured itself of a continuous supply of cane for a six-month grinding period. Although some of the cane would be cut slightly before or after its period of maximum sucrose, the greater part would be ground at maturity, thus maintaining a generally high sugar yield. Further they would complete the harvest before the clouds and rain of summer shut off primary production of sucrose within the planted canes. On the other hand, the cane's biological processes that required the company to restrict the grinding period to the six months when the cane's sucrose content made it economically viable. During this period managers could run their factories nearly around the clock, attempting to maximize efficiency and productivity in an operating window over which they had limited control. At Manatí, as the first harvest period approached, some 200,000 tons of perishable raw material stood in the fields, the value of which depended on the productive capacity of Rionda's "ideal sugar factory," then rising over the Cuban landscape.

The factory site of a modern central comprised much more than the mill buildings. The *batey*, as it was known in Cuba, was akin to an industrial community that served as a hub for the agricultural, manufacturing, and transportation operations. At Manatí, the *batey* encompassed 160 acres including the factory proper, residential areas, and space for the "laying of sufficient tracks to take care of all the incoming cane and outgoing sugar and molasses."[29] A main street nearly a half mile long separated the industrial section from the living quarters of the two thousand to three thousand workers employed in and around the mill. The site also included offices, stores, restaurants, a hotel, theater, post office, hospital, and school.

The focal point of the *batey* was, of course, the factory itself, including several adjoining buildings that housed the technology required to extract the liquid from the cane and produce the finished sugar. The buildings, known collectively as the "sugar house," covered over 100,000 square feet and rose to a height of 106 feet, not including the 200-foot-high brick and steel chimney. The Honolulu Iron Works designed the new factory using the most up-to-date equipment and advanced techniques for making sugar. A crusher, four three-roller mills, three vacuum pans, forty-six crystallizers, twenty-four centrifugals, and ten four-thousand horsepower boilers gave the sugar house the ability to extract and process the sucrose from some two thousand tons of cane each day.[30] Additional equipment would be added in subsequent years, making it possible for the factory to grind over 700,000 tons of cane in the 1922–1923 harvest.

In keeping with Rionda's goal of building a factory that could "be worked to the greatest efficiency,"[31] Manatí's operators sought to tweak the equipment to ensure the highest yield possible. As in other centrales, Manatí assessed its productivity and efficiency in several ways. First, the resident chemist determined the amount of sucrose in the juice as a measure of the cane's total sucrose content, ranging botanically from 7 to 20 percent. The measure of sucrose content reflects ecological and agricultural factors such as climate, soil conditions, and cane variety. Sucrose content can vary from year to year, depending on rainfall or on the proportion of new fields to old fields then in production. Yield, on the other hand, is the percentage of raw sugar obtained from the cane that entered the factory. In general, sucrose content of the canes at harvest, the speed with which growers delivered cane to the mill, and the efficiency of factory operations combined to determine the yield of sugar ultimately produced.[32] Differences between the sucrose content of the canes and the yield of sugar from the factory are considered "losses," and the company laboratory was constantly on the lookout for sucrose left behind in the bagasse, the molasses, and in the impurities filtered out of the juice. In 1916, the mill manager expressed dismay about the increasing "undetermined" losses and reported that they had completed "about four hundred experiments" to investigate their causes. The factory managers replaced motors, adjusted scales, washed filters, cleaned equipment more often, and shifted employees from one job to another in their effort to make the factory run more efficiently.[33]

An example of a factory refinement adopted by the company in its second year was a process known as maceration, in which freshwater was sprayed on the mat of bagasse as it emerged from individual milling apparatuses. The water "coming in intimate contact with the cells containing the sugar, soften[ed] them, and upon the pressing in the succeeding mill an increased fraction of sugar [was] obtained."[34] Maceration, in combination with a further pass through a milling machine, yielded an additional .5 to .75 percent of sugar. The factory analysts watched these figures closely as they translated them directly into corporate profits. As Rionda noted, a 1 percent wastage for a 375,000-ton harvest meant that the company was "throwing away" $225,000 worth of sugar values at current prices.[35]

Yet having a factory with a large productive capacity and high efficiency was of little value if the cane was not delivered promptly and in great quantity from the fields. Because of the sizable distances involved, sugar companies saw railroads as the only means of moving the cut cane rapidly from the fields to the factories and for shipping the 325-pound bags of raw sugar from the mills to markets. With some 150 square miles of land at the outset, Manatí needed a network of railways to ensure efficient transportation of the mill's raw material and finished product. The first rail line from the company's private shipping port on Manatí Bay through the *batey* and on to the fields covered about twenty-five miles. By 1919 the main line and the spur tracks extended 132 miles from the *batey* through the surrounding fields. Rolling stock included twenty-nine locomotives and 680 cane cars, in addition to various other types of train cars for supplies, passengers, and bagged sugar.[36] Efficient operation of the mill depended on the timely arrival of the trains bearing the cut cane. The cane cars transported their cargo directly into the factory where they were emptied onto conveyors that carried the cane to the grinding apparatus. As much as possible the company sought to keep the cane in constant motion from the time it was cut until it dropped into the maw (or mouth) of the grinding mills; any delays between field and factory would mean financial losses for the company and its shareholders. In many ways, then, the company's private railroad formed the framework for the biotechnological operation of making sugar. With the railroad in place, the fields, the factory, and the port became an integrated unit in which a continuous process incorporated the biology of the cane plant and the technology of the factory.

The Central Manatí ground its first cane on January 19, 1914, remarkably, fewer than twenty months after land clearing began. During the mill's first season, cane farmers and their laborers harvested 188,560 tons of cane, from which the mill produced nearly 22,000 tons of raw sugar. The amount of cane ground on most of the 160 days that the mill was in operation was about the same as that delivered from the fields, an important accomplishment that helped to avoid sucrose loss after harvest. Although the company had exceeded its production goals, there was enough cane still standing in the fields to have made an additional 2,437 tons of sugar when the summer rains brought grinding to a halt.[37] Generally, the condition of the cane determined when to begin and end the harvest and which fields to cut. The company relied on close observation of the growing plants as well as chemical analysis of sample canes to make such decisions. As fields reached maturity the company gave the order to cut canes, and when they determined that the sucrose content of the plants had declined to such a point that grinding became uneconomical the management called the crop finished. To a considerable extent, then, the factory is in a race with nature. Since storing cut cane is not an option, a factory with a higher capacity and greater productivity reduces losses in the field by providing the means to grind as much cane as the farmers grow.

While the mill managers sought to maintain high yields of sugar and raise the capacity of the factory, the company also increased the supply of its raw material. The amount of the cane area available to Manatí grew rapidly, from 10,156 acres for the first season to nearly 65,000 acres ten years later. While some centrals bought their cane exclusively from independent landowners, Manatí initially owned most of the land themselves and either leased fields to independent colonos or farmed the land as "administration" or company fields. Later the company increased its use of cane from land "controlled" because the owners had "no other present nor apparently imminent outlet for sugar cane" than the Central Manatí.[38] Their proximity to the company's railroad made these landowners "tributaries" to the central, obligated to sell their cane to Manatí if they wished to grow and sell cane at all.[39] Within a decade of its founding, the company had the potential to control agricultural production in an area encompassing four hundred square miles, giving the Manatí Sugar Company a " very cheap supply of cane."[40] Although only a fraction of this land was actually under cultivation, Manatí's sugar zone could provide a supply of cane "much greater even than the ultimate capacity of the factory."[41]

Matters of sugarcane biology influenced company operations from the field to the corporate headquarters. Forecasting labor requirements, budgeting new plantings, and timing factory operations all relied on precise assessments of both the quantity and the quality of the "inventory" of raw material as it grew in the fields. Further, the raw material effectively was part of the manufacturing process. The plant itself produced the sugar that was extracted by the company, which added to the sucrose's commercial value by turning it into a useful and profitable commodity. The purchase of cane from colonos accounted for the largest share of the cost involved in producing sugar. In the Cuban system, the price paid for cane varied along with the price received for the sugar; Manatí's colonos received the market value of five pounds of sugar for each one hundred pounds of cane delivered to the mill. During its first decade of production, the cost of cane for the Manatí Sugar Company averaged 41 percent of the price received for its sugar, or 1.93 cents for each pound of sugar sold between 1914 and 1923. Thus, as the cost of cane for the mill was relatively stable, linked, as it was, to the market price for the commodity, the company wanted to be sure that the cane they bought had a high content of sugar. In addition, since most of the cane used by the Central Manatí grew on company-owned land, they sought to ensure that cane farmers used the land productively.

The cane farmers worked independently but were tied to the company through contracts and close supervision to ensure that the fields were "properly cultivated . . . well drained and clear of weeds."[42] By contract the company had the right to inspect the cane farms leased to colonos to ensure that the farmers managed the fields efficiently and that the yields of cane and its sucrose content were high.[43] As noted above, the productivity of these fields had important implications for the success of factory operations. Rionda had made it clear that getting more cane per acre was essential to the success of the new venture, and fields with low sucrose content had little economic value to the company. Although they saw their company as an industrial firm, Manatí's directors noted that "a sugar factory differs from other factories in that it is dependent for its output upon the growth of sugarcane on tributary land."[44] Regardless of how large, modern, and efficient the factory, low productivity in the field would reduce the ultimate production of sugar and the company's profits. Field productivity is a highly complicated issue that depends on factors such as ecological conditions, cane variety, and cultural practices. The most important determinant for the company's overall cane supply seemed to be whether the cane was first-year plant cane or subsequent-year ratoon. Recall that sugarcane is a perennial, and after a newly planted field is cut, shoots develop from the original root stock to form ratoons; the cycle of cutting and growing in the same field can continue for years. Generally, the amount of cane produced in a newly planted field was significantly higher than in subsequent years. For example, the fields planted for Manatí's first harvest produced an average of about fifty-three tons of cane per acre. The ratoons from these same fields six years later produced only about nineteen tons per acre.[45] Nevertheless, while fields of ratoons are less productive than new fields, in the words of one sugarcane technologist, "ratoon crops are always cheaper than those grown from new plantings," given the high cost of clearing and sowing the initial crop.[46]

Equally important for the company was the fact that they developed their enterprise on new lands that they considered of exceptionally high fertility. The Manatí tract gave the company a "material advantage over the majority of the estates on the island, especially as the sucrose and purity of the juices of the canes grown on these lands have been materially higher than in most newly developed districts."[47] Problems that plagued other sugar areas, especially the older zones in western Cuba, seemed largely absent at Manatí. After about five years of production the central's manager expressed some concern about soil problems on the estate's land, but in general believed that lands could produce abundant cane for many years. The high fertility of the soils at Manatí allowed the company to avoid the use of chemical fertilizers that the old plantations in the West were already using. The augmentation of the soil at Manatí appears to have largely consisted of the return to the fields of cane trash, molasses, and the "mud" filtered from the cane juice in the mill.[48] The company's representative in Cuba reported in 1914 that "there is plant food enough in these virgin lands to permit the cutting of crops therefrom for at least ten consecutive years."[49]

The integration of plants, science, technology, and corporate management added up to substantial profit for the Manatí Sugar Company, although the commodity's price experienced wild fluctuations through the 1920s. In addition to assuring themselves a raw material supply, the company had to deal with the costs of production and sugar market considerations. Production costs included not only the manufacturing operations in the factory, discussed earlier, but the expense of shipping the raw sugar to buyers primarily in the United States. The company had an advantage over many other producers by having a private shipping port on Manatí Bay. The company's auditors believed the port offered a savings of fifty cents per bag of sugar over the cost of shipping by public railway to a noncompany port.[50] Equally important was the sugar tariff imposed on the company's product by the United States. Because Manatí sold almost all of its sugar to refineries in the United States, the tariff was a major consideration. At the time work began on Central Manatí, Cuban sugar had a 20 percent advantage over other foreign producers in the U.S. market due to the Cuban Reciprocity Treaty of 1902. The treaty set the tariff at 1.685 cents per pound for most foreign sugars but 1.348 cents for Cuban. Before Manatí's first output reached the market, however, the United States lowered the tariff on Cuban sugar to 1.0048 cents, a shift that came at a time of rising prices due to the outbreak of war in Europe. For the Manatí Sugar Company the war brought high sugar prices and a strong motivation for expansion and increased production through 1920. The war also led to a dramatic reduction in the production of Cuba's chief competitor, European beet sugar, which had accounted for 41 percent of world sugar in 1914. Prices climbed within weeks of the war's outbreak and remained high for the duration of the conflict and for some months afterward. Although sugar prices collapsed by the summer of 1920, the Manatí Sugar Company continued to turn a handsome profit during its first decade in business.

Sugarcane and Industrialization

J. R. Zell, a sugar grower in Cuba, once pointed out that sugar is made in the fields, not in the "so-called sugar factories."[51] His point was well taken. While this essay refers to the Manatí Sugar Company as being in the business of making sugar, Zell is quite correct in noting that the cane plant makes sugar, the technology of the factory extracts it and turns it into a useful commodity. Throughout the process the sugar remains chemically $C_{12}H_{22}O_{11}$. The goal of the industrial capitalists who developed the modern sugar industry was to add value to the naturally occurring product, making it profitable. Thus, we may question whether sugarcane is an industrial raw material, although the sugar companies clearly saw the plant that way. Sugarcane did not provide the basis for the creation of something else, the definition of a raw material. Cane, in fact, was part of the process of manufacturing the commodity, an example of what might be called "living technology,"[52] a plant that synthesized molecules of fructose and glucose to make sucrose.

Sugar production exemplifies the interaction of nature and technology, an especially intimate interaction in what might be called the biotechnological revolution that accompanied the Industrial Revolution. While sugarcane biology did not change in any fundamental way, sugar production advanced dramatically in the nineteenth century through the application of the new technologies and scientific knowledge of the industrializing era. As technology progressed, sugarcane came to be seen as a raw material for industrial capitalism, the basis for producing a commodity of mass consumption. More important, applying advanced technology and scientific analysis required large amounts of capital, something that could only be marshaled by well-connected entrepreneurs and the modern corporation. The era of the industrial sugar company was born, known popularly today as "big sugar."[53]

Corporations capitalized their sugar operations with millions, building huge, sophisticated factories, and controlling the agricultural production in vast areas. As Alan Dye notes in his study of Cuban sugar, the "adoption of continuous-process, high-throughput technologies introduced economies of scale into milling" and, consequently, the "sugar enterprise passed from being the traditional self-contained plantation into a modern business that managed an enormous industrial complex."[54] From another perspective, however, throughout this transformation sugar production continued not only to rely on basic biological processes, but also to be shaped by them. *Saccharum officinarum* determined the decisions made by the industrial capitalists on matters of mill size, technology employed, analytical techniques, transportation systems, and business practices. To be sure, the nature of the raw material sets conditions for every industrial operation, whether the raw material is organic or inorganic. But the role of sugarcane in determining the fundamental factors of production shows in the extreme how biology was a critical element in the development of industrial societies.

Notes

1. Sugar Research Foundation, *Sugar: An Illustrated Story of the Production and Processing of a Natural Food and Useful Chemical* (New York: Sugar Research Foundation, 1947).
2. M. Rionda to Alfred Jaretzki, March 27, 1912. Braga Brothers Collection, University of Florida Special Collections (BBC), Series 2, Letter Book, Volume 30.
3. Rionda to Czarnikow-Rionda Co., New York, February 21, 1912, BBC, Ser. 5, Internal Memoranda.
4. M. Rionda to Jaretzki, March 27, 1912, BBC, Ser. 2, Letter Book, Vol. 30.
5. F.S. Earle, *Sugar Cane and Its Culture* (New York: John Wiley and Sons, 1928), 206.
6. Noel Deerr, *The History of Sugar*, Volume II (London: Chapman and Hall, 1949), 532.
7. For an extended analysis of the adoption of agricultural and mechanical innovations see J. H. Galloway, *The Sugar Cane Industry: An Historical Geography from Its Origins to 1914* (Cambridge: Cambridge University Press, 1989).
8. Deerr, *History of Sugar*, Vol. I, 130.
9. Galloway, *Sugar Cane Industry*, 135.
10. Alexander Von Humboldt, *Personal Narrative of Travels to the Equinoctial Regions of the New Continent During the Years, 1799–1804* (London: Longman, Rees, 1829), 200.
11. Deerr, *History of Sugar*, Vol. II, 546.
12. M. Rionda to Jaretzki, March 27, 1912, BBC, Ser. 2, Letter Book, Vol. 30.
13. Deerr, *History of Sugar*, Vol. I, 130.
14. Czarnikow-Rionda Co., "The Cuban Colono System," *Cuba Review*, December 1925, 20.
15. Earle, *Sugar Cane*, 63.
16. Earle, *Sugar Cane*, 6.
17. Earle, *Sugar Cane*, 8.
18. "New Experiment Station," *Cuba Review*, December 1929, 21.
19. Liandro Rionda to Manuel Rionda, July 19, 1928, BBC, Ser. 10, Box 21, "Cane Seed for Francisco."
20. Earle, *Sugar Cane*, 254–324.
21. Vladimir P. Timoshenko and Boris C. Swerling, *The World's Sugar: Progress and Policy* (Stanford, Calif.: Stanford University Press, 1957), 131.
22. Sugar Research Foundation, *Sugar*.
23. T. H. P. Heriot, *The Manufacture of Sugar from the Cane and Beet* (New York: Longmans, Green, 1920), 354.
24. Sugar Research Foundation, *Sugar*.
25. Alan Dye, *Cuban Sugar in the Age of Mass Production: Technology and the Economics of the Sugar Central, 1899–1929* (Stanford, Calif.: Stanford University Press, 1998), 87.

26. Dye, *Cuban Sugar.*
27. "Appraisal of Properties," BBC, Ser. 118, Vol. 1, 1917.
28. "Datos sobre terrenos, montes, etc. de la Hacienda Dumanuecos," February 3, 1912, unsigned report, BBC, Ser. 10, Manatí Sugar Company.
29. "Report of M. E. Rionda," April 14, 1913, BBC, Ser. 115, Manatí Sugar Company minutes, 13.
30. *The Louisiana Planter and Sugar Manufacturer,* October 25, 1913, 291; "Amortization Report," BBC, Ser. 118, V. 2, Appendix A.
31. M. Rionda to Jaretzki, March 27, 1912, BBC, Ser. 2, Letter Book, Vol. 30.
32. For the central's first decade of production, the company's analysts determined that the cane contained from 11.93 to 14.49 percent sucrose, and yield ranged from 10.32 to 12.39 percent.
33. E. D. Ulzurrun to M. Rionda, May 12, 1916, BBC.
34. "Amortization Report," BBC, Ser. 118, Vol. 2, 43, Appendix A.
35. "Amortization Report," BBC. Ser. 118, Vol. 2, 43, Appendix A.
36. Manatí Sugar Company, "Annual Report, Fiscal Year 1919," 9, BBC.
37. *Louisiana Planter,* July 11, 1914.
38. "Appraisal," BBC, Ser. 118, Vol. 1, 1917.
39. Manatí Sugar Company, "Annual Report, Fiscal Year 1916," 2, BBC.
40. "Appraisal," BBC, Ser. 118, Vol. 1, 1917.
41. Manatí Sugar Company, "Annual Report, Fiscal Year 1917," BBC.
42. "Appraisal," BBC, Ser. 118, Vol. 1, 1917.
43. "Colono Contract," paragraph H, BBC.
44. "Amortization Report," BBC, Ser. 118, Vol. 2, October 9, 1923.
45. The lower production of ratoon fields is related primarily to the physiology of the cane plant, and, at least in the early years, was not a matter of soil fertility. See Earle, *Sugar Cane,* 223.
46. Earle, *Sugar Cane,* 19.
47. "Appraisal," BBC, Ser. 118, Vol. 1, 1917.
48. "Amortization," BBC, 48.
49. *Louisiana Planter,* June 6, 1914, 389.
50. "Appraisal of Properties," BBC, 1917.
51. J. R. Zell, "Abandoned Virgin Soils in Cuba," *Louisiana Planter,* July 1924, 74.
52. Edmund Russell suggested this term in his opening remarks at the "Rutgers Conference on Industrializing Organisms: Plants, Animals, and Technology," April 5–6, 2002.
53. See Michael Gunwald, "When in Doubt, Blame Big Sugar," *Washington Post,* June 25, 2002, A9.
54. Dye, *Cuban Sugar,* 254.

Manufacturing Green Gold
Industrial Tree Improvement and the Power of Heredity in the Postwar United States

WILLIAM BOYD and SCOTT PRUDHAM

> Why . . . should the forester be satisfied to gather his seeds from the wild, unimproved forms that are only partially adapted to his needs? Should he not take steps to develop better strains of trees, and especially those capable of more rapid growth?
> —Lloyd Austin, "A New Enterprise in Tree Breeding," 1927[1]

> Eventually, through the application of breeding methods, we may expect to produce high yielding and otherwise desirable genotypes at will.
> —Scott S. Pauley, "The Scope of Forest Genetics," 1954[2]

Although few foresters would likely have challenged the basic maxim of early inheritance studies that "like begets like," the science and practice of forestry was relatively slow in appreciating and incorporating the role of genetic variation in influencing the viability and success of trees. Despite the widespread use of improved seeds in agriculture, most American foresters in the 1930s and 1940s did not think twice about using unimproved seeds in their reforestation efforts. Unlike farmers, foresters typically had little appreciation for genetics and made almost no effort to match seeds to particular sites.[3] For some reason, foresters saw trees as different—beyond the scale of practical human manipulation.[4] Most operated on the working assumption that trees of a given species were genetically uniform and that intraspecific variation was entirely the result of environmental influences.

Of course, there were those who did not accept such assumptions. Writing in 1929, Aldo Leopold chastised the forestry community for what he called "the highly improbable assumption that 'all trees are born free and equal'."[5] Leopold, along with Lloyd Austin and a few other pioneers, felt that forestry was missing an important opportunity, and that both genetics and environment had to be taken seriously in forestry practice. By controlling both genetic and environmental variability, timber growing could achieve much needed calculability. In Leopold's view: "what we are trying to create is not timber, but confidence that valuable timber can be made to grow; not dividends, but confidence that dividends can be made to accrue."[6] Part of the difficulty, of course, lay in the long biological time lags involved in tree growth. As Austin himself noted, "It is, at the present time, hard to interest people in the planting of trees that they know will not reach merchantable size in their generation."[7]

For these and other reasons, the "rediscovery" of Mendel's work on inheritance at the turn of the century took almost fifty years to find its way into forestry practice, and industrial tree improvement did not emerge on an operational scale in the United States until the mid-twentieth century. Yet, when it did so, the landscape of American forestry shifted quickly from a logic of extraction to one of increasingly intensive cultivation. Drawing on previous work in Europe and Scandinavia, the considerable enthusiasm for genetics and breeding, stimulated by advances in agriculture, and several early experimental tree breeding programs in the United States, forestry professionals and industry leaders began to develop systematic tree improvement programs in the South and the Pacific Northwest during the 1940s and 1950s. Over the next fifty years, as much of the U.S. forest products industry came to depend on intensively managed timber plantations, intensive selection and breeding of forest trees emerged as a driving force behind the development of these industrial forests. Today, trees of improved genetic stock supply most of the seed for reforestation on commercial forestlands, both private and public. Through selective breeding and, more recently, the application of biotechnology, timber plantations in the United States are now being subjected to a sort of biological time-space compression, providing the foundation for a massive industry and opening new opportunities for innovation via the intensification of industrial cultivation.

This essay examines the historical development of industrial tree improvement programs in the United States, focusing specifically on two dominant timber producing regions—the South and the Pacific Northwest. In both of these regions, tree improvement has been marked by extensive cooperation among competing firms, as well as between the public sector, academia, and private industry—a reflection of the considerable obstacles facing any effort to establish intensive tree breeding on a commercial scale. At the same time, organizational strategies have diverged in important ways—owing to regional differences in land tenure, forest type and history, and industry organization. While industrial tree improvement has featured the appropriation of genetic resources into the realm of industrial innovation, the subordination of tree biology to the dictates of continuous-flow industrial production carries with it significant ecological and economic implications. Pressed into the service of industry, the ecology of the forest is being radically simplified, creating new environmental risks and raising a host of political concerns. At the same time, while renewing an exhausted industry, biological intensification has opened up new opportunities for the creation of proprietary technologies—a fact that puts considerable pressure on a cooperative approach that worked so well in the past. As industrial tree improvement enters into a more mature phase—marked by the application of recombinant DNA and other modern biotechnologies—new economic actors may come to wield considerable influence over the nature and pace of innovation in the sector. In an industry where access to cheap timber once provided much of the basis of competitive advantage, control over genetic resources and proprietary technologies may soon hold the key to creating the new industrial ecologies necessary to compete in an increasingly globalized industry.

Viewed in the abstract—and in keeping with a key theme of this volume—the story of industrial tree improvement is one of biological systems being appropriated by and incorporated into industrial systems. Yet, also consistent with other chapters here, the industrialization of these organisms has required extensive involvement of state actors. It is compelling to see the state-centered dimensions of tree improvement in light of threads in contemporary critical theory as a modernizing project aimed at inscribing the gospel of efficiency into biophysical landscapes, rendering nature more "legible" via a governmentalizing vision exercised in the realm of "biopower."[8] Yet, while the modernizing view of the state is central to the history of tree improvement, this is hardly the whole story. Rather, it is critical to appreciate the fundamental role of capitalist firms in seeking to harness the power of heredity as an instrument of accumulation, and thus the layering of governmentalization and commodification in propelling industrial tree improvement. Indeed it is the combination of these projects previously in harmony but seemingly destined for conflict that underscores the tree improvement story in America.

The chapter is organized as follows. Section I briefly discusses some of the biological and economic challenges confronting industrial tree improvement. The following section then elaborates on some of the key institutional and intellectual precursors to industrial tree improvement. Sections III and IV focus on the story of industrial tree improvement in the South and the Pacific Northwest. The final section then discusses how some of the recent developments in molecular biology and biotechnology are impacting the organization and practice of industrial tree improvement.

Biological and Economic Challenges of Tree Improvement

For those interested in applying classical breeding techniques to tree improvement, a number of obstacles stood in the way.[9] Compared to crop breeders, tree breeders had to deal with very large, immobile organisms with long nonreproductive periods and generation cycles that often exceeded human life expectancy. In contrast to most commercial crops, basic knowledge of forest and tree biology was quite thin in the early twentieth century, particularly regarding the relationship between the genetic structures of trees and tree characteristics.[10] Given the large heterozygosity of forest trees, finding superior trees and seeds, matching them to particular sites, and controlling the selection process in the context of open-pollination posed numerous logistical challenges not found in crop breeding. Furthermore, the relative difficulty of controlling the biotic and abiotic environment of a particular stand of trees over the long time frame needed to judge breeding successes strongly limited the application of conventional breeding techniques.[11]

Of all the obstacles facing tree improvement, though, it was the time constraint—time to achieve phenotypic stability, time to reach reproductive maturity, and time to harvest—that presented the greatest challenge. Maximizing the rate of return on investments in tree improvement would thus depend fundamentally on optimizing the amount of genetic gain per unit of time.[12] As gains accrued, the incentives to invest in tree improvement would increase. The problem, however, was that even under the best conditions, such gains would not be apparent until a decade or more after the initiation of a formal tree improvement program. Given the risk and uncertainty associated with such long time lags, combined with the tremendous logistical and institutional challenges involved in such an effort, systematic investment in tree improvement exceeded the scope of most individual firm's investment horizons. Developing a tree improvement program in regions such as the South or the Pacific Northwest would thus have to be based on cooperation among firms, as well as between private industry and the state—all of which would take time. As Harvard professor Scott Pauley indicated in the mid-1950s, direct manipulation of forest genetics on a commercial scale was still a rather distant prospect.[13]

In addition to the various biophysical challenges confronting tree improvement, there were also a number of economic and institutional challenges. Any successful tree improvement program depended fundamentally on the creation and maintenance of a stable environment for long-term investment in forestry. Until the landscape of investment was made amenable to rational economic calculation, in other words, tree improvement would not be a viable technological opportunity. Such measures included most prominently sufficient protection from the threat of forest fires, particularly in regions such as the U.S. South, where extensive customary woodsburning practices and other activities created the potential for widespread destruction. In addition, in areas with extensive private timberland ownership, credit and tax systems needed to be made amenable to the kind of long-term investments necessary to sustain industrial tree improvement programs.[14] Finally, from an operational standpoint, any successful tree improvement program would also have to produce improved seeds in commercial quantities sufficient to meet the needs of forest regeneration—a daunting task in a region such as the South where almost two million acres per year were being regenerated in the late 1950s.[15]

Institutional Foundations and Knowledge Networks

The industrial tree improvement programs initiated in the mid-twentieth century drew on a variety of different intellectual and institutional lineages. Although much of the basic knowledge came from the larger world of agricultural genetics and breeding, the distinctive challenges facing tree breeding and the distinctive experiences associated with early efforts to recognize the role of "parentage" made it difficult (if not impossible) to simply transfer the agricultural model wholesale to the forestry context. In order to realize tree improvement on an industrial scale, therefore, new institutions and knowledge networks would have to be constructed. Although there is considerable work to be done on the intellectual and institutional precursors to industrial tree improvement, three areas in particular stand out. First, European insights into the role of seed source or provenance in tree growth provided the basic foundation for future tree improvement programs. Second, the American Breeders Association (1903–1913) served as a conduit for transmitting the insights of Mendelian genetics to the forestry community and, more important, as a network for forestry professionals interested in breeding. Third, early experiments in forest tree breeding in the United States and elsewhere, combined with the development of seed orchard techniques, provided the overall framework for intensive selection and seed production for the regeneration of industrial forests.

Geography, Race, and Seed Certification
Concern with the influence of "provenance" or seed source on the success of trees first emerged in Europe in the early nineteenth century. Early provenance studies by Vilmorin in France highlighted the role of geography in determining the differential behavior of trees that were ostensibly members of the same "species."[16] With the growing recognition that plantation failures in several European countries stemmed from the use of nonlocal seed not adapted to the conditions of particular sites, the importance of provenance research grew. In 1892, the International Union of the Forest Research Organizations (IUFRO) established a cooperative program of provenance research for major forest tree species.[17] These early studies focused primarily on determining the genetic variation of individual tree species due to their geographic origins through the use of provenance tests. Such tests involved planting trees of the same species but from different geographic regions side by side to determine the existence and extent of "racial" differences.[18] Early European provenance tests focused on Scotch pine and dated from the late nineteenth century. In the United States, the earliest provenance tests were conducted during the 1910s on Ponderosa pine and Douglas fir in the West and on loblolly pine in Louisiana.[19] As discussed below, however, systematic research on provenance did not really take off in the United States until the mid-twentieth century.

Intensive focus on "racial" variation and the importance of provenance took on added urgency in countries such as Sweden where the land area was relatively small, the forest products industry provided a major source of income, forest management techniques were advanced, and extreme climate placed serious limits on the number of viable species and the adaptability of local "races." During the late nineteenth and early twentieth centuries, Swedish reforestation efforts often utilized seed purchased from Germany. Large-scale plantation failures, particularly of Scotch pine, led to careful attention to the seed source and the development of strict standards regarding provenance. During the 1930s, Swedish researchers began to develop a set of basic principles and rules that foresters could follow in selecting seed for reforestation efforts. Based on extensive provenance testing, these researchers developed regression lines that correlated the differences in growth rate, physiological form, and other characteristics with changes in climate, latitude, and elevation, providing simple rules to govern seed selection. By mid-century, Swedish reforestation efforts progressed largely on the basis of such rules and their influence soon spread to other countries.[20]

Integrating the new knowledge regarding seed source and geographic "races" into reforestation efforts also required the development of seed certification programs aimed at verifying "racial" classifications and matching them to particular sites. Developing more precise and functional "racial" classifications for major forest tree species, in other words, had to be matched with trusted certification programs that would allow foresters to quickly determine whether a particular batch of seed was appropriate. Early seed certification programs emerged in Scandinavia and Europe, and by the 1950s most seed used in reforestation programs in these regions had been carefully evaluated for its "racial" fitness to a particular environment.[21]

Taken together, provenance testing, "racial" classification, and seed certification provided the foundation for industrial tree improvement efforts.[22] As discussed below, the major industrial tree improvement programs initiated in the South and the Pacific Northwest during the middle decades of the twentieth century could only go forward on the basis of adequate knowledge and appreciation of provenance. In order to succeed, selection and breeding of forest trees would have to take place within an overall framework of provenance studies.

The American Breeders Association and the New Science of Genetics

Stimulated by the "rediscovery" of Mendel's work on inheritance, the American Breeders Association provided an important forum for early discussion of genetics in relation to tree improvement. Founded in 1903 and composed of commercial breeders, scientists from agricultural colleges and experiment stations, USDA researchers, and other people interested in inheritance and breeding, the American Breeders Association (ABA) proved instrumental in the assimilation of Mendel's work into the practice of breeding.[23] ABA members were among the first to appreciate the predictive value of Mendelian ratios and set to work on applying the "fundamental laws of breeding" to agricultural improvement. Here was an early illustration of the marriage of science and industry in an effort to press biology into the service of commercial gain. In the vision of Willet Hays, the ABA would bring together scientists and practical breeders "in a grand cooperative effort to improve those great staple crops and magnificent species of animals." Only on the basis of such cooperation between "the breeders and the students of heredity," Hays argued, could the "wonderful potencies of heredity be harnessed and placed under the control and direction of man as are the great physical forces of nature."[24]

Although the ABA directed the majority of its attention and efforts to plant and animal breeding, the organization did include a committee on tree breeding attended by various forestry professionals, including Gifford Pinchot. During the ABA's ten-year existence (1903–1913), the tree breeding committee provided an important vehicle for disseminating the new science of genetics to the forestry community and developing a program for forest tree breeding. In large part, however, this consisted of a recognition of the pervasive lack of appreciation for the importance of heredity in forestry practices and the considerable challenges facing tree improvement.

At the 1907 meeting of the ABA, for example, the Committee on Breeding Nut and Other Forest Trees reported on its preliminary inquiry into the state of tree breeding, concluding that "very little has been attempted, and practically nothing has been accomplished in the selection and breeding of better forms, varieties or species of the commercial trees."[25] But the committee was optimistic about the potential for using "the modern knowledge of breeding as a basis for the development of methods specially applicable in the breeding of forest trees." To this effect, the committee reviewed existing work in the field, particularly in the provenance area, and recommended a series of practical steps toward a tree improvement program, including systematic attention to seed source and selection of elite trees for breeding experiments.[26]

The following year, with Gifford Pinchot as chair, the committee focused specifically on the institutional requirements for a systematic tree breeding effort, reflecting the institution building proclivities of its chairman.[27] According to Pinchot, the government had a critical role to play in tree improvement, largely because of the long biological time lags involved in tree growth.

> The life of one investigator is ... too short to produce any appreciable change in trees whose natural age may be from three to five hundred years.... To accomplish any results in breeding forest trees it is absolutely essential that there should be long-lived institutions or organizations such as the Government Forest Experiment Stations ... throughout Europe, which would carry on the experiments continuously for a long period, recording most minutely the accumulated experience. In this country, no matter how desirable such work may seem, no tangible results can be expected, unless it is undertaken by institutions which can guarantee continuity of work.[28]

Pinchot, of course, was a major proponent of a more active federal role in forestry research, establishing a Section on Silvics within the forest service in

1903 and supporting the creation of early experiment stations before his dismissal in 1910.[29] As he noted in his 1908 ABA committee report, moreover, the Forest Service had already initiated "investigations for the improvement of useful qualities of forest trees."[30] For the most part, such efforts were restricted to seed source studies and provenance testing. In his view, "[m]ethodical selection [was only] just beginning to be recognized in forestry."[31]

Later meetings of the ABA committee devoted to tree breeding reflected an increasingly sophisticated effort to develop a research agenda for the growing community of tree improvement professionals. Seed source occupied pride of place in the discussions, as most participants were beginning to realize just how enormously challenging an industrial-scale tree breeding program would be.[32] The importance of the ABA thus lay not so much in the actual progress it made in tree improvement but in the forum that it provided for collective learning. Through its efforts to bring together those interested in tree breeding and in its role in surveying the state-of-the-art in tree improvement (both nationally and internationally), the ABA facilitated the growth of an emerging knowledge network. As a result, forestry professionals and others interested in tree improvement gained a certain institutional legitimacy in the public and private sectors that would prove critical in the years to come.

Early Experimental Programs in Forest Tree Breeding

As with provenance and seed source, attention to the possibilities of breeding forest tree species dates from the nineteenth century. In 1845, for example, the German forester Johann Klotzsch reported on successful crosses between different species of pine, alder, and cedar.[33] In the United States, Luther Burbank, the so-called wizard of Santa Rosa, produced successful walnut crosses and a variety of fruit tree crosses in the late nineteenth century, some of which led to the creation of the famous Burbank plum.[34] By the early twentieth century, growing attention to experimental tree breeding led some to suggest that considerable gains were within reach.[35] Reflecting such enthusiasm, two foresters urged the forestry community in 1912 to consider the use of hybrid seeds as a basis for reforestation efforts. "[W]e have no doubt," they concluded, "that with many good lumber trees crossing would be found easy and hybrid seed could be sold with a wide margin of profit both to producer and to forester."[36]

Twelve years after this optimistic assessment, the Oxford Paper Company of Rumford, Maine, initiated the first industrial-scale breeding program aimed at producing rapid growing poplar hybrids for pulpwood reforestation. Managed by the New York Botanical Garden, the project produced some thirteen thousand hybrid seedlings from one hundred different cross combinations between thirty-four different species or varieties of poplar.[37] When the depression hit, however, the company could no longer afford to finance the program. In 1936, it transferred all hybrids and breeding records to the U.S. Forest Service Northeastern Forest Experiment Station. Such ambitious tree breeding programs, it seemed, could only be justified in the context of a prosperous and stable economic environment.[38]

In addition to the Oxford program, several other forest tree breeding efforts were initiated in the United States during the 1920s, the most important of which was the Eddy Tree Breeding Station in Placerville, California. Established in 1925 by James G. Eddy, the new program sought to replicate Burbank's breeding successes in the area of commercial forest trees.[39] Under the direction of Lloyd Austin, the Eddy station focused its attention in pines and launched an ambitious program of "geographical race tests," selection of superior trees, progeny tests, and hybridization.[40] By 1931, the station contained the most complete arboretum of pines in the world.[41] In 1932, it was reorganized as the Institute for Forest Genetics, and three years later it was donated to the U.S. Forest Service.[42] By this time, the institute had become the leading international center for research on pine breeding.

Reflecting the influence of Burbank and the general enthusiasm surrounding hybrid corn in the 1930s and 1940s, the new institute focused much of its energies on hybridizing pine species. The prospect of generating heterosis or hybrid vigor in forest trees captivated the imaginations of many of the early professionals working at the institute. By the end of World War II, the prospects for ramping up the institute's hybridization experiments to a commercial scale seemed quite favorable. Reporting on the progress made at the institute, two foresters predicted in 1947 that

> Soon there will be hybrid forest trees that may grow to harvesting size in one-half or one-third the time required for a good, nonhybrid timber tree to reach the same size.... Already we have a hybrid pine, that at three years is more than twice as high and three times as heavy as the better of its two parents. So a new era is beginning in reforestation and, although the hybridizing of forest trees seems to be at about the same stage that hybridization of corn had reached in the mid-1920s, and failures and disappointments are to be expected along the way, we confidently predict that in another 20 years, the forester will be using and discussing hybrid tree strains as casually as midwest farmers now discuss their hybrid corn.[43]

The ultimate commercial success of hybrid forest trees, however, rested on finding a cheap and practical method of vegetative propagation,[44] a challenge that turned out to be far more difficult, particularly in the case of pines, than early tree breeders recognized.[45] Because hybrids do not "breed true," seed-based regeneration strategies could not be used to maintain hybrid vigor in the progeny.[46] Use of hybrids in commercial reforestation efforts also ran the risk that the new hybrid might not be successfully adapted to the complex set of ecological factors that come into play during the long-term growth of an industrial timber plantation.[47]

Given these operational challenges, hybridization was limited largely to experimental programs. Tree improvement professionals would thus have to find alternative techniques if they hoped to realize the goal of producing improved seed in quantities necessary for commercial planting. Although not as glamorous as hybridization, selection of superior trees, combined with the development and use of seed orchard techniques, emerged as the most viable method for such a program. The key advances here came from Scandinavia. Beginning in the 1930s, the Danish forester Carl Syrach Larsen, one of the pioneers of industrial tree improvement, developed a technique for grafting cuttings from superior trees onto previously existing root stock. The resulting "ramets," as they were known, reflected the genotypes of the grafted cuttings without any residual genetic influence from the root stock. During the 1940s, Swedish foresters used Larsen's technique to select superior individuals and establish seed orchards with the grafted ramets of these selected trees. Although these practices were dismissed as unrefined by those interested in hybridization, the practical implications proved too powerful to ignore. Here was a relatively cheap and easy method to begin producing commercial-scale quantities of "superior" seed selected from elite individuals of the appropriate geographic "race." By 1950, Sweden had initiated a national tree improvement program based on elite clonal seed orchards. Following Sweden's lead, similar programs were initiated in Germany, Australia, Japan, the United Kingdom, Hungary, and the United States.[48] Industrial tree improvement had arrived.

Industrial Tree Improvement in the American South

Systematic efforts to develop tree improvement programs on an industrial scale first emerged in the U.S. South in the early 1950s. During this time, a particular constellation of factors—the growth of the pulp and paper industry and its appetite for wood, the rising value of timberlands, the increased demand for seedlings to furnish regeneration efforts, and the development of practical methods for implementing tree improvement programs—created the opportunity for those interested in tree improvement to move toward operationalizing a program for the region. One of the first formal steps in this direction came in early 1951, when a group of southern foresters convened in Atlanta to develop a tree improvement strategy for the South. This meeting, which led to the formation of the Southern Forest Tree Improvement Committee (SFTIC) later that year, underscored the growing concern among some in the forestry profession that southern regeneration efforts then under way were missing important opportunities by giving only passing attention to forest genetics.[49]

Provenance Studies

In one of its first and most important undertakings, the SFTIC established the Southwide Pine Seed Source Study under the direction of Philip Wakeley.[50] Up until this time, disregard for seed source was the rule rather than the exception among seed collectors in the South as well as the rest of the country. Most southern state nurseries obtained loblolly pine seed (the most widely used planting stock) from the cheapest source available, without making any real effort to match seed source to the particular region where seedlings would be planted. This was a major problem, leading to unnecessary planting failures.[51]

Wakeley's seed source study, which involved provenance research for the four major pine species of the South (loblolly, slash, longleaf, and shortleaf) drew on earlier European work in an effort to explicitly identify those genetic factors within tree species responsible for variation.[52] It was a massive undertaking, requiring the cooperation of federal, state, industry, and university actors in spreading the costs and dividing the labor. Wakeley's study, however, proved vital not only to efforts to plant "unimproved" seedlings in areas to which they were adapted, but also to designing future tree improvement programs for the region. Without attention to provenance, planting stock could all too easily end up on sites to which it was ill suited. Likewise, without knowing the geographic origin of particular seeds, forest geneticists would be unable to make accurate assessments of breeding successes and develop reliable pedigree analyses.[53] In Wakeley's words:

> genetic differences between individuals or local strains within a geographic race ... seem likely to be overshadowed by the genetic unsuitability of the race as a whole when stock is transferred to a less favorable place. The inescapable conclusion is that selections and hybrids must be made separately region by region, within the framework of existing geographic races. To the extent that this is true, provenance studies designed to identify such races and define their territorial boundaries are fundamental to other phases of tree improvement.[54]

Put crudely, geography mattered. Provenance studies would henceforth provide the foundation for future efforts in tree improvement throughout the region. As a result of Wakeley's work, the importance of seed source was widely accepted and southern states adopted formal seed certification programs. By the end of the 1950s, virtually all state and industry efforts in artificial regeneration drew seed from local sources.[55]

Extensive Improvement

As artificial regeneration efforts increased in the South during the 1950s, early tree improvement supporters also began to argue for selecting seed stock from superior trees. Impressed by the productivity gains being recorded in agriculture, particularly with hybrid corn, some suggested that an intensive breeding program might also bring large benefits to industrial forestry. If you were going to plant, they argued, why not plant the best. Intensive breeding with forest trees, however, represented a substantially greater challenge than that involving agricultural crops, and it was far more complicated than provenance studies. Procuring seed from phenotypically superior trees throughout the South and using this seed as the basis for the commercial production of seedlings represented a major hurdle facing those who wanted to operationalize tree improvement on an industrial scale. Such a program would have to be carefully planned and, due to the biological time lags involved, would necessarily take more than a decade to achieve any tangible results. Writing in 1954, Clemens Kaufman, the director of the School of Forestry at the University of Florida, reminded his fellow foresters that they were still operating in "what might be termed the fire protection-planting stage of forestry practice." "We must remember," he continued, "that at this time we have neither super trees nor seed from trees of proven quality, or even stands of average quality or better, phenotypically, which have been selected to supply the quantity of seed required annually." Kaufman thus suggested that the principles of genetics could be employed immediately in an extensive manner by selecting and leaving phenotypically superior trees as seed trees for natural regeneration.[56] The intensive application of genetics to tree improvement, however, would have to wait until reliable volumes of improved seed could be produced for artificial regeneration—a challenge that, more than any other single factor, led to the creation of the first industrial tree improvement cooperatives in the South during the 1950s.

Intensive Improvement and the Cooperative System

The first university-industry tree improvement cooperative in the United States was established at Texas A&M University in 1951. Initially, the cooperative included eight companies, which together provided most of the financing, while the university provided facilities and staff. Bruce Zobel, who had just received his Ph.D. in forest genetics from the University of California, was hired as the first director. Influenced by Scandinavian research in forest genetics and tree improvement, most of the early work focused on developing an applied breeding program oriented to the members' areas of operation in Texas, Louisiana, and Oklahoma. Because of the relative lack of basic knowledge on tree biology and forest genetics, however, participating firms saw the whole effort as something of a gamble.[57] These companies were investing in a long-term effort that no one could say for sure would actually result in tangible gains.

Shortly after the establishment of the Texas cooperative, industry leaders, in concert with members of the forestry community, initiated two more tree improvement cooperatives—one at the University of Florida (1955) and one at North Carolina State University (1956).[58] In 1956, Zobel moved to Raleigh and became the director of the new North Carolina State Cooperative as well as a professor of forestry at the university. The cooperative at North Carolina State, which focused primarily on loblolly pine, quickly emerged as the largest and most well known of the three. Its principal objective was to develop strains of trees with desired characteristics—yield, quality, and adaptability—and to produce seeds of these strains on a commercial scale. Reducing turnover time was the critical challenge. As Zobel put it: "The objective of our tree improvement program is to get as much improvement as possible as quickly as possible.... We are interested in gain per unit of time."[59]

As noted, the major operational objective of the cooperative was to provide commercial quantities of improved seed to meet the annual planting needs of members (approximately 400 thousand acres or 300 million seedlings in the early 1970s). At the outset, the tree improvement cooperatives promised members a 5 percent gain in volume from improved seed. Member firms would select seeds from superior trees from their own lands and exchange them with the cooperative. These "superior trees" (approximately three thousand for each cooperative) provided the genetic base from which gains could be realized in both the long-run (through advanced generation breeding) and the short-run (through intensive selection). Based on selected trees, the cooperatives established breeding or research orchards to maintain diversity and support long-run breeding efforts as well as first-generation clonal seed orchards (based on the Swedish model) to begin producing improved seed in commercial quantities.[60] In the process, researchers performed progeny testing to establish pedigree lines of improved trees, which in turn became the genetic foundation for industrial forestry in the South.

Given the turnover times associated with trees, however, applying the principles of quantitative genetics to select for and produce commercial quantities of improved seed was a long and laborious process. Indeed, even though the southern tree improvement cooperatives began operating in the 1950s, it was not until the early 1970s that commercial quantities of improved seed became available. By 1973, the North Carolina State Cooperative was only producing about half the planting stock needed by its members. Not until the end of the decade did the cooperative meet all its members' needs. For the region as a whole, the proportion of improved seedlings produced in southern nurseries (public and private) increased from about one in four in 1976 to more than 90 percent in 1986 (out of a total 1.6 billion seedlings).[61] By this time, North Carolina State Cooperative members had established some four thousand acres of seed orchards which yielded some 630 million seedlings per year (almost 40 percent of the regional total)—enough planting stock to regenerate 900 thousand acres annually.[62]

As for the actual gains from the program, the first-generation of improved trees developed through the North Carolina State Cooperative showed an average increase of 7 percent in height growth, 12 percent in stem volume, and 32 percent in harvest value. Real after-tax returns from the total investment in tree improvement were estimated to be between 17 percent and 19 percent—quite healthy by any standard. By the mid-1990s, the North Carolina State Cooperative was moving into its third generation of selection and breeding. In roughly twenty-five years of seed production, the member organizations harvested sufficient seed to plant more than thirteen billion genetically improved loblolly pines—enough to cover ninteen million acres.[63] Biological intensification had become the driving force of industrial forestry in the region.

Along with this transition to intensive forest breeding and management has come the increased risks and vulnerabilities associated with any process of ecological simplification. Though loblolly pine has been growing in parts of the South in even-aged stands for centuries, the program of intensive selection and breeding undertaken in the South since the 1950s combined with the extensive practice of artificial regeneration in the region has rendered southern timberlands more vulnerable to various insect and disease problems. As with high-input monocrop agriculture, maintaining high productivity on intensively managed timberlands has required increased applications of chemicals to combat pathogens, control competing vegetation, and make up for nutritional deficiencies.[64] Consequently, both the U.S. Forest Service and the cooperative system have allocated considerable resources to the ongoing effort to combat these problems.[65]

At the same time, major firms in the industry have also developed their own tree improvement research programs, raising new questions about ecological risk and about the long-term viability of the cooperative system. Indeed, in the 1980s and 1990s, as some of these firms began experimenting with techniques derived from molecular biology, the forestry community found itself confronting a host of difficult issues regarding ecological risk and proprietary control over genetic resources and technologies. As discussed in more detail below, although wide-scale use of the "new" biotechnologies is still in the future as far as industrial forestry is concerned, there is ample reason to suspect that as particular processes or products become strategic resources and subject to the laws governing intellectual property, the cooperative arrangement that has worked so well in the past may undergo substantial change. As one industry executive put it: "Now when you begin to look at the role of biotechnology and gene insertion and cloning your best materials you step farther away from the cooperatives. You're getting into intellectual property at that point."[66]

For the time being, however, the cooperative system continues to provide much of the foundation for tree improvement efforts in the South. Looking back over the past half century, one is struck not only by the remarkable success of these efforts in transforming large acreages of the South into highly productive industrial timber plantations, but also by the hybrid public-private character of the enterprise and, perhaps most surprising of all, by the extensive cooperation between competing firms. Part of this may stem from what some have identified as the "cooperative culture" of the forestry profession in the South during the middle decades of the twentieth century. A more important factor has surely been the complex and long-term nature of the enterprise that militated against any single firm embarking unilaterally on tree improvement.[67]

Industrial Tree Improvement in the Pacific Northwest

The development of tree improvement in the Pacific Northwest parallels the southern tree improvement story in many ways. In particular, while early scientific, experimental work on provenance variation among commercial forest trees was initiated prior to World War I and spearheaded by the federal government, it was not until the 1950s that serious industrial efforts to capture genetic gains in industrial forestry began. In addition, and in parallel with the South, willingness among rival firms to cooperate in the production of improved seed, assisted by public science, has been critical to sustaining regional tree improvement—the lone exception being Weyerhaeuser's proprietary initiative in tree improvement. In later years, and continuing to the present, state support became more important in assisting private cooperation, as public and private contributions came together under the auspices of cooperative research institutions housed at research universities in the region.

Experimental work on the geography of genetic variation in Douglas fir in western Oregon and Washington was undertaken by a U.S. Forest Service scientist named Thornton Munger beginning in 1913.[68] Munger's arrival heralded a geographic shift in the U.S. forest industry, as lumbermen turned their eyes from the cutover lands of Northeastern, Great Lakes, and Southeastern forests to the heavily timbered lands along what Earl Pomeroy termed the Pacific Slope.[69] Munger's research reflected a growing interest (public and private) in the last great stands of old-growth forests in the continental United States, and the staple tree by which the region has become known in the forest industry, the Douglas fir.[70] Moreover, as a federal scientist, Munger signified the growing profile of the federal government in the contemporaneous emergence of scientific forestry in the United States,[71] including the incipient field of forest genetics. It was Munger's work that brought West to the Douglas fir region enthusiasm for the new genetics in forestry, building on the work of the ABA.

Yet, despite Munger's early interventions, it took more than forty years for industrial tree improvement to come to life in the region. All of the same challenges to industrial tree improvement in the Southeast confronted the industry in the West. Moreover, the relatively untapped Douglas fir forests that rose to greet Munger provided a geographic fix[72] for the nation's hungry timber industry and sustained essentially extractive practices for several decades.[73] Interest in reforestation on an industrial scale—a precursor for tree improvement—only took hold during the 1940s, propelled by mounting concern about timber depletion on industrial lands in western Oregon and Washington. These concerns propelled passage of the Oregon Forest Conservation Act of 1941, the nation's first set of legislated forest practice regulations.[74] That same year, Weyerhaeuser established the nation's first tree farm near Grays Harbor, Washington. And the West Coast Lumbermen's Association, under the direction of William Greeley, established a cooperative nonprofit nursery to supply five million seedling trees per year for reforestation purposes. Seven years later, in 1948, the Industrial Forestry Association (IFA) was formed as a collaborative effort among industrial forest landowners to coordinate nursery and planting operations in the Douglas fir region.[75] It was the IFA that spearheaded regional tree improvement as a strategy for capturing genetic gain to serve burgeoning reforestation efforts.

The Northwest Tree Improvement Cooperative

In 1954 the IFA hired the region's first forest geneticist, Jack Duffield, to coordinate an applied, albeit experimental program of Douglas fir tree improvement involving multiple landowners. Duffield held a Ph.D. in genetics from the University of California at Berkeley and came to the IFA by way of the U.S. Forest Service's Institute of Forest Genetics in Albany, California, where he had been exposed to the legacy of early tree improvement research on the genetics of pine.[76] Two years later, the IFA invested $40,000 in a research facility at Nisqually, Washington, in order to support Duffield's work. At Nisqually, Duffield also drew on the work of Syrach Larson, developing a system for establishing desirable genotypes by grafting scions of selected Douglas fir onto established root systems in seed orchards.[77] Using this technique, Duffield set up the first three seed orchards in the Douglas fir region, two in western Oregon and one in western Washington.[78]

Duffield's initiative led to establishment of the region's first operational tree improvement cooperative in Vernonia, Oregon, in 1966. This cooperative was spearheaded by the Crown Zellerbach Corporation, Longview Fiber, and International Paper, with assistance from staff in the Oregon State Department of Forestry. Roy Silen of the U.S. Forest Service's Forest Science Laboratory at Oregon State also provided assistance. Using Silen's "Progressive System" of tree improvement, the Vernonia cooperative established an institutional model for formal collaboration of several firms with lands in a given area.[79] Under the program, firms selected superior-looking trees from their lands and harvested seeds from their cones. These seeds were then used to test the performance of progeny in experimental plots to identify desirable lines. A second selection process based on these plots generated so-called "plus" trees, from which scions were grafted onto established root systems in seed orchards, as per Larsen's practice.[80]

Initially, each firm in the Vernonia coop had its own seed orchard, but as the program grew in size and scope, firms began to develop cooperative seed orchards as well. As more and more firms with land in western Oregon and Washington expressed interest, Vernonia became the template for a network of local cooperatives that grew to a total of twenty-two. In turn, the members of these local coops formed an umbrella organization called the Northwest Tree Improvement Cooperative (NWTIC) to provide coordination and guidance. By the late 1980s, the NWTIC network was generating more than 80 percent of the total seed required by coop members for Douglas fir reforestation.[81]

Public Science Research Cooperatives

Despite striking parallels, the NWTIC represents a departure from institutional tree improvement practices in the South insofar as involvement by public research universities in the Northwest remained limited and informal into the 1980s. The NWTIC model, while drawing on advice provided by forestry scientists from the Forest Service and OSU's College of Forestry, was and remains an entirely private cooperative effort. However, beginning in the early 1980s, operational tree improvement under the NWTIC was supplemented by cooperative tree improvement research institutions housed at public research universities in the region, most importantly OSU in Corvallis, and the University of Washington in Seattle. Of these cooperatives, the most important to tree improvement is the University of Washington's Pacific Northwest Tree Improvement Research Cooperative (PNWTIRC).[82]

The PNWTIRC was founded in 1983 to provide research support for the network of Douglas fir tree improvement conducted under the NWTIC. That is, while the NWTIC manages actual cooperative tree improvement activities among members, the PNWTIRC acts as its research arm, providing technical assistance based on ongoing research programs. The organization's membership is composed of some of the largest integrated forest products companies in the region, together with public forest land management agencies. The extensive geographic reach represented by this membership, including participation by public and private institutions from British Columbia, Canada, attests to the existence of a common research agenda among timberland managers along the West Coast attempting to undertake systematic tree improvement. This is not least attributable to the relatively short history of regional commercial tree improvement, and thus the existence of a host of scientific and technological uncertainties common to multiple landowners across this wide region.

The research undertaken by the PNWTIRC provides a good indication of the generally commercial orientation of cooperative research, again, no surprise given the membership, funding, and governance structures of such coops. For example, major emphasis in the PNWTIRC has been placed on the early selection of Douglas fir for cold and drought resistance, identifying the genetic basis of these traits, and the relationship between these traits and growth characteristics with more direct commercial advantages (e.g., wood density).[83] Research has also been directed at breeding trees across wider geographic ranges than is practiced in the spatially fragmented structure of the NWTIC. The impulse here is clearly to capture greater genetic gains by selecting from a wider range of stock, and at the same time increase the efficiency of tree improvement by reducing institutional fragmentation.[84] The PNWTIRC is also working on techniques to more tightly control the lineage of crosses produced in the seed orchards, given the uncertainties in gene flow introduced by relying on open, wind pollination.[85]

Despite some regional variations in institutional practices, one of the remarkable aspects of the history of tree improvement in both the South and the Douglas fir region is its cooperative character. The ability of individual private firms—most of which are in some form of direct competition with one another for access to timber and markets for wood products—to cooperate extensively with one another and with public science in sharing genetic resources and underwriting research and operational tree improvement programs has allowed them to collectively address the uncertainties and delays endemic to the undertaking. The relatively open property regimes of research and operational tree improvement in the Northwest parallels the institutional organization of southern tree improvement, reflecting common underlying issues of scientific and economic uncertainty, biological delays and other challenges that encourage firms to pool their efforts. While efficient, the disadvantage of this strategy is that it does not allow firms to compete on the basis of their tree improvement programs. In this context, one would expect that with decreasing levels of uncertainty and associated increases in levels of investment in tree improvement, firms might seek to pursue more exclusive forms of association and control over improved varieties of trees.

Proprietary Tree Improvement

The lone exception to the Douglas fir region's cooperative rule—Weyerhaeuser's intensive forestry research and development program—is the exception that would seem to prove the rule. While Weyerhaeuser participates in public science research cooperatives doing work relevant to its operations, including the PNWTIRC, the company does not and has never been a member of the NWTIC's operational tree improvement program in the Pacific Northwest. Instead, Weyerhaeuser embarked on its own entirely proprietary program of tree improvement as part of a broad based-effort to intensify forest tree growth on its lands in the Douglas fir region (and elsewhere). Initiated in 1969 and dubbed the "High Yield Forestry Program," the program represented an escalation in the company's reforestation efforts coincident with the progressive exhaustion of its holdings of old-growth.[86] Under the program, Weyerhaeuser attempted to improve timber quality and yield in its timberlands through integrated stand management techniques (e.g., fertilizing, thinning, and so forth) and tree improvement. At its peak, the program employed on the order of thirty Ph.D.s in various aspects of forest science and genetics at its headquarters in Centralia, Washington, all working toward Weyerhaeuser's goal of doubling yields within four generations.[87]

Like other forestry firms embarking on tree improvement, Weyerhaeuser faced uncertainty and delays in tree improvement. However, the company's method for selecting phenotypes emphasized the establishment of immediate operational seed orchards to reduce delays in the progeny testing. The firm leaped ahead by pursuing progeny testing and operational tree improvement in parallel, not in sequence, reasoning that the tree improvement program could do no worse on average than wild breeding processes. As a result, Weyerhaeuser very quickly advanced to the point where all of its requirements for reforestation stock in western Oregon and Washington could be met with genetically improved lines. The company has predicted the first harvest from trees planted under the high yield program sometime in the next decade.[88]

This would be a considerable achievement given that Weyerhaeuser is by far the region's largest forest land owner, retaining on the order of two million acres of timberlands in the Northwest. More than half of this land was purchased in a single transaction of 900,000 acres negotiated between Frederick Weyerhaeuser of the Weyerhaeuser Timber Company of Minnesota and Weyerhaeuser's St. Paul neighbor, James J. Hill, head of the Northern Pacific Railway in December 1899, with a further 380,000 acres purchased in August 1901.[89] With this tremendous amount of land, most of it in large contiguous blocks (and thus not as fragmented by the proliferation of breeding zones interspersed among different land owners), the company has been able to undertake on its own what other firms in the region have only undertaken cooperatively. In other words, the economies of scale available from a relatively large and largely contiguous acreage has allowed Weyerhaeuser to justify its in-house approach and opt out of cooperative tree improvement.

Yet Weyerhaeuser's proprietary approach begs the question as to how firms can protect their investments as tree improvement advances. What will happen if improved trees become competitive, strategic assets? At the same time, as discussed in the following section, the advent of the "new" biotechnologies also poses difficult issues regarding ecological risk attendant with genetic modification. The combination may well lead to significant institutional reorganization in U.S. tree improvement.

Tree Improvement and the New Biotechnologies

Biotechnology has become a more significant feature in industrial forestry in recent years, including avenues by which these new techniques have been employed in cooperative tree improvement.[90] At present, there are no commercial genetically engineered (GE) forest trees in commercial cultivation in the United States. Moreover, work with conifers, staple trees of both the South and Northwest, has generally lagged other species. In fact, while the first successful regeneration of a transgenic tree (a poplar) occurred in 1987, the first successful regeneration of a transgenic conifer (a spruce) did not take place until 1993.[91] One of the main problems with the regeneration of transgenic conifers is that they are largely resistant to tissue culture techniques; thus, the first successful experimental regeneration of a non–transgenic conifer—specifically Norway spruce (*Picea abies*)—didn't take place until 1985.[92] Resistance to tissue culture and clonal propagation is a principle reason that genetic engineering of forest trees is still only routine in poplars.[93] As one forest geneticist framed it, "Producing a transgenic plant with Douglas-fir is technically very possible. Clonally propagating that [plant] and using it on a broad scale is much more expensive and technically challenging than it is in poplar, for instance."[94]

Tissue Culture and Somatic Embryogenesis

In this context, it is no surprise perhaps that some of the most ambitious private sector research in forestry biotechnology in North America has been undertaken on the development of a technology called somatic embryogenesis. This is a tissue culture technique involving in vitro production of cloned embryos. Suitable fertilized embryos are typically used (as opposed to mature cells) to supply the original plant material. These are cultured and multiplied using processes that may involve several iterations to generate enough embryos. The resulting cloned embryos can then be further cultivated in the lab before being transferred to soil to grow as cloned trees. Since its first successful use on Norway spruce in 1985, the technology has been the subject of attempts to overcome difficulties in cloning improved varieties of other conifers.[95]

While considerable research has been conducted in numerous countries on the development of somatic embryogenesis for use with conifers,[96] Weyerhaeuser and Westvaco have most aggressively pursued the technology in the United States. Weyerhaeuser has focused on Douglas fir, while Westvaco, with its extensive operations in the American southeast, has emphasized techniques suitable for use with pine. Together, these firms accounted for thirteen of the twenty-one patents for somatic embryogenesis that were issued between 1989 and March 1998.[97] Weyerhaeuser also holds nine patents related to the production of manufactured seed, a related technology for encasing the cultured embryos. Somatic embryogenesis and other micropropagation techniques may greatly facilitate the further intensification of tree improvement programs using conventional techniques as well as genetic engineering by enabling the mass production of advanced varieties. Noting exactly this appeal, Weyerhaeuser's general manager of western regeneration Stephen Hee described somatic embryogenesis as "a means to efficiently propagate selected individuals possessing the growth and quality attributes we desire."[98]

Genome Mapping and Transgenic Trees

Tissue culture is not the only challenge facing the deployment of biotechnology in tree improvement. Work with at least some forest trees has been slowed by the fact that their genomes are relatively complex and take longer to map; thus, work with trees having simpler genomes (e.g., poplars—see below) has progressed faster.[99] Increasingly, however, approaches including quantitative trait loci (QTL) techniques have been used with commercial conifer species in order to map their genetic sequences and to identify the association between particular DNA sites and commercially relevant traits. This includes cooperative research at the University of Washington, and federally sponsored research on conifer genomes has been carried out at the Institute of Forest Genetics (IFG), in Albany, California. In addition, researchers at the IFG have been experimenting with genetically engineered forest trees, including conifers. The IFG achieved a first in the late 1990s by successfully regenerating a transgenic pine tree.[100] The fact that this was a publicly funded effort underscores the significance of public research in forestry biotechnology, confronted by some of the same technical and economic challenges that have always been central to conventional tree improvement.

In fact, it is a public and cooperative institution, building on the tradition of conventional tree improvement, that is likely to become the first source of commercially deployed GE forest trees. Since its formation in 1995, the Tree Genetic Engineering Research Cooperative (TGERC) at Oregon State University in Corvallis, Oregon, has been developing GE trees for commercial forest plantations in the Northwest; prototypes of these trees are now in field trials being evaluated under regulatory guidelines prior to their use in commercial operations.

Though located in Oregon, the TGERC does not work with Douglas fir nor indeed any species of conifer. Instead, the TGERC has been engineering crosses of a hybrid between black (*Populus trichocarpa*) and eastern (*P. deltoides*) cottonwood, hybrids that are used in a relatively small cumulative acreage of pulpwood plantations, primarily located in eastern Oregon and Washington. At the TGERC, researchers successfully developed prototypes of GM cottonwoods now in field trials under the regulatory review processes of the Animal and Plant Health Inspection Service of the U.S. Department of Agriculture and the U.S. Environmental Protection Agency. Three basic types of GE trees have been developed thus far: (1) varieties with engineered sexual sterility;[101] (2) varieties with engineered insect resistance;[102] and (3) varieties with engineered herbicide resistance.[103]

In many ways the TGERC builds on and extends the cooperative tradition of tree improvement research and development. Based at OSU in Corvallis, the TGERC is a joint undertaking relying on public and private contributions and contributions of research staff and facilities from the university. Members originally included major owners and operators of cottonwood plantations in the region, and interested public agencies, including the Department of Energy and OSU. In addition, subsequent additions have included forest products giants International Paper, Georgia-Pacific, Weyerhaeuser, and Westvaco who, though without significant investments in cottonwood operations in the region, are interested more generally in potential applications of the coop's research in forestry.[104]

New Economic Spaces and Ecological Risks
The deployment of genetic engineering in tree improvement raises at least two major concerns for future tree improvement efforts. The first of these pertains

to the management of genetic resources as formal property. The second pertains to the challenges of ecological risk and its social regulation. In the realm of property, it bears repeating that the institutional foundation for cooperative tree improvement in both the South and the Northwest has been relatively fluid and open in its approach to the sharing of genetic material. Biotech may change this. The TGERC, for example, departs from this model in significant ways. Consider, for instance, that Monsanto and Mycogen joined as associate members of the TGERC several years ago, reflecting the financial stake they have in the research given that both firms have formally licensed gene constructs to the TGERC for use in the research. Specifically, both the insect resistant and the herbicide tolerant varieties have gene constructs owned by one or both of these firms. Any commercially deployed GE trees containing the genes they own would generate licensing royalties to the firms.[105] The TGERC thus suggests a more complete form of commodification of improved trees, the long-term result of which may be more exclusive partnerships between the private and academic sectors, or involving exclusively private actors.[106] In fact, the potential of more exclusive forms of partnerships involving commercial biotechnology in tree improvement was recently underscored by the announcement of a wholly private joint venture in the production of GE trees called Arborgen in January 2000.[107] To what extent this may actually lead to more fundamental shifts in the political economy of tree improvement in the United States is contingent on a number of influences, including commercial viability, regulatory approval, and wider social norms.

In many respects, these concerns turn on the question of ecological risk introduced by commercially deployed GE trees. Key concerns include the effects of genetically engineered species on wild populations, either through breeding with wild varieties or through competition with wild populations. There is also the potential for engineered genes to leak to other species, notably from crops to weeds. While these issues are significant enough to warrant a more complete treatment than we can give them here, several observations are pertinent.

First, there are good reasons to worry that introduced genes will spread to wild populations of poplar.[108] In fact, despite a lower level of public controversy, GE trees may introduce greater risk of engineered genes spreading into nontarget populations than engineered agricultural crops. The risk of gene transfer in commercial GE crop deployments is generally proportional to how closely related the donor and recipient species are,[109] and industrially cultivated trees are more closely related to noncultivated or "wild" trees than most crops. This is because, as we have noted in this essay, forest tree cultivation is a relatively recent phenomenon (compared with agricultural crop breeding), and tree breeding programs typically rely on "wild" genetic resources. Indeed, this points more generally to the potential impact of tree improvement on tree genetic diversity, since the vast majority of improved trees are fertile and may breed freely with surrounding "wild" trees.[110] What if tree improvement, however inadvertently, breeds into or out of populations via conventional or genetic engineering traits that are, respectively, detrimental or essential to their perpetuation? No one, it seems, has a very good idea of the magnitude of such risks. In the case of the TGERC, engineered sterility for controlling the risks of commercially released GE trees offers a potential solution.[111] Yet, given the close relationship between wild and cultivated trees, the margin of error in induced infertility must be very low.[112]

Second, there is also growing concern associated with the interaction of engineered species with wild populations of other species. Interspecific gene flow is a controversial topic, and one about which there is little consensus. However, recent research seems to indicate that the possibility is real and more pervasive than was previously thought.[113] One of the most significant concerns in this respect is that genes for herbicide resistance, or other competitively advantageous traits, might be passed from crop species to nuisance plants, producing superweeds. If glyphosate resistance were transferred to weeds in cottonwood plantations for example, this could prove extremely damaging to the plantations and to other crops in the region.

Finally, it is important to recognize the significance of the length of deployment of GE trees in discussions of risk. Rotations in cottonwood plantations (six to eight years) are long by agriculture's standards, but these in turn are dwarfed by the rotation ages of most commercial forest tree plantations. Not only do long rotations potentially increase the risk of unintentional gene transfer, they also introduce time lags. That is, with long delays in the maturation of forest trees, widespread deployment of potentially damaging or disruptive varieties could take place well before the negative consequences are even apparent.

Conclusions

At the Fifth World Forestry Conference held in 1960, Carl Syrach Larsen, the man who arguably invented industrial tree improvement, reflected on the progress made in the preceding decades and laid out his vision for the future. Echoing much of the enthusiasm that had animated plant and animal breeders since the turn of the century, Larsen emphasized the importance of taming the power of heredity to serve the needs of industrial forestry:

> Just as water in rivers and streams must be tamed by engineers before it can be put to work for the benefit of mankind, so genetics, one of the mightiest powers in nature, must be regulated and controlled before it can be utilized. The heritable factors, the genes in forest trees, must be studied intensively, regardless of expense, and "sorted" so that later we can make them work free of cost to build up the castle of our dreams.[114]

Such optimism, of course, had to be tempered by the massive logistical challenges involved in operationalizing tree improvement on an industrial scale.

As the preceding study of tree improvement has demonstrated, it took decades after the commencement of tree improvement programs for significant gains to be realized.

But the gains did come, and over the past thirty years the acceleration of biological productivity achieved through tree improvement has been nothing short of remarkable. As illustrated in the case studies discussed above, the subordination of tree biology to the dictates of continuous-flow industrial production required a cooperative approach to harness the power of heredity—one that drew extensively on regional, national, and international knowledge networks. In some respects, the story of industrial tree improvement is as much a story about the development of knowledge networks in resource-based industrialization as it is about the transformation of trees into industrial organisms.

One might also suggest that the story of industrial tree improvement is best seen in light of a broader narrative, one involving the modernization of nature aimed at rationalization and control. Indeed, the critical role played by state institutions, particularly in early forest genetics and tree improvement, suggests the relevance of the governmentality literature, and theories of biopower, as modern state administrations have sought to expand their scope via what James Scott has framed as rendering nature "legible."[115] Although not discussed in detail in this essay, it is clear that much of the early enthusiasm for tree improvement emerged out of and drew sustenance from the progressive conservation movement, a largely expansionist period for state administration of American lands and resources. What better illustration of the gospel of efficiency than a rational, scientific program aimed at accelerating tree growth and enhancing desirable characteristics? But to cast tree improvement simply as one facet of a broader canvass of state legibility and simplification would be a mistake. Indeed, while the state did play an important role in establishing many of the background conditions necessary for tree improvement, it was seldom if ever the prime mover in the industrial tree improvement programs that flourished in the second half of the twentieth century. This was (and is) ultimately a story about how capital seeks to harness reproductive biology as a productive force.[116]

Where all of this is heading remains in question. As biological intensification has come to be the driving force behind industrial forestry in North America and much of the rest of the world, new challenges and vulnerabilities have emerged. Industrial tree improvement programs inevitably carry with them new forms of environmental risk attendant with dramatic ecological simplification, including susceptibility to new pathogens and other largely unknown consequences from reducing the genetic diversity of "natural" forests. These risks are only heightened by the seemingly imminent introduction of genetic engineering into commercial tree improvement programs. At the same time, as the "new" biotechnologies are employed in forest tree breeding, proprietary issues will very likely alter the cooperative institutional arrangements that served the tree improvement enterprise so well in the past. As in agriculture, the extension of tree improvement into the realm of genetic engineering and the new biotechnologies poses at least as many challenges as it does opportunities, the struggle over which will likely shape the future of tree improvement, and indeed, the entire forestry sector.

Notes

This chapter draws on William Boyd, "The Forest is the Future? Industrial Forestry and the Southern Pulp and Paper Complex," in *The Second Wave: Southern Industrialization from the 1940s to the 1970s*, Philip Scranton, ed. (Athens: University of Georgia Press, 2001); and Scott Prudam, "Taming Trees: Capital, Science, and Nature in Pacific Slope Tree Improvement." Forthcoming, *Annals of the Association of American Geographers* (September 2003).

1. Lloyd Austin, "A New Enterprise in Forest Tree Breeding," *Journal of Forestry* 25 (1927): 928.
2. Scott S. Pauley, "The Scope of Forest Genetics," *Journal of Forestry* 52 (1954): 644.
3. As one forester noted in 1950, "Interest and activity in forest genetics . . . had hardly begun to make appreciable inroads into the rank and file of the profession." Ernst J. Schreiner, "Genetics in Relation to Forestry," *Journal of Forestry* 48 (1950): 37.
4. Bruce J. Zobel, "Forest Tree Improvement—Past and Present," in *Advances in Forest Genetics*, ed. P. K. Khosla (New Delhi: Ambika Publications, 1981), 13.
5. Aldo Leopold, "Some Thoughts on Forest Genetics," *Journal of Forestry* 27 (1929): 710.
6. Leopold, "Forest Genetics," 710.
7. Lloyd Austin, "A New Enterprise in Forest Tree Breeding," *Journal of Forestry* 25 (1927): 928. To put things in perspective, one molecular biologist noted in 1996 that "[t]ree breeding is like reinventing agriculture. Maize and wheat have gone through 5 to 10 thousand generations of cultivation and selective breeding, while the most advanced forest tree-breeding programs are in their third generation." Ronald Sederoff of North Carolina State University, quoted in Anne Simon Moffat, "Moving Forest Trees into the Modern Genetics Era," 971 *Science* 760 (February 9, 1996).
8. For discussion of the notions of governmentality and biopower, see G. Burchall, C. Gordon and P. Miller, eds., *The Foucault Effect: Studies in Governmentality* (Chicago: University of Chicago Press 1991), 87–104; Michel Foucault, *Discipline and Punish: The Birth of the Prison* (New York: Vintage Books 1979). See also James Scott, *Seeing Like a State: How Certain Schemes to Improve the Human Condition Have Failed* (New Haven, Conn.: Yale University Press. 1998), esp. chapters 1 and 8. On applications of these notions to scientific forestry and particularly sustained yield regulation, see D. Demeritt, "Scientific Forest Conservation and the Statistical Picturing of Nature's Limits in the Progressive-Era United States." *Environment and Planning D-Society & Space* 19 (2001): 431–459.
9. According to Bruce Zobel, tree improvement involves the combination of genetic selection and improved silvicultural practices to increase the productivity of forest trees. Major objectives include: (1) adaptability; (2) resistance to pests and diseases; (3) growth rate; (4) tree form and quality; and (5) wood qualities. Bruce J. Zobel, "Increasing Productivity of Forest Lands Through Better Trees," *The S.J. Hall Lectureship in Industrial Forestry*, April 18, 1974 (Berkeley: University of California, Berkeley School of Forestry and Conservation, 1974), 4–5.
10. G. Muller-Starck and H. R. Gregorius, "Analysis of Mating Systems in Forest Trees"; and J. P. van Buijtenen, "Quantitative Genetics in Forestry," both in *Proceedings of the Second International Conference on Quantitative Genetics*, ed. B. S. Weir, E. J. Eisen, M. M. Goodman, and G. Namkoong (Sunderland, Mass.: Sinauer Associates, 1988), 573–574; C. Gaston, S. Globerman, and I. Vertinsky, "Biotechnology in Forestry: Technological and Economic Perspectives," *Technological Forecasting and Social Change* 50 (1995): 80; and Bruce J. Zobel and Jerry R. Sprague, *A Forestry Revolution: The History of Tree Improvement in the Southern United States* (Durham: Carolina Academic Press, 1993), 24–29
11. Tree breeders often have to wait as much as twenty to thirty years (depending on the species) before mature traits can be evaluated. See William L. Olsen, "Molecular Biology in

Forestry Research: A Review," in *Forest and Crop Biotechnology: Progress and Prospects*, ed. Frederick A. Valentine (New York: Springer-Verlag, 1988), 315–316.

12. S. E. McKeand and R. J. Weir, "Economic Benefits of an Aggressive Breeding Program," *Proceedings of the 17th Southern Forest Tree Improvement Conference* (1983): 100.
13. Scott S. Pauley, "The Scope of Forest Genetics," *Journal of Forestry* 52 (1954): 644.
14. For a discussion of efforts to rationalize the landscape of timberland investment in the American South, see William Boyd, "The Forest Is the Future? Industrial Forestry and the Southern Pulp and Paper Complex," in *The Second Wave: Southern Industrialization from the 1940s to the 1970s*, ed. Philip Scranton (Athens: University of Georgia Press, 2001), 175–185.
15. For a discussion of regeneration efforts in the South, see Boyd, "The Forest Is the Future?" 185–193.
16. Vilmorin gathered seeds from Scotch pine growing all over Europe and grew them under the same set of conditions, demonstrating that some of the important differences among the various "races" of Scotch pine resulted from hereditary factors. See Jonathan W. Wright, "The Role of Provenance Testing in Tree Improvement," in *Advances in Forest Genetics*, ed., P. K. Khoslà, (New Delhi: Ambika Publications, 1981). Wright, "Provenance," 103.
17. IUFRO expanded its provenance program in 1908 and then again in 1938, focusing on Scotch pine, European larch and Norway spruce. During World War II, however, some of the test plantations and records were destroyed and the overall project unraveled. See Wright, "Provenance," 105. See also Wolfgang Wettstein-Westersheim and Hans Hufnagl, "Improvement Through Racial Selection and Planting," *Proceedings of the Fifth World Forestry Conference*, Vol. 2 (1960), 775.
18. Although we realize the problematic character of terms such as "race" and "racial" and the obvious links to eugenics, an inquiry into that contested history is beyond the scope of this essay.
19. See Wright, "Provenance," 103; Philip C. Wakeley, "Importance of Geographic Strains," in *Report of the First Southern Conference on Forest Tree Improvement*, January 9–10, 1951 (Atlanta, 1951), 1–9; and Henry I. Baldwin, "Seed Certification and Forest Genetics," *Journal of Forestry* 52 (1954): 654–655.
20. For the most part, this meant selecting seed from habitats of similar elevation and soil type to the planting area. In the case of Scotch pine seed, for example, planting was restricted to areas that were less than one hundred meters higher or lower than the original habitat of the parent trees. Joseph H. Stoeckeler, "Science on the March: Bigger and Better Forest Trees for Sweden," *Scientific Monthly* 70 (1950): 329.
21. See Baldwin, "Seed Certification and Forest Genetics," 654; Ernest Rohmeder, "Problems and Proposals for International Forest Tree Seed Certification," *Proceedings of the Fifth World Forestry Conference*, Vol. 2 (1960), 685–689; Leo A. Isaac, "Problems and Proposals for International Certification of Tree Seed Origin and Stand Quality with Particular Reference to Western North American Species," *Proceedings of the Fifth World Forestry Conference*, Vol. 2 (1960), 690–695.
22. Baldwin suggests that close attention to "local races" or what he calls forest ecotypes illustrated the importance of ecological knowledge to forestry. See Henry I. Baldwin, "The Application of Ecological Knowledge to Forestry," *Scientific Monthly* 72 (1951): 225–229.
23. Founded three years after the rediscovery of Gregor Mendel's work on inheritance and two years after the publication of the first volume of Hugo de Vries *Die Mutationstheorie*, the ABA was the first nationally organized, membership-based institution promoting genetic and eugenic research in the United States. During its ten-year existence (1903–1913), the ABA's research program shifted gradually from a primary focus on agricultural improvement to one focused on eugenics. For historical treatment of the ABA and its relation to genetics and eugenics in the United States, see Barbara A. Kimmelman, "The American Breeders' Association: Genetics and Eugenics in an Agricultural Context, 1903–1913," *Social Studies of Science* 13 (1983): 163–204. Charles Rosenberg discusses the reception of Mendel's work among plant and animal breeders in the United States in comparison with the reception among biologists and members of the medical profession. See Charles E. Rosenberg, *No Other Gods: On Science and American Social Thought* (Baltimore: Johns Hopkins University Press, 1997), 211–224. See also Diane B. Paul and Barbara A. Kimmelman, *Mendel in America: Theory and Practice, 1900–1919*, in *The American Development of*

Biology, ed. Ronald Rainger, Keith R. Benson, and Jane Maieschein (Philadelphia: University of Pennsylvania Press, 1988), 281–310.
24. In 1904, Hays became undersecretary of agriculture, but continued on as secretary and "guiding force" of the ABA. See Willet M. Hays, "Address to the First Meeting of the American Breeders' Association," in *American Breeders' Association* I (Washington, D.C.: ABA, 1905), 9–10. Hays often compared heredity or "breeding power" to electricity and always alluded to its potential as an economic force: "As electrical energy must be harnessed, so these investigations are showing that the peculiar breeding potencies of the rare plant or animal must be singled out and given opportunity to work. Both in practical breeding and in evolutionary studies the individual with exceptional breeding power is gaining respect.... The world is learning to seek the 'Shakespeares' of the species with the same avidity that it seeks gold mines." Willet M. Hays, *American Work in Breeding Plants and Animals,* in American Breeders' Association II (Washington, D.C.: ABA, 1906), 158. For further discussion of Hays and his role in the ABA, see Kimmelman, "The American Breeders Association," and Paul and Kimmelman, *Mendel in America.*
25. George B. Sudworth and A. D. Hopkins, "Report of the Committee on Breeding Nut and Other Forest Trees," American Breeders' Association III (Washington, D.C.: ABA, 1907), 224. See also the statement of J. Russell Smith: "In these days of intense activity in plant breeding, one might suppose that forestry literature would give due weight to the subject of tree breeding, but the most progressive foresters do not seem to have recognized that there is a great field for the producer or discoverer of improved varieties of forest trees. The long time required for the maturity of a saw log is no doubt the chief reason why foresters have ignored this usually fascinating field. The patient, painstaking plant breeder may render a future valuable service to mankind by devoting part of his attention to forest trees." Sudworth and Hopkins, "Committee on Breeding Nut and Other Forest Trees," 227.
26. Sudworth and Hopkins, "Committee on Breeding Nut and Other Forest Trees" (1907), 225.
27. See American Breeders Association, Committee on Breeding Forest and Nut Trees, "Report of the Committee on Breeding Forest and Nut Trees," *American Breeders Association* IV (Washington, D.C.: ABA, 1908), 304. Other illustrious forestry professionals and academics on the committee included Raphael Zon, J.W. Toomey, and Willis L. Jepson.
28. "Report of the Committee on Breeding Forest and Nut Trees" (1908), 305.
29. The section on Silvics was established "to contribute to ordered and scientific knowledge of our forests." The major push for forest experiment stations came from Raphael Zon, who upon visiting several European countries in 1908, recommended that the Forest Service create separate branches for research and administration. Zon was also the head of the Section on Silvics. In 1908 he established the first forest experiment station in Cocinino National Forest in Arizona. In 1915, the Forest Service formally created a Branch of Research. For a discussion, see William Robbins, *American Forestry: A History of National, State, and Private Cooperation* (Lincoln: University of Nebraska Press, 1985), 17.
30. "Report of the Committee on Breeding Forest and Nut Trees" (1908), 305.
31. "Report of the Committee on Breeding Forest and Nut Trees" (1908), 306.
32. Thus, the 1912 Report of the Committee on Breeding Nut and Forest Trees concluded that "[t]his committee firmly believes in the importance of the source of seed and in the extension of the range of valuable species as the most immediate possible means of adding to or of enriching our different regions and thus increasing the productivity of our forests." *American Breeders Association* VIII (Washington, D.C.: ABA, 1912), 522.
33. Johann Klotzsch's work dates from 1845 and is considered the earliest European attempt at deliberate crossing of forest trees. See Ernst J. Schreiner, "Improvement of Forest Trees," *Yearbook of Agriculture 1937* (Washington, D.C.: GPO/USDA, 1937), 1249.
34. For a brief discussion of Burbank's work and his influence on forest-tree breeding, see Austin, "A New Enterprise," 929–930.
35. Schreiner discusses a number of other early breeding efforts in the United States. See Schreiner, "Improvement," 1249–1253.
36. E. M. East and H. K. Hays, *Heterozygosis in Evolution and in Plant Breeding,* USDA Bureau of Plant Industry Bulletin 243 (Washington, D.C.: GPO, 1912).
37. Schreiner, "Improvement," 1249–1250.
38. Ernst Schreiner, "Genetics in Relation to Forestry," *Journal of Forestry* 48 (1950): 34. Given the changing dynamics of the pulp and paper industry, moreover, there was less incentive

39. to continue an intensive breeding program with poplars in New England. Indeed, by 1950 it was clear that the South had become the dominant region for pulp and paper production while the PNW was emerging as the leader in solid wood products. Tree improvement efforts focused on pine and Douglas fir thus offered the greatest possible return.

39. Eddy was deeply influenced by Burbank and consulted extensively with him in developing the project. See Lois C. Stone, "The Institute of Forest Genetics: A Legacy of Good Breeding," *Forest History* 12, no. 3 (1968): 21–22. See also Austin, "A New Enterprise," 928–930.

40. The details of the program are discussed in Austin, "A New Enterprise."

41. Stone, "Institute of Forest Genetics," 24.

42. See Austin, "A New Enterprise" and Ryookiti Toda, "An Outline of the History of Forest Genetics," in *Advances in Forest Genetics*, ed. P. K. Khosla (New Delhi: Ambika Publications, 1981).

43. Palmer Stockwell and F. I. Righter, "Hybrid Forest Trees," *Yearbook of Agriculture 1943–1947* (Washington, D.C.: USDA, 1947), 465. See also the remarks of Duffield and Stockwell, "Today pine breeding has progressed to the point that promising pine hybrids exist for each of the three major timber-producing regions in the United States." J. W. Duffield and Palmer Stockwell, "Pine Breeding in the United States," *Yearbook of Agriculture 1949* (Washington, D.C.: USDA, 1949), 147.

44. Vegetative propagation involves the use of rooted cuttings and grating (and more recently tissue culture techniques) to essentially clone a particular tree. The chief advantages of vegetative propagation in forestry, as opposed to using a seed-based regeneration strategy, include the potential to capture greater genetic gain; the potential to capture greater genetic uniformity; and the potential to accelerate tree improvement activities. The primary uses of vegetative propagation in forestry include preservation of selected genotypes through clone banks; multiplication of desired genotypes for special uses such as seed orchards and breeding orchards; evaluation of specific genotypes through clonal testing; and as a basis for regeneration in industrial-scale planting programs. For a discussion, see Bruce J. Zobel and John Talbert, *Applied Forest Tree Improvement* (New York: Wiley, 1984), 310–344.

45. As Schreiner noted in 1937, "Few forest trees . . . can be commercially propagated by cuttings, and although many species can propagated vegetatively by grafting, budding, or layering, these methods are not feasible at present because of their comparatively high costs. It is essential to find cheaper methods of vegetative propagation if select hybrids or strains are to be multiplied and utilized immediately for forestation planting." Schreiner, "Improvement," 1253–1254. See also Schreiner, "Genetics in Relation to Forestry," 36–37.

46. Righter suggested in response that allowing hybrid plantations to reseed, even though it would mean that the original hybrids (F1 trees) would not "breed true," would still be advantageous in the silviculture context because there would be enough offspring resembling the parents to crowd out the inferior F2 trees before the stand reached maturity. See F. I. Righter, "New Perspectives in Forest Breeding," 104 *Science* 2688 (1946): 1–3.

47. See Toda, "Outline of the History of Forest Genetics," 7.

48. In 1955, the term "plus-tree" was attributed to the selected phenotypes and the overall method came to be known as the "plus-tree" method of selective improvement. See Toda, "Outline of the History of Forest Genetics," 7–8. On the place of Sweden as a leader in industrial tree improvement, see Joseph H. Stoeckeler, "Science on the March: Bigger and Better Forest Trees for Sweden," *Scientific Monthly* 70 (1950): 328. One observer estimated that for 1947 the total amount spent on tree improvement in Sweden exceeded the U.S. total by a factor of two. See Schreiner, "Genetics in Relation to Forestry," 35.

49. For more details on the committee's early efforts, see U.S. Forest Service, *Report of the First Southern Conference on Forest Tree Improvement*, January 9–10, 1951 (Atlanta: USFS, 1951); U.S. Forest Service, *Report of the Second Southern Conference on Forest Tree Improvement*, January 6–7, 1953 (Atlanta: USFS, 1953); and U.S. Forest Service, *Report of the Third Southern Conference on Forest Tree Improvement*, January 5–6, 1955 (New Orleans: USFS, 1955).

50. Wakeley was known throughout the southern forestry community as "Mr. Southern Pine." Zobel and Sprague, *Forestry Revolution*, 59–61. See also Jonathan W. Wright, "The Role of Provenance Testing in Tree Improvement," in *Advances in Forest Genetics*, ed. P. K. Khosla (1981) and Philip C. Wakeley, "The Relation of Geographic Race to Forest Tree Improvement," *Journal of Forestry* 52 (1954): 653.

51. Zobel and Sprague, *Forestry Revolution*, 4. In 1981, Zobel estimated "that over 30 percent of all tree improvement programs have been failures or have been only marginally successful

because geographic variability within the species was ignored." Zobel, "Forest Tree Improvement," 16. See also Zobel and Talbert, *Applied Forest Tree Improvement*, 75–93.
52. Wakeley's study involved collecting seeds from each of the four southern species of pine in various areas of their natural range. A dozen or so plantations were then established in these areas to test for "racial" differences in survival, growth rate, and disease resistance within each species. See Philip C. Wakeley, "Importance of Geographic Strains," in *Proceedings of the First Southern Conference* (1951), 1–9; E. L. Demmon and P. A. Briegleb, "Progress in Forest and Related Research in the South," *Journal of Forestry* 54 (1956): 674–682, 687–692.
53. Baldwin, "Seed Certification," 655; J. W. Duffield, "The Importance of Species Hybridization and Polyploidy in Forest Tree Improvement," *Journal of Forestry* 52 (1954): 645.
54. Wakeley, "Geographic Strains," 653.
55. The West Virginia Pulp and Paper Company developed the first industry seed selection program in 1949. By 1956, the company was planting all seedlings from selected stock. See L. T. Easley, "Loblolly Pine Seed Production Areas," *Journal of Forestry* 52 (1954): 672–673; Zobel and Sprague, *Forestry Revolution*, 84.
56. Clemens M. Kaufman, "Extensive Application of Genetics by the Silviculturist," *Journal of Forestry* 52 (1954): 647–648.
57. Zobel and Sprague, *Forestry Revolution*, 17–21.
58. Thomas O. Perry, "The Cooperative Genetics Program at the University of Florida," in *Proceedings of the Third Southern Conference* (1955). Originally, the North Carolina State Cooperative had twelve charter members. In 1966 it was divided into two programs, one of which focused on hardwood and one of which focused on loblolly pine, with the pine program receiving the bulk of the attention and financial support. See Zobel, "Increasing Productivity," 17. In addition, a host of other cooperatives focusing on various aspects of industrial forestry have been established in the South, and more recently, in other parts of the country.
59. Zobel, "Increasing Productivity," 4. From an organizational standpoint, the program was unique. There were no formal contracts or written agreements. Members (about thirty-five in the early 1970s: thirty-two from industry and three state agencies) paid yearly dues and participated on an equal basis. Any member could withdraw at any time, and any member could be asked to withdraw if its contribution was considered inadequate. Members who withdrew, however, would not be allowed to return. The cooperative's university-based managers designed, analyzed, and interpreted field tests, while professional foresters employed by the member firms performed fieldwork under the guidance of cooperative staff. More fundamental research, which was not a primary focus of the cooperative, enlisted associated graduate students, some of whom would later become employees of member firms. The entire effort was multidisciplinary—drawing on a number of fields outside of forestry, including pulp and paper science, genetics, botany, statistics, soil science, plant breeding, and plant pathology. See Zobel, "Increasing Productivity," 17–18.
60. Zobel, "Increasing Productivity," 17, and Zobel and Sprague, *Forestry Revolution*, 50–55.
61. One hundred percent of the seedlings produced in Georgia, Alabama, and Louisiana in 1986 were genetically improved. See Clark W. Lantz, "Overview of Southern Regeneration," in *Proceedings of the Annual Meeting of the Southern Forest Nursery Association* (1988) 5; Clark W. Lantz, "Role of State Nurseries in Southern Reforestation—An Historical Perspective," in *National Proceedings: Forest and Conservation Nursery Association—1996* (Portland: United States Forest Service, 1997).
62. This was roughly double the output of the other two tree improvement cooperatives. See A. E. Squillace, "Tree Improvement Accomplishments in the South," in *Proceedings of the 20th Southern Forest Tree Improvement Conference* (1989), 10.
63. Gains from the second and third generations are expected to exceed those of the first. See J. B. Jett, "Thirty-five Years Later: An Overview of Tree Improvement in the Southeastern United States," in *Proceedings of the Annual Meeting of the Southern Forest Nursery Association* (Charleston, S.C.: Southern Forest Tree Improvement Committee, 1988); Steve McKeand and Jan Svensson, "Loblolly Pine: Sustainable Management of Genetic Resources," *Journal of Forestry* 95, no. 3 (1997): 5, 8–9.
64. For discussions of some of the risks and vulnerabilities associated with forest tree improvement and the spread of pine monocultures in the South, see the collection of articles in Duke University School of Forestry, "Topic 1: Possible Consequences of Southern Pine

Monoculures," *Proceedings of the Fourth Conference on Southern Industrial Forest Management* (Durham, N.C.: Duke University Press, 1960). For a more general discussion, see John W. Duffield, "Forest Tree Improvement: Old Techniques and the New Science of Genetics," *H. R. MacMillan Lectureship Address* (Vancouver: University of British Columbia, 1960). See also James Scott, *Seeing Like a State*, 11–22.

65. The Southern and Southeastern Forest Experiment Stations of the U.S. Forest Service, for example, undertook research on forest disease and insect problems. See L. W. Orr and R. J. Kowal, "Progress in Forest Entomology in the South," *Journal of Forestry* 54 (1956): 653–656; George H. Hepting, "Forest Disease Research in the South," *Journal of Forestry* 54 (1956): 656–660; and A. J. Riker, "Opportunities in Disease and Insect Control Through Genetics," *Journal of Forestry* 52 (1954): 651–652.

66. Interview with author (Boyd), November 17, 1997.

67. In the words of Bruce Zobel: "The genetic improvement of forest trees is a long-term, expensive undertaking. It can be done best through Cooperative efforts because most organizations cannot afford a team of highly trained specialists. In a Cooperative, one trained man can oversee a great deal of research. The need to keep the genetic base sufficiently broad is almost impossible for a single organization but is easily achieved in a cooperative effort. The funds and manpower required for a tree-improvement program for each member is minimal in a cooperative, yet enables maximum genetic gains. Plant materials, methods, equipment and even manpower are exchanged amongst members to the benefit of all. In my view, naturally biased, it would seem wasteful and even foolish for each organization to strike out on its own with an expensive, inadequate, and inefficient program when faster and greater gains are assured through joint action. The Florida, Texas, and North Carolina Cooperatives are convincing demonstrations of how well combined action has succeeded." See Zobel, "Increasing Productivity," 4.

68. T. T. Munger and W. G. Morris, *Growth of Douglas-fir Trees of Known Seed Source* (Washington, D.C.: USDA, 1936), 40; M. A. Bordelon, "Genetic Improvement Opportunities," in *Assessment of Oregon's Forests 1988*, ed. G. J. Lettman (Salem, Oregon, Ore.: State Department of Forestry, 1988): 231–239.

69. E. S. Pomeroy, *The Pacific Slope: A History of California, Oregon, Washington, Idaho, Utah, and Nevada* (Lincoln: University of Nebraska Press, 1991).

70. The Douglas fir region comprises a strip about 125–150 miles wide along the western portions of Oregon and Washington beyond the Cascade Mountains, together with the Northwest corner of California excluding the redwood forests of the California coast. Within this region, Douglas fir is ubiquitous, and is by far the most important commercial species. For example, Douglas fir accounted for 75 percent of timber volume removed from western Oregon lands in Oregon Department of Forestry, *1995 Annual Reports* (Salem, Ore., Oregon Department of Forestry, 1997).

71. Demeritt, "Scientific Forest Conservation," 431–459.

72. David Harvey, *The Limits to Capital* (Oxford, B. Blackwell, 1982). As expressed by former chief forester William B. Greeley in the 1920s, "It is written in the immutable law of commerce that industries seek their cheapest source of raw material." Quoted in W. G. Robbins, "Lumber Production and Community Stability: A View from the Pacific Northwest," *Journal of Forestry* 31, no. 4 (1987): 187–196, quote on 189.

73. M. Williams, *Americans and Their Forests: A Historical Geography*. Cambridge, UK: Cambridge University Press, 1989. Coupled with the ruinous exhaustion of forest resources in other regions of the country, Pacific Northwest forests were estimated to contain more than half of the nation's standing timber resource at the turn of the century. United States Bureau of Corporations, *The Lumber Industry* (Washington, D.C.: GPO, 1913); C. R. Howd, *Industrial Relations in the West Coast Lumber Industry . . . December, 1923* (Washington, D.C.: U.S. Bureau of Labor Statistics/GPO, 1924); M. Hibbard and J. Elias, "The Failure of Sustained-Yield Forestry and the Decline of the Flannel-Shirt Frontier," in *Forgotten places: uneven development in rural America*, ed. T. A. Lyson and W. W. Falk (Lawrence, Kans.: University Press of Kansas, 1993), 195–215.

74. On the broader politics of emerging scientific forestry, see Demeritt, "Scientific Forest Conservation," 431–459; see also W. G. Robbins, *Lumberjacks and Legislators: Political Economy of the U.S. Lumber Industry, 1890–1941* (College Station: Texas A&M University Press, 1982) on these same issues, as well as on the specific circumstances of the Oregon legislation.

75. W. D. Hagenstein, *Growing 40,000 Homes a Year* (Berkeley, Calif.: School of Forestry and Conservation, University of California, 1973).
76. Interview with a regional forest geneticist, December 19, 1997. On early work at the Institute of Forest Genetics, see Zobel, "Forest Tree Improvement," 11–22.
77. C. S. Larsen, *Genetics in Silviculture* (London: Oliver, 1956) and Toda, "An Outline of the History of Forest Genetics," 4–12. It was Larsen who pioneered the technique of cloning "elite" or "plus" trees by grafting scions from different genotypes onto host root stocks for the purposes of controlled experimentation and breeding. This work is key in Douglas fir tree improvement since the species is difficult to propagate from shoots.
78. Hagenstein, *Growing 40,000 Homes a Year*.
79. R. R. Silen, *A Simple Progressive Tree Improvement Program for Douglas-Fir* (Portland, Ore., USDA Forest Service, Pacific Northwest Range and Experiment Station, 1966).
80. M. A. Bordelon, "Genetic Improvement Opportunities," in *Assessment of Oregon's Forests 1988*, ed. G. J. Lettman (Salem, Ore.: Oregon State Department of Forestry, 1988), 231–239. My characterization is slightly different than Bordelon's, which seems to exclude the grafting stage. I included this stage based on an interview conducted December 19, 1997, with a regional forest geneticist whose background includes detailed knowledge of regional tree improvement programs. This is also in line with practices described elsewhere. Of course, since the orchards are typically located in small clearings surrounded by wild forests, a persistent problem with this model is that open pollination in the seed orchards runs the risk of contamination from pollen contributed by surrounding, unimproved trees. This has proved a very difficult challenge to contain, and is a prime example of the logistical difficulties in breeding wild forest trees. See W. T. Adams, V. D. Hipkins, J. Burczyk, and W. K. Randall, "Pollen Contamination Trends in a Maturing Douglas-Fir Seed Orchard," *Canadian Journal of Forest Research* 27 (1997): 131–134.
81. Bordelon, "Genetic Improvement Opportunities," 231–239.
82. The confusing combination of acronyms NWTIC and PNWTIRC is unfortunate, but unavoidable.
83. Pacific Northwest Tree Improvement Research Co-operative, *Annual Report 1995–1996* (Corvallis, Ore.: Forest Research Laboratory, Oregon State University, 1996).
84. Interview conducted December 17, 1997, with a forest geneticist; interview conducted October 2, 1997, with a forest geneticist.
85. Adams et al., "Pollen Contamination Trends," 131–134.
86. In 1964, recognizing its old-growth days were over, the firm had reorganized all timberland-related functions, including "logging operations, forestry and research, procurement and sales, log sales, log measurement, timberland purchasing, and other related functions" into a single timberlands division. The firm then developed a planning approach based on a highly abstracted model "Target Forest," and set about rationalizing all aspects of production, including timberland management. High Yield Forestry was an extension of this rationalization project. H. E. Morgan, *The Environment of High-Yield Forestry* (Berkeley, Calif.: School of Forestry and Conservation, University of California, 1969).
87. See Morgan, *High Yield Forestry*; also S. Hee, *Intensive Silviculture Stewardship: The Weyerhaeuser Experience* Presented at the Silviculture Conference: Stewardship in the New Forest, Vancouver, B.C., Forestry Canada, 1992. Information also taken from an interview with a Pacific Slope forest geneticist with knowledge of Weyerhaeuser's program, and regional tree improvement more generally, December 1997.
88. This claim is made by a Weyerhaeuser executive in B. Virgin, "Weyerhaeuser's Forestry Gamble Pays Off," *The Oregonian*, Portland, Oregon, January 1, 1997, E2, and is corroborated by statements made in an interview with one of the authors by a regional forest geneticist with expert knowledge of Weyerhaeuser's tree improvement program.
89. See C. E. Twining, *George S. Long, Timber Statesman* (Seattle: University of Washington Press, 1994). As Frederick Weyerhaeuser told George Long, manager of the company's West Coast operations "the less [timber] we sell and the more we buy, the more money we make" (Twining, *George S. Long*, 29).
90. For a useful albeit dated overview of applications of biotechnology in tree improvement and their potential, see R. Haines, *Biotechnology in Forest Tree Improvement: With Special Reference to Developing Countries* (Rome: Food and Agriculture Organization of the United Nations, 1994).
91. A. Séguin, G. Lapointe, and P. J. Charest, "Transgenic trees," in *Forest Products Biotechnolog*, ed. A. Bruce and J. W. Palfreyman (Bristol, Penn.: Taylor and Francis, 1998), 287–304.

92. S. von Arnold, D. Clapham, U. Egertsdotter, I. Ekberg, H. Mo, and H. Yibrah, "Somatic Embryogenesis in Norway Spruce (*Picea abies*)," in *Somatic Embryogenesis and Synthetic Seed I: Biotechnology in Agriculture and Forestry*, ed. Y. P. S. Bajaj (New York: Springer-Verlag, 1995), 415–430.
93. Séguin et al., "Transgenic Trees," 287–304.
94. Interview with one of the authors conducted September 1997 with a forest research geneticist.
95. Séguin et al., "Transgenic Trees," 287–304.
96. For a survey, see Y. P. S. Bajaj, ed., *Somatic Embryogenesis and Synthetic Seed I*.
97. Patent information obtained on CD-ROM, accessed at the San Francisco Public Library, U.S. Patent Office, Patent and Trademark Center, July 24, 1998.
98. Hee, *Intensive Silviculture Stewardship*.
99. Haines, *Biotechnology in Forest Tree Improvement*; Séguin et al., "Transgenic Trees," 287–304.
100. The IFG is home to the national pine genome mapping project, initiated by David Neale and Claire Kinlaw, and dedicated to mapping out the genome of important commercial conifer species, including loblolly and sugar pine, Douglas fir, and Pacific yew. Mapping the genomes of these trees feeds into the Dendrome Project, a database for forest tree genomic information and itself part of the larger USDA effort to map and make genomic information available for forest and crop species (the Plant Genome Project). See U.S. Department of Agriculture, "Conifer Molecular Genetics and Genomics," Institute of Forest Genetics, Albany, California, retrieved January 31, 2003, from http://www.psw.fs.fed.us/ifg/genomic.htm.
101. Engineered reproductive sterility in the trees is primarily of interest in order to facilitate commercial deployment of GM varieties, since regulatory and wider public sanction depends on assurances that the trees will not interbreed with wild populations of cottonwoods. S. H. Strauss, A. M. Rottmann, and L. A. Sheppard, "Genetic Engineering of Reproductive Sterility in Forest Trees," *Molecular Breeding* 1 (1995): 5–26.
102. Engineered insect resistance makes use of an increasingly common technique in GM food crops involving the insertion of a gene controlling the production of *Bacillus thuringiensis* toxicity extracted from a common soil bacterium giving the toxin its name. The result is a plant that produces the Bt toxin itself. This toxin is commonly used as an agricultural pesticide in conventional applications, including control of the cottonwood leaf beetle in existing cottonwood plantations in Oregon and Washington. Tree Genetic Engineering Research Cooperative, *TGERC Annual Report 1996–97* (Corvallis, Ore.: Forest Research Laboratory, Oregon State University, 1997), 37. The chief commercial advantage of Bt cottonwoods—similar to Bt crops now deployed in agricultural fields—would be that they could produce their own toxin. This would then eliminate the need for aerial spraying, something that may have ecological benefits if it means less of the toxin is dispersed into the surrounding environment. S. Krimsky and R. P. Wrubel, *Agricultural Biotechnology and the Environment: Science, Policy, and Social Issues* (Urbana: University of Illinois Press, 1996).
103. Glyphosate or Roundup resistant varieties have an introduced gene that makes the trees tolerant of glyphosate, a common herbicide. These trees, as their name suggests, tie their use directly to Monsanto, the producer of Roundup, an herbicide used commonly in plant cultivation. Prototypes of such Roundup Ready cottonwoods are also in TGERC field trials, and would be of interest in existing plantations because glyphosate is toxic to conventional hybrid cottonwoods. Although effective as a chemical herbicide, glyphosate is now of limited use in the plantations because it kills or severely damages nonresistant cottonwoods. See S. H. Strauss, S. A. Knowe, and J. Jenkins, "Benefits and Risks of Transgenic Roundup Ready® Cottonwoods," *Journal of Forestry* (May 1997): 12–19.
104. Membership has changed over time. The TGERC entered its second five-year research plan in 1999 with membership consisting of Aracruz Cellulose, Alberta Pacific, the Department of Energy, International Paper, the National Science Foundation, Oregon State University, Potlatch, Westvaco, and Weyerhaeuser. Retrieved January 14, 2002, from http://www.fsl.orst.edu/tgerc/hist.htm.
105. Moreover, under the auspices of the Oregon State University Office of Technology Management, formal patent claims have been filed with the U.S. Patent Office listing the TGERC and OSU as claimants for a range of products and processes, including transgenic trees produced by the coop. Under the terms of a Memorandum of Agreement signed by all coop

members upon joining the TGERC, members will receive priority in the negotiation of formal rights to deploy proprietary transgenic trees in the event that such a deployment takes place. Nevertheless, they would pay royalties.
106. See M. Sagoff, "On Making Nature Safe for Biotechnology," in *Assessing Ecological Risks of Biotechnology*, ed. L. Ginsburg (Stoneham, Mass.: Butterworth-Heineman, 1991), 341–365. Sagoff notes his expectation that biotechnology and genetic engineering in forestry will lead to what he refers to as a more vertically integrated silviculture. See also M. Kenney, *Biotechnology: The University-Industrial Complex* (New Haven, Conn.: Yale University Press, 1986), 31, who notes that "One reason the commercialization of biotechnology is such an interesting phenomenon is the possibility that biotechnology is the cutting edge for the creation of new social relationships, though it is still unclear what the final stable configuration of these social relationships will be."
107. Involving Westvaco, International Paper, and Fletcher Challenge together with a New Zealand genomics company called Genesis Research and Development.
108. S. R. Radosevich, C. M. Ghersa, and G. Comstock, "Concerns a Weed Scientist Might Have About Herbicide-Tolerant Crops," *Weed Technology* 6 (1992): 635–639; L. C. Duchesne, "Impact of Biotechnology on Forest Ecosystems," *Forestry Chronicle* 69, no. 3 (1993): 307–313, Strauss et al., "Benefits and Risks," 12–19.
109. P. J. Regal, "Scientific Principles for Ecologically Based Risk Assessment of Transgenic Organisms," *Molecular Ecology* 3 (1994): 5–13; Strauss et al., "Genetic Engineering of Reproductive Sterility," 5–26.
110. Adams et al., "Pollen Contamination Trends," 131–134; S. T. Freidman and G. S. Foster, "Forest Genetics on Federal Lands in the United States: Public Concerns and Policy Responses," *Canadian Journal of Forest Resources* 27 (1997): 401–408. For discussions of some of the risks and vulnerabilities associated with forest tree improvement and the spread of pine monocultures in the South, see the collection of articles in Duke University School of Forestry, "Topic 1: Possible Consequences of Southern Pine Monocultures," *Proceedings of the Fourth Conference on Southern Industrial Forest Management* (Durham: Duke University Press, 1960). For a more general discussion, see John W. Duffield, "Forest Tree Improvement: Old Techniques and the New Science of Genetics," *H. R. MacMillan Lectureship Address* (Vancouver: University of British Columbia, 1960). See also Scott, *Seeing Like a State*, 11–22.
111. Strauss et al., "Benefits and Risks," 12–19.
112. There are also risks "stemming" from the persistence of GE cottonwoods in the environment since they are easily propagated from cuttings. See R. R. James, S. P. DiFazio, A. M. Brunner, and S. H. Strauss, "Environmental Effects of Genetically Engineered Woody Biomass Crops," *Biomass and Bioenergy* 14, no. 4(1998): 403–414.
113. Radosevich et al., "Concerns," 635–639.
114. Carl Syrach Larsen, "Progress in Genetics and Tree Improvement," in *Proceedings of the Fifth World Forestry Conference*, Vol. 2 (Seattle, Wash.: World Forestry Congress, 1960), 680.
115. Scott, *Seeing Like a State*, passim.
116. For elaboration on this theme, see W. Boyd, S. Prudham, and R. Schurman, "Industrial Dynamics and the Problem of Nature," *Society and Natural Resources* 14 (2001): 555–570.

II
Animals, Aggression, Arrogance, and Analysis

War Horses
Equine Technology in the American Civil War

ANN N. GREENE

The American Civil War was the first industrialized war. Conventional descriptions of industrialization have focused on the adoption of mechanical devices and inanimate forms of energy, and this applies to many aspects of the Civil War. Mechanized production made it possible to feed and equip enormous armies, steam-powered railroads expanded the war's geographical size, and mass-produced weaponry with increased range and accuracy caused staggering casualties and altered battle tactics. However, the Civil War was also a war of animal power that used horses on an unprecedented scale. Thousands of horses accompanied Civil War armies. In four years, the Union army alone employed between 650,000 and 1,000,000 horses.[1]

The extraordinary reliance on animal power during the Civil War suggests a more complicated relationship between industrialization and animal power than is usually included in the story of industrialization. The Civil War was a war of extensive animal power because it was an industrialized war, for newer technologies of transportation and weaponry made it possible to employ horses more broadly and extensively. This essay attempts to rethink our understanding of industrialization by examining the Union army's use of horses during "the first industrialized war." In particular, it analyzes horses as a war technology by exploring the following questions: Why did the Union army need so many horses? Who was responsible for acquiring army horses? Where did the horses come from? What did using them require? What did it take to keep them in working condition? How did the Union's horses factor into the nature and outcome of the war? These questions lead to a fuller picture of the biotechnological environment of the war.[2]

Antebellum Animal Power

The Union army's use of horses and mules grew out of the economic and technological changes that had occurred during the antebellum years. One of the effects of industrialization was to expand the use of animal power. Historians have not always recognized this relationship, but people in the nineteenth century did. In the census of 1810, Tench Coxe included animal power in a list of the young republic's resources, noting that "the people of America shall

143

continue too pursue, to the utmost of their power, the use of water, steam, horses, cattle machinery, dexterity, and various modern processes and devices." By the census of 1860 the impact of industrialization on animal power was clear. Work animals provided the majority of the power in antebellum America, complementing many of the newer industrial technologies appearing at that time.[3]

The series of developments known as the "transportation revolution"—road construction, the development of steamboats, and the construction of canals and railroads—required large numbers of horses to haul stagecoaches, canal boats, streetcars, and wagons. Road construction and improvement made possible extensive stagecoach service and wagon traffic among the cities of the Northeast. Regarded as one of the most modern innovations of the time, stagecoaches were regular, reliable, and fast, achieving speeds of almost ten miles per hour. Boston alone had seventy-five different companies offering one hundred arrivals and departures daily. By 1833, "nonstop" coaches to New York City took only thirty-three hours. Heavily subsidized by postal contracts, stagecoaches, by the 1850s, provided passenger service and mail delivery throughout the country and across the Trans-Mississippi West to California.[4]

Water transportation depended on animal power as well. The completion of the Erie Canal in 1825—a thrilling technological event at the time—stimulated a flurry of canal building that resulted in three thousand miles of canals by 1840. These canals employed thousands of horses, as many as eight thousand on the Erie Canal alone at mid-century. Freight traffic on canals continued to expand even as passenger traffic moved to the railroads. Most canals remained horse-powered, because steamboat wakes eroded their banks. Interest in developing a steam-powered canal boat led the New York State Legislature to offer an award for "the profitable introduction of some motor for canal boats other than animal power," but no feasible solution appeared. Thus into the 1880s less than 1 percent of the approximately four thousand canal boats on the Erie were steam-powered. The absence of bridges made ferries an important link in the transportation network; teamboats, or horse ferries, powered by horses walking on treadmills, operated at short crossings of harbors, rivers, and lakes. The first ran between Brooklyn and New York City in 1814, and by 1819 there were eight. The Niles *Weekly Register* noted that "We have many teamboats at different ferries in the United States." Horse ferries largely disappeared from the Northeast by the Civil War, but remained in use in the South, Midwest, and West, some well into the twentieth century.[5]

The invention of mechanical agricultural implements made it possible to apply animal power to tasks previously done by hand. The most significant of these were lightweight, lower-draft plows, and reapers and mowers for harvesting hay and grain, but there was a plethora of others: seed drills, hay rakes, harrows, hayforks, and tedders. Animals also provided stationary power by means of "horse-powers," widely employed mechanisms that converted the linear motion of the horse into rotary motion to drive grain threshers, wood saws, mills, cotton gins, and other machinery.[6]

Rapid urban growth expanded the use of animal power in public transportation, industry, construction, shipping, and commercial hauling. Omnibuses, or large passenger wagons, first operated in New York in 1829 and were common in cities by mid-century, often with multiple lines providing frequent service over many routes. By the 1850s, cities were constructing horse railways—horse-drawn streetcars on rails—to replace omnibuses. These increased commuting distance by half, doubled the residential area of cities, and changed the social geography of the nineteenth-century city. Rising industrial production and commercial activity clogged the streets with burgeoning numbers of wagons and carts, and horses provided stationary power in small manufacturing establishments, on construction sites, and at harbors.[7]

Perhaps the most important complementary relationship between work animals and newer technologies was the one between horses and railroads. Railroads provided cheap, fast, efficient hauling over long distances, but by creating new commercial networks and expanding the volume of freight and passenger traffic, they increased the demand for horses as well. Horse-drawn transportation provided the flexible, short-distance hauling needed for local distribution and travel that railroads could not. Antebellum railroads were not an integrated system, but a fragmented network containing many gaps that necessitated the transfer of passengers and freight. Horse-drawn transportation flourished in the interstices of the railroad network and made it work. Tracks of different lines often converged without connecting, because they used rails of different gauges, or because they wanted to protect rolling stock and monopolies over local business. Nor did railroads coordinate their timetables to facilitate through traffic, resulting in long layovers. Cities often banned steam locomotives, and had horses pull the cars into downtown depots or located depots outside the city limits. Many major rivers had ferries but no bridges. All these gaps necessitated many transfers of passengers and freight. Conversely, the ability of railroads to carry bulky items like hay and grain and to increase the distribution of industrial products like wagons, farm implements, horseshoes, and harnesses, made it possible to use horses in a wider range of places and to maintain concentrated horse populations in cities and along canals, railroads, and turnpikes. This complementary relationship between railroads and horses was an important aspect of antebellum development. As the Agricultural Census of 1860 noted, "Horses have multiplied more rapidly since the introduction of locomotives than they did before.... Three fourths of the miles of railroad have been made since 1850; and we see that since then the increase of horses has been the greatest.... Railroads tend to increase [horses'] number and value. This is now an established principle."[8]

Table 5.1 Estimated U.S. Horse and Mule Populations, 1840–1860

Year	Horses and Mules	% Change	Horses	% Change	Mules	% Change
1840	4,335,669		N/A		N/A	
1850	4,896,050	12.9	4,336,719	N/A	559,331	
1860	8,753,029	78.8	7,434,688	71.4	1,318,341	135.7

Sources: *Compendium . . . as obtained from the Returns of the Sixth Census* (1841; reprint, New York: Arno Press, 1976), 359; *Statistical View of the United States . . . being a Compendium of the Seventh Census. By J. D. B. De Bow, Superintendent* (Washington, D.C.: Beverley Tucker, 1851), Table 185, 170, Table 186, 174; *Preliminary Report on the Eighth Census*, compiled by Joseph C. G. Kennedy, Superintendent of the Census (Washington, D.C.: GPO, 1862), Table 36, 198, 210.

As industrialization expanded the use of animal power, the demographics of the work animal population changed. Census tables show that the populations of both horses and mules grew dramatically between 1840 and 1860 (see Table 5.1). Particularly in the North, horses and mules constituted the majority of work animals. The amount of power generated by work animals increased dramatically as well (see Table 5.2).[9]

Army Horses

The Civil War made the complementary relationship between horses and railroads particularly salient, adding to the numbers of horses required by the army and altering the ways in which they were used as well. Railroads widened the geographical scope of the war by making it possible for armies to operate over greater distances, freeing them from traditional dependence on water transportation. Railroads moved enormous numbers of soldiers and huge quantities of matériel overland. However, this required using wagon trains containing hundreds of teams and wagons to enable field armies to operate away from railheads. For example, in 1864 the Army of the Potomac traveled

Table 5.2 Horsepower in the United States, 1850–1860

Year	Total Horsepower (in Thousands)	% Change	Horsepower from Draft Animals	% Change	Draft Animals as Percent of Total Horsepower
1850	8,495	N/A	5,960	N/A	70.2
1860	13,763	62.0	8,630	44.8	62.7

Source: U.S. Bureau of the Census, *Historical Statistics of the United States, Colonial Times to 1970*, Part 2 (Washington, D.C.: GPO, 1975), Series S1–14, 818.

with 4,000 wagons for 125,000 men. Since the standard army wagon used a team of six mules or horses, such a wagon train required at least 24,000 animals. Not surprisingly, wagon horses accounted for the largest percentage of all army horses. For example, of the 43,422 horses attached to General William S. Rosecrans's army in 1862, approximately two-thirds were wagon horses.[10]

In addition to extending the geographical size of the war, railroads changed the terrain by creating new strategic territory. Roadbeds, grades, cuts, tunnels, bridges, junctions, and stations were important locations to control. It was possible to defend whole sections of states without securing every inch, but railroads, with their long exposed lines, were vulnerable at every point. Reconnoitering railroad lines and attacking their strategic points became the special job of cavalry. One of the tactical innovations of the Civil War was the independent use of cavalry for large-scale raids. The Confederate army led the way in this area, honing raiding to an art form in the eastern and western theaters and creating a continual headache for Union commanders. In 1861, and again a year later, General "Jeb" Stuart rode completely around the Union army, attacking garrisons, cutting rail lines, seizing supplies, stealing horses, gathering intelligence, fighting mounted or on foot, and evading capture. Nathan Bedford Forrest used the same tactics in the West. By 1863, the Union Cavalry—a branch of the army that many generals had thought unnecessary and frivolous in 1861—had achieved a similar level of strength and effectiveness, as innovative officers like Philip Sheridan persuaded Ulysses Grant to let the cavalry fight independently. Furthermore, the geographical scale of the war, the need for intelligence about enemy location and strength, and the lack of maps made cavalry the eyes of the army without which commanders were operationally blind.[11]

Also salient were developments in weaponry that created new configurations of firepower and equine mobility. Rifles with greater accuracy and range reduced the efficacy of the traditional saber charge, but allowed cavalry and mounted infantry to combine considerable firepower with fast strike capability. Artillery horses maneuvered new light field guns on the battlefield. A commander could call up his artillery, then reposition it as the battle evolved. One of the most widely used artillery pieces was the 12-pounder Napoleon. This gun weighed 1,200 pounds, traveled on a wheeled carriage that adjusted the angle of fire, and had a range of 1,100–1,600 yards. It was attached to a wheeled device called a limber, with the whole rig pulled by a team of six horses. Six horses, more than enough to pull the gun, provided speed and endurance. A wheeled ammunition chest called a caisson, also pulled by six horses, accompanied each gun. A full battery consisted of eight guns, a traveling forge, a battery wagon, several supply wagons, and almost one hundred horses. After wagon horses (usually mules), artillery horses (almost never mules) were probably the most sizeable category of army animals. Thus the advent of industrialization, which contributed to Civil War armies being substantially larger than any previous American force, created new technological complementarities that required the Union to mobilize hundreds of thousands of horses and mules to serve as wagon horses, artillery horses, and cavalry mounts.

Supplying Horses

With the outbreak of war in 1861, the army began buying horses immediately and horse purchases escalated. As the army expanded from its prewar level of 16,000 to 187,000 by the end of 1861, the number of horses grew proportionately. The army needed one horse for every two men, and thus there was an immediate need for almost 100,000 horses.[12]

Managing the Union horse supply was the responsibility of Quartermaster General Montgomery C. Meigs. A native of Philadelphia and an 1832 graduate of West Point, Meigs had spent his early career as a military engineer. Based on his success and probity in managing two enormous projects—the construction of an aqueduct to supply Washington, D.C., with fresh water from the upper Potomac River and the completion of the Capitol—he was appointed Quartermaster General in 1861 shortly after the outbreak of war. In his own words, "his command extended from the Atlantic to the Pacific, from the Great Lakes to the Gulf." The Quartermaster department supplied everything except ordnance and subsistence. Within two years it had sole authority over the process of supplying Union forces and did not have to compete with state officials. The department spent $8 million in the fiscal year ending in July 1861; this figure quintupled to $40 million in the next year, tripled to $120 million by 1863, and peaked at $285 million for fiscal 1863–1864. Horses constituted one of the largest single expenditures in the budget. To this immense task, Meigs brought formidable organizational skills, an unrivaled capacity for detail, a talent for military planning, high standards of personal honor, and a fierce commitment to the righteousness of the Union cause. He often consulted with Abraham Lincoln, Secretary of War Edwin M. Stanton, General Henry W. Halleck, and field commanders in the planning of military operations.[13]

Horse purchases began immediately. In June 1861 Meigs ordered that 3,030 horses be purchased, 4,950 in July and 19,320 in August. Between July 1861 and September 1862, the Union government acquired 146,453 horses and 101,135 mules. Where did all these animals come from and how did the government obtain them? According to the census of 1860, the Union possessed 5.5 million horses and 400,000 mules at the beginning of the war, approximately twice as many as the Confederacy. However, these horses and mules were widely scattered across the Union states on farms and in towns. There were no breeding operations, organized markets, or large suppliers of any kind from which the Quartermaster department could purchase large numbers of animals. Furthermore, not all these horses were "serviceable," or suitable for military labor. Some were the wrong age, some were the wrong sex, some were unsound, and some were intractable or otherwise unusable. Meeting the immense horse requirements of the Union army while ensuring that the horses purchased were serviceable was an ongoing challenge for Meigs.[14]

Meigs relied on horse dealers to facilitate horse purchasing and to accumulate large groups of horses quickly. These dealers agreed to accumulate and deliver a certain number of horses at a prearranged price. They then would comb the countryside gathering and purchasing horses. These contracts could range from small lots of a dozen horses to groups of several hundred at a time. One horse dealer, Levi Straw of Boston, agreed to deliver one thousand horses, though this was an unusually large contract. Horse dealers were important middlemen in a market consisting of thousands of small suppliers, and they served to organize that market for the Quartermaster department. Furthermore, competitive bidding helped keep horse prices down as well.[15]

Quality Control

A wave of complaints followed the government's first horse purchases because so many proved to be unserviceable. All horses presented to the government for purchase had to pass inspection. Inspectors had to determine each horse's age, health, and sex, and assess its strength and soundness. This was a task requiring considerable skill and experience on the part of the inspector, including a working knowledge of equine health, physiology, mechanics, and movement. Determining a horse's age is difficult, let alone assessing its soundness. Few inspectors possessed skill and experience, but all came under immense pressure from the army to purchase horses and not discourage the horse dealers, and from horse dealers at inspection sites to approve the horses they presented. An incident in the fall of 1861 illustrates the difficulties of the situation.

Captain E. C. Wilson, a Quartermaster in Harrisburg, Pennsylvania, was ordered to inspect seventeen hundred horses in Harrisburg, as well as a thousand horses ninety-eight miles to the west in Huntingdon, and another three hundred in Chambersburg, fifty-six miles to the south. Wilson hired agents to handle the inspections in Huntingdon and Chambersburg. Wilson's first inspector was strict in his inspections and caused a complete uproar among the horse dealers when he rejected many of their horses. Wilson replaced this man with General James, a trusted friend, who did somewhat better, but fell ill, forcing Wilson to hire a third agent, Sherbaker, to finish the inspections.[16] In November 1861, twenty citizens of Huntingdon wrote an outraged letter to Meigs alleging fraud by Sherbaker. They claimed many of the horses Sherbaker had approved were ones they had known for years and knew to be unsound, diseased, or otherwise unsuitable. Their letter concluded,

> Corruption so apparent deals a blow as fatal as treason to the life of our country. The citizens almost doubt whether a government so beset by the base and unprincipled and so used by the knavery and cupidity of the vile is worth preserving. Patriotism is sorely tried, because those who coldly support this war have

such occasion to talk only of the corruptions which disgrace its conduct.... The honest and faithful citizen ... must aid in every service where his hands can help. Duty demands that we expose and strike this monstrous evil.... In the name of a bleeding and suffering country we appeal ... for prompt relief from such unmitigated disgrace as has befallen us.[17]

Meigs sent Major R. Jones to Huntingdon to investigate the allegations. When Jones reinspected the Huntingdon horses, he approved less than 200 cavalry horses, about 100 good workhorses, 30 artillery horses, and another 150 he described as "indifferent"—less than 480 horses all told. Among the horses approved by Sherbaker he found 164 mares, 35 of which were pregnant, 120 horses over nine years old, 86 less than three years of age, 60 that were too small, 8 that were blind, and at least 5 with permanent debilitating conditions. Jones concluded that the situation in Huntingdon provided "abundant evidence of the determination of the dealers ... to make as much as possible, regardless of the means used."[18]

Another difficulty was that many men were proud of their ability to judge horseflesh and touchy about it being challenged. This made the job of inspecting horses that other men verified as serviceable a demanding and delicate business. Wilson, by his own admission, had "no experience in any such business" and "no knowledge whatever of horses," making him vulnerable to pressure from horse dealers and unable to supervise agents like Sherbaker. As one Quartermaster observed, "The ability to inspect a horse ... is a source of great pride. When horses they deem sound are rejected, it's an affront to honor.... It involves both ability to judge horseflesh and human character."[19]

Meigs responded to criticisms of horse purchases from citizens and from Congress by issuing specifications to guide inspectors and standardize army horses in terms of age, size, weight, and sex. All horses purchased had to be between five and ten years of age. Team horses had to be equivalent in size and strength so they could be used interchangeably and so that harnesses could be one standard size. For example, artillery horses had to be of compact build, but strong and fast, in order to maneuver field guns during a battle. Mares were unacceptable, nor were stallions desirable, except for the occasional officer's horse. The majority of army horses were geldings. Early in the war, the specifications included color, but these requirements seem to have disappeared as the war consumed horses. Mule specifications differed in that mules were considered mature at only two years and were to be purchased primarily as wagon animals. The Quartermaster's rules read in part, as follows:

> Cavalry Horses must be sound in all particulars, well broken, in full flesh and good condition, from fifteen (15) to sixteen (16) hands height, from five (5) to nine (9) years old, and well adapted in every way to Cavalry purposes. Horses between nine (9) and (10) years of age, if still vigorous, sprightly and healthy, may be accepted.

Artillery Horses must be of dark colors, sound in all particulars, strong, quick and active, well broken and square trotters in harness, in good flesh and condition, from six (6) to ten (10) years old; and less than fifteen and one half (15-1/2) hands high; each horse to weigh not less than ten hundred and fifty (1050) pounds.

Mules must be over 2 years of age; strong, stout, compact, well-developed animals, in full health, free from any blemish or defect which would unfit them for severe work, and must have shed the four front colts teeth and developed the corresponding four permanent teeth, two in each jaw.

The foregoing specifications must be rigidly adhered to. No discretion is allowed to an Inspector to accept any animal which these specifications would otherwise reject.[20]

Meigs conflicted with field generals who were unsatisfied with the horses he provided or who claimed that Meigs was undersupplying them. In the fall of 1862, General George B. McClellan complained to General Halleck that Meigs would not give him enough horses. Embarrassed by Confederate General "Jeb" Stuart's second uncontested ride in a year around the entire Army of the Potomac, including a raid on Chambersburg, Pennsylvania, McClellan blamed it on a "deficiency" of horses from Meigs that made it "impossible to prevent the rebel cavalry raids." McClellan complained to Halleck that he had only received 150 horses per week; however, Meigs reported to Halleck that he had supplied McClellan with 1,459 horses a week, and 10,254 in the six weeks prior to Stuart's mid-October ride-around.[21] Meigs had a similar dispute with General Rosecrans of the Army of the Cumberland. In late 1862, Rosecrans complained that Meigs was sending him unusable horses and suggested that Meigs was refusing to pay enough to get good horses. Rosecrans claimed that due to lack of serviceable horses, "I have lost the control of the country between my infantry and that of the enemy . . . for want of an adequate mounted force. The fruits of victory have been wrested from me." When Meigs improved the inspection process at the Louisville horse depot, Rosecrans then complained that Meigs's standards were too high and had constricted his supply of horses. "What I learn is that only 29 horses per day are coming in since the new inspector began to be vigorous. We must have speed of delivery as well as quality."[22]

Equine Soldiers

Army horses had needs that resembled those of human soldiers for food, clothing, equipment, and medical care. Meigs had to provide transportation, forage, horseshoes, blacksmithing tools, all manner of horse furniture (saddles and bridles for cavalry and artillery, saddle blankets, halters, girths, bits, breast straps, harnesses for wagons and artillery), accoutrements such as currycombs, horse brushes, buckets, and nose bags, routine upkeep, and medical care.[23] The most important of these were forage and horseshoes. Underfed horses tired quickly and broke down, and unshod horses went lame. Requests for and concerns about forage supplies were a constant refrain in military reports. The

Quartermaster department assembled immense supplies of forage, by purchasing from many small dealers and producers. The daily ration for horses was fourteen pounds of hay and twelve pounds of grain, and for mules fourteen pounds of hay and nine pounds of grain. In comparison, human soldiers required three pounds of food daily. During the winter of 1861–62, the horses of the Army of the Potomac ate 400 tons of forage each day; in January 1864, they needed 74 railroad cars of grain and 375 cars of hay daily. Horses needed grain to maintain their strength and condition in army service, but they needed the bulk of hay or grass as well. They could not forage for themselves. Not only did the army travel with so many animals that finding pasture for them was impossible, but in uncultivated areas of the South there was no forage at all. The army often relied on delivered forage, and if it ran short, the cause centered on delivery rather than supply. For example, when Sherman campaigned along the coast in 1864, Meigs informed him that it would be difficult to supply forage for all his animals in that location (though he then managed to do it). Meigs used the railroads to transport this heavy and bulky freight to supply depots. Equally, a large portion of army wagon trains was devoted to carrying forage for the horses. If the army moved too far away from railheads where they could get supplies, the number of wagons needed to transport forage for artillery horses, cavalry horses, and the horses pulling the forage wagons soon became so huge as to reach the point of diminishing returns. On these grounds, Meigs questioned Rosecrans's demands for more horses:

> The large numbers of horses you have sent back to Louisville . . . over 9,000, shows that you have had more horses than your troops have been able to take care of. You say that there has been great mortality, for want of long forage, which could not be furnished for want of transportation. Were there then so many animals in the department that they could not transport their own food? When our army reaches this limit, what is the remedy? Is not every additional horse another subject for starvation? . . . You say that the rebels outnumber you five to one, and this I do not take to be a careless expression, for I find it repeatedly used in your dispatches too the General-in-Chief and to myself. Have they 60,000 mounted men? How do they find food for them?[24]

Army horses used thousands of horseshoes and needed frequent shoeing. A cavalry regiment of one thousand horses wore four thousand shoes (and many more nails) at a time and needed to carry spares. The Quartermaster department ordered different sizes of shoes for the front and hind hooves of horses and mules, often placing orders for fifty thousand kegs of them at a time. No single manufacturer of iron products could meet this demand, and Meigs purchased shoes from many manufacturers. Quality control was maintained through Boards of Survey, committees of officers convened to examine and report on the quality and fitness of purchased articles. Except for some shortages reported in late 1863, the Union had plenty of horseshoes and nails. A machine that mechanized horseshoe production had been patented and put to work in 1857, making the mass production of horseshoes possible.[25]

Equine Management

Despite efforts to procure good horses and keep them supplied, army horses did not receive very good care. Meigs was constantly frustrated at the apparent inability or unwillingness of soldiers to keep their horses serviceable and effective for military use. Few soldiers knew much about horse care or were willing to do the extra work that it required. Although the effectiveness of artillery and cavalry depended on having serviceable horses, many army horses were neglected, overworked, and ill-used. Field conditions made attending to horses difficult. Charles Francis Adams, Jr., eloquently described his struggles as a cavalry officer to spare his horses as much as possible and to make sure his men kept them in good condition. But horses were treated little better away from the field. Army horses often lived in large depots where they received little attention. Horses are already susceptible to a variety of skin and hoof conditions that are aggravated by poor care, but the army's large herds suffered epidemic diseases as well. A "violent and destructive" epidemic of hoof-and-mouth disease put four thousand of the Army of the Potomac's horses out of commission after Antietam, and there were periodic outbreaks of glanders, a highly contagious and incurable disease. One of the worst occurred in 1864 when glanders swept through the thirty thousand horses at the Giesboro depot outside Washington. These incidents also alarmed nearby farmers and other civilian horse owners who feared contagion from army horses.[26]

Meigs lectured officers on the condition of their horses, investigated complaints, established depots for rehabilitating wounded and worn out horses, and issued regulations about maintenance. He reminded field commanders that "extraordinary care [should be] taken of the horse, on which everything depends." When commanders complained they did not have enough horses, Meigs reminded them to preserve the condition of the horses they already had. In a letter to Rosecrans that was a virtual manual on horse care and tactical use, Meigs said:

> You report to General Halleck that you have received, since December 1, [1862] 18,450 horses and 14,607 mules. . . . You had on hand March 23, 19,164 horses and 23,859 mules—43,023 animals in all . . . about one horse or mule to every two men in your army. You have broken down and sent off as unserviceable, in addition to these, over 9,000 horses and report that one-fourth or one-third of the horses on hand are worn out. Now, all this, it seems to me that the horses are not properly treated. They are either overworked, or underfed, or neglected and abused . . . Compel your cavalry officers to see that their horses are groomed; put them in some place where they can get forage, near the railroad, or send them to your rear to graze and eat corn . . . never move off a walk unless they see an enemy before or behind them; to travel only so far in a day as not to fatigue their

horses; never to camp in the place in which sunset found them, and to rest in a good pasture during the heat of the day.[27]

Another way that Meigs tried to keep army horses from being wasted was by attempting to limit the length of wagon trains by restricting the amount of soldiers' equipment. Baggage trains were hundreds of wagons long; at six horses or mules per wagon, such trains required many animals and consumed immense quantities of forage. The more wagons there were for equipment, the more wagons there had to be to carry forage for wagon horses, in an unending upward spiral. The length of the wagon train determined how many horses were needed, but the availability and condition of the horses determined the length of the wagon train. In addition, heavier loads, poor roads, or bad weather could increase the size of the team needed for each wagon. On the muddy mountain wagon road into Chattanooga in the fall of 1863, wagons sometimes required sixteen mules, plus a soldier per mule to help the mule push.[28]

Meigs wanted the army to adhere to the standard prescribed by Napoleon of twelve wagons per thousand men; to this end he told McClellan in the fall of 1862 that six thousand wagons were too many, and he had to restrict his wagons to one for every eighty men, including officers. The Union army averaged *at least* twenty-six wagons for every thousand men, but this figure rose as the war went on. In the spring of 1864 Grant's army averaged thirty-one wagons per one thousand men. Napoleon, however, had campaigned across densely settled and cultivated Europe where his armies could live off the land, while the Union army often campaigned in uncultivated areas of the South where foraging was difficult if not impossible. Meigs repeatedly issued orders that specified the exact number and kind of wagon permitted for each organizational unit of the army, but his success in limiting the length of wagon trains was more limited than the trains proved to be.[29]

Dead Horses

There was constant attrition of army horses in the never-ending skirmishes and campaigning that defined much of the Civil War military action. On long marches, horses remained harnessed or saddled for hours or days at a time, went unshod, and had no regular water, forage, or rest while traveling as much as forty miles a day. Under these conditions horses wore out, some dying of fatigue or becoming so unable to work that they were abandoned along the way or even shot to keep them out of enemy hands. Commanders frequently described their horses as "jaded" after military operations and in need of days of rest, forage, and care before being serviceable again. Despite their size, strength, and often-formidable appearance, horses are surprisingly vulnerable and even fragile creatures.

Transporting or relocating horses often exposed them to hazards. Horses were lost on overland drives and drowned during river crossings. For example, in 1863, the army had to bring 2,176 horses and 345 mules overland from Aquia Creek in Virginia to Washington, D.C., employing nearly 200 herders and a company of cavalry. Near Dumfries, the herd stampeded into a salt marsh, and almost 200 horses drowned. Once out of the marsh the herd entangled itself with a passing army train, and soon there were army animals scattered across northern Virginia. Railroad transport was risky as well. There were few cars properly constructed to protect horses from injury. The long delays and frequent layovers typical of railroad travel kept horses confined in railroad cars without adequate food, water, or care for hours and sometimes days. Few railroad depots and personnel were equipped to offload and care for large numbers of horses. Army horses routinely needed days of rest and care after railroad travel. Transporting horses by steamboat was much the same.[30]

Major battles caused immense equine casualties along with human casualties. There were many horses on battlefields—the horses of officers commanding troops, of couriers carrying messages so field commanders could communicate with armies scattered across large battlefields, and of the artillery and cavalry. The size, bulk, and build of horses made them excellent targets, and battle reports frequently describe officers having horses shot out from under them. Artillery horses suffered particularly high casualties. When a field commander ordered up the artillery, six-horse teams galloped the guns from the rear and wheeled them into position—as many as eight guns and eighty-four horses for a full battery. Once there, soldiers detached the guns from the limbers and were supposed to then move the horses and caissons well back from the gun into a protected location. However, artillery soldiers often wanted access to the ammunition carried in the caisson and limber, and they wanted the horses available in case they needed to "limber up" and move. Field guns were placed to bring the enemy into range, but this brought them into the enemy's range as well. Chickamauga battle reports from September 1863 paint a picture of artillery horses under fire:

> Here [the rebel] battery attempted to get into position, but their horses and men were shot down as often as attempted.
>
> A section of [Confederate] battery was in our front. The regiment killed all the horses belonging to one of the guns.
>
> I immediately gave orders for the battery to limber up, but it could not be done as the horses as they were brought up to the guns were shot down.
>
> The battery was hardly in position before, the troops giving way on the right, it was exposed to a most terrific fire of musketry from the front and flank. General King ordered the battery to limber to the rear, but it was impossible to execute the order, since many of the cannoneers were killed and wounded, and the horses shot at the limbers.

> Three pieces were limbered up with much difficulty, under the most galling fire, and got away. The horses had been shot belonging to the other limbers.
> A loss of 13 men killed and wounded, and 25 horses killed.
> My loss was . . . 16 horses permanently disabled.
> Horses wounded, 15; and in consequence of not unharnessing for six days and the hardship they have undergone, I will lose 25 more horses.
> At this point the enemy had returned to our front with a battery of two 12-pounder guns . . . at the first volley from my regiment every horse was killed.
> On the afternoon of Sunday, when the final retreat was made, our loss in horses was 30.[31]

Artillery soldiers tried to carry on with the remaining horses. They freed live horses from the wreckage and attached them to other guns and removed guns from the field. Guns needed less than six horses to move; the extra horse power provided maneuvering speed. However, often guns were abandoned "for want of animals to bring them off."[32]

Experienced and trained artillery and cavalry horses became accustomed to being in battle, especially if part of an established team and handled by skilled drivers. However, horses could panic if wounded, frightened, or disoriented. This made driver skill paramount, but artillery drivers were not always skilled and not always able to control six horses under battle conditions, as these battle reports show:

> Just as the howitzer entered the road the horses took fright and started off at full speed up the road. The driver of the horses (who was a volunteer and not accustomed to the team) informs me that he attempted to halt just as he got into the road, and that the dashing by of the troop which accompanied us caused his horses to become unmanageable and to run off. . . .
> Their percussion shells were bursting in quick succession among us. One of them knocked off the two drivers of the limber of the 12-pounder, and the horses ran entirely away. . . .
> In the meantime, shell and bullets were rained on the 12-pounder so fast that the limber was broken and the horses so repeatedly wounded that they could not be held to their places but ran away with both it and the caisson. . . .
> My horses [became] unmanageable when the firing commenced. . . . My horses are too green to be serviceable.[33]

Stampeding horses were a factor on the battlefield, disrupting lines and inflicting injury, especially if still harnessed to guns and caissons. "The horses became restive and gave way to the rear, breaking the lines of Company A." "Horses without riders . . . dashed through our ranks with great speed . . . our lines were broken several times by horses and mules running away."[34]

Moreover, horses became interchangeable parts in battle. Officers whose horses were killed or disabled jumped on strange horses; cavalry and wagon horses were commandeered to pull artillery; horses extricated from the wreckage of one gun were commandeered for another. Under battle conditions, any horse was preferable to none. This was why Meigs tried to standardize the size and the equipment of army horses, so all harnesses would fit all horses, and viable teams could be assembled on the spot.

Wartime Union equine casualties have been estimated at a million and a half. Army animals broke down from hard service; drowned during river crossings; died of disease, poor care, and battle wounds; and were captured or killed in raids. In the fall of 1863, the army lost more than ten thousand animals during the siege of Chattanooga, when the only supply route was a sixty-mile mountain road and forage ran low. Soldiers described dead mules stacked high in the streets. Meanwhile, Confederate General Joseph Wheeler's Cavalry intercepted an eight-hundred-wagon Union supply train, burned the wagons, and sabered the mules. Dead horses were a ubiquitous part of any battlefield. The *Atlantic Monthly*, describing the aftermath of Antietam, noted that "at intervals, a dead horse lay by the road side, or in the fields, unburied. . . . At the edge of the cornfield lay a gray horse. . . . Not far off were two dead artillery horses in their harness."[35] This description, written for civilian consumption, did not convey the reality of most battlefields. Battle reports routinely described landscapes of dead horses; photographs taken after Gettysburg show hundreds of carcasses piled for burning.

Union Horse Supply

Union commanders could count on having large numbers of horses at their disposal. After the battle of Gettysburg, Meigs had almost seven thousand horses ready to send the Army of the Potomac to replace losses. In 1864, when General Grant opened his campaign against General Robert E. Lee, his 125,000 soldiers crossed the Rapidan River accompanied by almost 30,000 horses, 23,000 mules and 4,000 officers' horses. Later that year, General Sherman departed Atlanta for the coast with sixty-four thousand soldiers and thirty-four thousand horses. Between January and August of 1864, Meigs remounted the entire cavalry of the Army of the Potomac twice. General Sheridan reported receiving two hundred horses a day by the summer of 1864, and by the end of the war, Meigs was sending replacements for horse casualties that approached five hundred per day.[36]

The Union contained 5.5 million horses and 400,000 mules, twice as many equines as compared to the Confederacy's 2 million horses and 900,000 mules. While the Union had a generous horse supply, it did not have an unlimited horse supply. Like any living population, not every horse was available or appropriate for military use, due to age, sex, or physical condition. In addition, the industrialized North used horses in all aspects of its economy, so that the army competed with the needs of the home front. The South used much less animal power by comparison. Moreover, the horse supply was somewhat inelastic. Horses could not be bred to meet war demand. The gestation period for horses is eleven months, and horses take four years to reach maturity. Thus, any prescient individuals who began breeding horses in 1861 would not have had horses ready to sell to the government until the spring of 1865. Although the demand for war horses did stimulate the horse market, people began breeding horses for postwar rather than wartime sales. The government augmented the horse supply by importing some horses and by capturing Confederate horses, but for the most part, Union horse supply consisted of the horses on hand in 1861, some of which matured and become available for use during the war. In 1863, Lincoln became so concerned about the horse supply that he signed an order forbidding horse exports. For Meigs, the question of production did not involve increasing the actual number of horses, but creating a system that would elicit as many serviceable horses as possible from the northern economy.[37]

However, having an adequate horse supply meant having horses that were in use, not just numbers in census tables. That the Union had such an ample horse supply by the last two years of the war was due to Meigs's skillful management. Meigs did not simply acquire horses for the army: he developed an integrated system of procurement and supply for this idiosyncratic, high-maintenance war commodity, a system he then tried to standardize as much as possible through purchase specifications and inspection standards. Some problems—inspection and horse care—were never fully solved, but effective purchasing procedures and adequate forage and horseshoe deliveries kept the army supplied with enough horses.

By the last two years of the war, Meigs was acquiring horses and their supplies from all over the North and from captured territory in the South. Using the railroads, he funneled them to a three-tier system of supply depots—principal depots located around Washington, D.C., and St. Louis, Missouri; regional advance depots; and temporary depots located near field armies. He tried to maintain the condition of army horses by ensuring adequate forage and horseshoes, issuing regulations on care, establishing rehabilitation depots for jaded horses, and establishing an army veterinary corps. For example, to address the problems associated with moving horses by rail, Meigs developed special cars for horses, instituted shipping policies that included mandatory rest stops, built facilities for feeding and watering horses at railroad yards, and had horse trains accompanied by experienced personnel. Though he constructed a bureaucracy focused on horse supply, he remained focused on the goal of delivering to field commanders the horses they needed—or in the case of McClellan and Rosecrans, thought they needed. His disputes with those generals centered on the numbers of horses supplied, but Meigs never refused to send horses and would break his own rules if necessary to meet urgent demands. If Meigs needed horses quickly, he would purchase on the open market even if it proved costly. He would even buy inferior horses, stating that in some situations any horse was worth more than no horse at all. If he needed horses for a month's work and needed them immediately, then Meigs would settle for horses that would only last a month.[38]

The difference between the Union and Confederate horse supply was more than numerical. Confederate soldiers supplied their own artillery and cavalry horses. For example, Confederate Quartermaster General Abraham C. Myers and later Alexander R. Lawton had to compete with state governors and field commanders for supplies. They resorted to impressment to acquire animals for the army, eroding civilian support and the home front economy. They also suffered much criticism and interference from politicians and generals, which dissipated their efforts and fragmented their goals. Neither was able to construct a system around the use of horses. By contrast, the Union government purchased and owned all army horses, except for the private horses of officers. Meigs benefited from the centralized political structure of the Union government and from his personal relationships with politicians and generals that for the most part were collegial rather than contested. He did not have to compete for resources with state governments and private institutions, nor did he have to resort to impressment. Able to use market mechanisms to obtain horses and related commodities without resorting to a command economy, Meigs generated demand and managed supply, moving horses through the system, negotiating at one end with horse dealers who wanted to sell as many of their horses as possible, and at the other end with field commanders who demanded more horses, complained about the quality of the ones they received, and didn't maintain the condition of those they already had.[39]

Railroads made a critical difference for both sides. Neither Myers nor Lawton was able, or even willing, to get control over rail transportation. The Confederate government received little cooperation from southern railroads on moving military supplies and had no authority to make them do so. The chronic lack of forage suffered by Confederate horses was as much a problem of distribution as a problem of supply. In the spring of 1865, Lee had to send many of his cavalry and artillery horses away to be foraged, leaving him without the resources he needed to hold off the oncoming Yankees. Because of authority granted in legislation such as the Railroad Act of 1862, Meigs could count on cooperation from northern railroads, even for goods they did not like to transport such as forage. The Union was able to reap the strategic benefits of a complementary relationship between railroads and horses by using railroads effectively to deliver supplies and even by having army horses ride iron horses to war. A particular dramatic incident occurred in 1863 when northern railroad executives and the War Department mounted a rescue mission to relieve the siege of Chattanooga. Over 20,000 men, thousands of horses, and their equipment moved 1,233 miles in less than two weeks, a distance it would have taken months to cover traveling overland. The Confederacy rarely shipped horses by rail and even forbade it, but consumed time and energy by moving horses overland instead.[40]

Organic Machines

Meigs's statement that horses were the factor "on which everything depends" illustrates the centrality of horses in the technological environment of the Civil War. Horses were components of war technology. Just as a steam boiler was part of a locomotive, not a separate or independent artifact, horses were part of supply wagons, field guns, and cavalry. All of these were technological ensembles. Yet, horses are largely absent from discussions of nineteenth-century technology, despite their importance as prime movers. As animals, they fall on the wrong side of a long-standing conceptual divide between technology and nature and are treated as transhistorical objects. Differentiating between horses and technology in this fashion is contrary to how nineteenth-century people viewed horses. Fascinated with machinery, people frequently employed mechanical metaphors to describe and understand horses as prime movers and to find ways to improve their size, speed, and power. As one agricultural report stated: "Considered in reference to utilitarian purposes, the horse may be called a machine." What counts as a technology changes over time; understanding Civil War technology requires recognizing how the nineteenth century understood technology.[41]

Army horses were the central component in a network of relationships between horses, mules, forage, horseshoes, railroads, iron manufacturers, horse dealers, farmers, railroad executives, and cavalry officers, and other artifacts, individuals, institutions, and practices. Meigs's success lay in constructing a comprehensive system of horse supply that engaged many aspects of the industrial economy. The Civil War was embedded in the process of industrialization, and it illustrates that industrialization was a process that created new configurations of organic and inorganic components. An analysis of the Union horse supply suggests a different relationship between horse power and industrialization than has been traditionally described.[42]

The topic of horse supply establishes Civil War logistics as an important connection between industrialization and military history. While many historians nod to Meigs's accomplishments, they rarely stress logistics in their narratives of the first industrialized war. Alfred Chandler has linked the emergence of business management to sectors and industries characterized by new technologies and expanding markets and focused on the middle management of railroad companies as providing a "visible hand" in nineteenth-century economic development. According to Chandler, "Increasing specialization must, almost by definition, call for a more carefully planned coordination if volume output demanded by mass markets is to be achieved." Did the war effort demand specialization and resemble a new industry? As the builder of a national system, was Meigs not a better example of the operation of the "visible hand" than the railroad executives who operated smaller, regional operations?[43]

An ongoing question about the Civil War is the role played by the Union's superior resources in Union victory. Charts displaying comparative resources of the Union and Confederacy remain a staple of textbooks and lectures on the Civil War, showing the Union possessing twice the population, ten times the factory production, twice the railroad mileage, sixteen times the iron production, and three times the farm acreage. But as recent events like the Vietnam

War demonstrate, superior resources and technological supremacy do not always guarantee military victory. In the context of nineteenth-century industrialized warfare, new combinations of horses and inorganic technologies required a new concept of organization and made management of the horse supply vitally important. It was a task at which the Union proved more successful than the Confederacy. One of the keys to Union victory lay in its ability to incorporate a traditional war animal into the technological and logistical framework of large-scale warfare and not simply in the numbers of horses at its disposal. The Civil War was, in Russell Weigley's words, "perhaps the last great war of animal power," but it was a war of animal power precisely because it was an industrialized war.[44]

From ancient times to the present, horses have been military instruments. In the recent war in Afghanistan, camels, donkeys, and horses went to battle alongside sophisticated military machines; and in the mountains, American army officers coordinated from horseback complex operations combining satellites, planes, ground troops, and artillery. In the Civil War, as in the present, horses proved to be integral components of the most modern technological systems.

Notes

1. Joseph Glatthaar, "Battlefield Tactics," in *Writing the Civil War: The Quest to Understand*, ed. James M. McPherson and William J. Cooper, Jr. (Columbia: University of South Carolina Press, 1998); Walter Licht, *Industrializing America* (Baltimore: Johns Hopkins University Press, 1995), xv; Russell Weigley, *History of the United States Army* (New York: Macmillan, 1967), 220; James M. McPherson, *Battle Cry of Freedom* (New York: Oxford University Press, 1988), 306 n41, 325; Philip M. Teigen, "The Zoogeography of Horses and Mules in the United States, 1860–1920," paper presented at the annual meeting of ICOHTEC, Lisbon, Portugal, August 1998, p. 3. The precise number of horses used in the war is unknown, but Union Quartermaster General Montgomery C. Meigs estimated that the Union army used one horse or mule for every two men. *The War of the Rebellion: A Compilation of the Official Records* (Washington, D.C.: GPO, 1880–1901), series III, vol. 4, 1212 (henceforth designated as *OR*).

2. People in the nineteenth century used "horses" to refer to both horses and mules unless they needed to be more specific, and this essay follows this convention. Mules are "half-horses"—the sterile offspring of male donkeys and female horses.

    These questions follow the systems theory of Thomas Hughes, who analyzes technology as networks of relationship between artifacts, institutions, individuals, practices, and values. See Thomas P. Hughes, *Networks of Power* (Baltimore: Johns Hopkins University Press, 1983). These questions also follow the model provided in John Singleton, "Britain's Military Use of Horses, 1914–1918," *Past and Present* 139 (May 1993): 179. Alan Nevins, in his four-volume work on the Civil War, stated "Most historians have neglected giving proper attention to the supply of horses and mules in Civil War days." *The War for the Union*, vol. 3 (New York: Scribners, 1959), 234.

3. *A Statement of the Arts and Manufactures of the United States of America for the Year 1810. Digested and Compiled by Tench Coxe* (Philadelphia: n.p., 1814), xxiv; *Agriculture of the United States in 1860, Compiled from the Original Returns of the 8th Census . . . by Joseph C. G. Kennedy, Superintendent of the Census* (Washington, D.C.: Government Printing Office, 1864), xii, xv, clxiv; Dolores Greenberg, "Energy Flows" in *An Emerging Independent American Economy, 1815–1875*, ed. Joseph R. Frese and Jacob Judd (New York: Sleepy Hollow Press and Rockefeller Archive Center, 1980), 35.

4. The classic work on antebellum transportation is George Rogers Taylor, *The Transportation Revolution* (New York: Rinehart and Company, 1951). On stagecoaches, see Richard John, *Spreading the News* (Cambridge, Mass.: Harvard University Press, 1995), 98–102; John F. Stover, "Canals and Turnpikes" in *An Emerging Independent American Economy, 1815–1875*, ed. Frese and Judd, 63. The Pony Express did not invent western mail delivery but provided express service in addition to existing delivery by stagecoach.
5. Ronald E. Shaw, *Canals for a Nation* (Lexington: University Press of Kentucky, 1990); *Scientific American* 13 (August 28, 1857): 405; *Scientific American* 27 (July 13, 1872), 15; 27 (October 19, 1872), 246; 27 (September 21, 1872), 185; Kevin J. Crisman and Arthur B. Cohen, *When Horses Walked on Water: Horsepowered Ferries in Nineteenth Century America* (Washington, D.C.: Smithsonian Institution Press, 1998). There was a horse ferry on the Tennessee River in Chattanooga at the time of the Union siege in 1863.
6. Leo Rogin, *The Introduction of Farm Machinery* (Berkeley: University of California Press, 1931); R. Douglas Hurt, *American Agriculture* (Ames: Iowa State University Press, 1994); Paul Gates, *Agriculture and the Civil War* (New York: Alfred A. Knopf, 1965) and, idem., *The Farmer's Age* (New York: Holt, Reinhardt and Winston, 1960).
7. Clay McShane and Joel Tarr, "The Centrality of the Horse in the Nineteenth-Century American City," in *The Making of Urban America*, ed. Raymond A. Mohl (Wilmington, Del.: SR Books, 1997), 105–130; Joel Tarr, "A Note on the Horse as an Urban Power Source," *Journal of Urban Affairs* 25 (March 1999): 434–449; and Clay McShane, "Gelded Age Boston," *New England Quarterly* 74 (June 2001): 274–302.
8. Conventional descriptions of the relationship between horses and railroads focus on locomotives replacing horses for motive power, rather than looking at the context of the entire transportation system. John F. Stover's statement that "the steam locomotive soon relegated the horse to a reserve and substitute status" is a typical example. John F. Stover, *American Railroads* (Chicago: University of Chicago Press, 1961, reprinted 1997), 24. However, Stover's book and the following provide a comprehensive picture of antebellum railroads: Taylor, *Transportation Revolution;* Robert C. Black, III, *The Railroads of the Confederacy* (Chapel Hill: University of North Carolina Press, 1952); John C. Clark, *Railroads in the Civil War* (Baton Rouge: Louisiana State University Press, 2001); George Rogers Taylor and Irene D. Neu, *The American Railroad Network, 1861–1890* (Cambridge, Mass.: Harvard University Press, 1956); George E. Turner, *Victory Rode the Rails* (Indianapolis, Ind.: Bobbs-Merrill, 1953; Westport, Conn.: Greenwood Press, 1975); Thomas Weber, *The Northern Railroads in the Civil War, 1861–1865* (New York: Columbia University Press, 1952). See also *Agriculture of the United States in 1860*, clxiv, clxv.
9. *Compendium . . . as Obtained from the Returns of the Sixth Census* (1841; reprint, New York: Arno Press, 1976), 359; *Statistical View of the United States, . . . Being a Compendium of the Seventh Census. By J. D. B. DeBow, Superintendent* (Washington, D.C.: Beverley Tucker, 1851), Table 185, 170; Table 186, 174; *Preliminary Report on the Eighth Census*, compiled by Joseph C. G. Kennedy, Superintendent of the Census (Washington, D.C.: Government Printing Office, 1862), Table 36, 198, 210; U.S. Bureau of the Census, *Historical Statistics of the United States, Colonial Times to 1970*, Part 2 (Washington, D.C.: GPO, 1975), Series S 1–14, 818.

Population numbers for horses and mules are from census returns for 1840, 1850, and 1860. Percentages are my own calculations based on those figures. The superintendent of the census relied on data supplied by U.S. marshals, but how the marshals collected data is unclear. These census returns are what Union and Confederate government officials possessed, and they provide a generally accurate picture of antebellum equine populations.

For an excellent discussion of energy and power in the nineteenth century, see Dolores Greenberg, "Energy Flows" in Frese and Judd, *An Emerging Independent American Economy*. Based on her reading of the census figures, Greenberg notes that work animals provided the majority of power until 1880. This fact belies the traditional narrative of industrialization. While the census figures in Table 2 have the same limitations as the ones for horse and mule populations, they accurately depict that the use of all power was rising, and that animal power continued to rise even as its percentage of total power declined relative to that of inanimate power sources. It is possible that humans produced the majority of the power consumed in the United States and that a more accurate picture of the energy landscape of antebellum America, and especially that of the antebellum South, would result if the power provided by both free and enslaved labor were included in these figures.

10. *OR* I, vol. 33, pt 1, 853–854; Russell Weigley, *Quartermaster General of the Union Army: A Biography of M. C. Meigs* (New York: Columbia University Press, 1959), 264, 269.
11. *OR* I, vol. 4, 340–341, 353; *OR* I, vol. 29, pt. 2, 254; *OR* I, 31, pt. 1, 737; Glatthaar, "Battlefield Tactics," 74; Archer Jones, *Civil War Command and Strategy* (New York: Free Press, 1992), 84–85.
12. U.S. Bureau of the Census, *Historical Statistics of the United States, Colonial Times to 1970.*
13. Sherrod East, "Montgomery C. Meigs and the Quartermaster Department," *Military Affairs* 25 (Winter 1961–62): 183–196, Meigs, quoted on p. 186; Weigley, *Quartermaster General*, 317.
14. Records of the Office of the Quartermaster General, Consolidated Correspondence File, Box 839, RG 92, National Archives; *Preliminary Report on the Eighth Census*, 199, 210. Weigley cites 4,688,878 horses from the 1860 Census; I have added almost a million horses from the addendum to Table 36 on p. 210. Weigley, *Quartermaster General*, 256.
15. John A. McKim to Capt. James A. Ekim, Boston, MA, February 22, 1864 and March 5, 1864; Potter to Ekim, March 21, 1864. RG 92, CCF, Box 838.
16. Letter Captain Edward Wilson to Major R. Jones, December 9, 1861. RG 92, Box 839.
17. Letter Citizens of Huntingdon, PA to Montgomery C. Meigs, November 2, 1861, RG 92 Box 839.
18. Letter, Major R. Jones to Brig. General L. Thomas, Adjutant General of the U.S. Army, December 18, 1861, RG 92, Box 839.
19. Captain G. W. Lee to Meigs, August 14, 1863, RG 92, CCF Box 839, file "horses 1863."
20. General Order No. 43, Quartermaster's Office, September 23, 1864. "Rules and Regulations relating to the purchase, procurement and disposition of horses and mules for the Army," CCF, Box 838, RG 92. Specifications for horses, dated July 2, 1863, included color. Artillery horses could be bays, browns, and blacks; cavalry horses could be sorrel as well; gray horses were only permitted for buglers. CCF, Box 840. There are cases where mules were used to mount cavalry and infantry, but they were considered too panicky for artillery unless they were the only animals available.
21. David W. Miller, *Second Only to Grant* (Shippensburg, Penn.: White Mane Books, 2000), 171; George B. McClellan, *Report on the Organization and Campaigns of the Army of the Potomac* (New York: Sheldon, 1864), 418; *OR* I, vol. 19, pt. 2, 422–424. Jeb Stuart's first ride around the Union Army had been in October 1861. Both rides around the Union Army garnered newspaper headlines in the North and the South.
22. *OR* I, vol. 23, pt. 2, 271, 289.
23. See *OR* I 30, pt. 1, 366, 407 for a partial list of horse accoutriments.
24. *OR* I, vol. 23, pt. 2, 301. Russell Weigley notes that in World War II the army faced the same situation of being immobilized by the size of its transport when General Patton's armored divisions outran their gasoline supplies. Weigley, *Quartermaster General*, 264.
25. *OR* I, vol. 3, 148; *OR* I, vol. 5, 117–118; *OR* I, vol. 10, pt. 1, 64; Proceedings of a Board of Survey, February 18, 1863, CCF, Box 840, RG92. In contrast, the Confederate army got so desperate for horseshoes that they stripped them from dead horses. *OR* I, vol. 5, 780, 783; *OR* I, vol. 31, pt. 1, 480.
26. Worthington C. Ford, ed., *A Cycle of Adams Letters 1861–1865*, 2 vols. (Boston: Houghton Mifflin, 1920); Stephen Z. Starr concluded in his magisterial work on the Union Cavalry that unwillingness to take proper care of the horses was "well-nigh universal." Stephen Z. Starr, *The Union Cavalry in the Civil War*, vol. 3, *The War in the West, 1861–1865* (Baton Rouge: Louisiana State University Press, 1985), 590. Nor were conditions for Confederate horses any better. Despite a persistent mythology that Southern soldiers were natural horsemen, they took no better care of their horses than did Yankees, though they had fewer horses to waste. For example, cavalry genius Jeb Stuart was known for being particularly hard on the horses in his command. McClellan, *Report*, 419; "GLANDERS!—Too Late," *American Agriculturalist* 24 (September 1865): 269. For a full account of Civil War glanders epidemics, see G. Terry Sharrer, "The Great Glanders Epizootic, 1861–1866: A Civil War Legacy," *Agricultural History* 69 (1995): 79–97.
27. *OR* I, vol. 10, pt. 2, 270.; Colonel W. Halstead to Meigs, February 2, 1862, Private Peter Lanks to Secretary of War Stanton, April 14, 1863, Private A. D. Still to Abraham Lincoln, November 2, 1863, CCF, Box 839, RG92; Nevins, vol. 2, 476; *OR* I, vol. 23, pt. 2, 303.
28. *OR* I, vol. 30, pt. 1, 218, 242; Jerry Korn, *The Fight for Chattanooga: Chickamauga to Missionary Ridge* (Alexandria, VA: Time-Life Books, 1985), 78.

29. *OR* I, vol. 31, pt. 3, 70, 130, 179; *OR* I, vol. 33, 417, 853–854, 920–921; *OR* I, vol. 52, pt. 1, 702; *OR* III, vol. 4, 888, 1212; Nevins, *War for the Union* vol. 2, 476–477.
30. *OR* I, vol. 30, pt. 3, 51; *OR* I, vol. 30, pt. 1, 544; *OR* I, vol. 30, pt. 2, 53–57, 670–673, 668; *OR* I, vol. 10, pt. 1, 820, 915; Report of Captain L. T. Peirce to the Deputy Chief Quartermaster, Army of the Potomac, June 26, 1863, CCF, Box 839, RG92; David J. Gerleman, "Unchronicled Heroes: A Study of Union Cavalry Horses in the Eastern Theater, Care, Treatment and Use, 1861–1865," Ph.D. diss., Southern Illinois University, 1999, 101–111, 123.
31. *OR* I, vol. 30, pt. 1, 285, 292, 309, 324, 414, 427, 437, 554, 575, 577.
32. This particular quote is from *OR* I, vol. 3, 281 but this phrase is used repeatedly in the *Official Records*.
33. *OR* I, vol. 2, 189–190; *OR* I, vol. 3, 227, 234; *OR* I, vol. 29, pt. 2, 221.
34. *OR* I, vol. 10, pt. 1, 223, 226–227.
35. Drew Gilpin Faust, "Equine Relics: Stuffed Horses, Articulated Skeletons and the Civil War," working paper, University of Pennsylvania, Philadelphia, 1999, 1; *OR* I, vol. 30, pt. 2, 723; *OR* I, vol. 30, pt. 1, 221; Korn, *The Fight for Chattanooga*, 79–81; Oliver Wendell Holmes, "My Hunt After 'The Captain,'" *Atlantic Monthly* (December 1862): 745, 748.
36. *OR* I, vol. 27, pt. 2, 543, 590; Journal of William Bolton, 51st Pennsylvania. Civil War Library and Museum, Philadelphia: *OR* I, vol. 33, 853–854; *OR* I, vol. 52, pt. 1, 702; "Horses Wanted," *American Agriculturalist* 23 (1864): 102; *OR* I, vol. 36, pt. 1, 797–798; *OR* III, vol. 4, 1212.
37. *Preliminary Report on the Eighth Census,* Table 36, 198, 210. The agricultural press encouraged farmers to breed horses because the war demand had eliminated the surplus horse supply and the horse market looked promising for many years. "Horses Wanted," *American Agriculturalist* 22 (1863): 231; "Horses Wanted," *American Agriculturalist* 23 (1864): 102; "The Prospective Demand for Horses and Mules," *American Agriculturalist* 23 (1864): 141; "Breeding Horses," *American Agriculturalist* 23 (1864): 174. It is unclear if the government imported horses. There are no horse imports in the census of 1870, or in the Reports of the Commissioner of Agriculture for the years 1861–1868. Nor does the Consolidated Correspondence File on Public Animals for Record Group 92 contain mention of imports beyond references to "Canuck" horses as a desirable type of horse to acquire. Horses were more often referred to by place of origin than by breed at this time, so there were New York horses, Maine horses, Michigan horses, and so forth. There may have been some movement of horses across the border from Canada. If the Quartermaster department made significant purchases of imported horses, they are not recorded with their purchases of domestic horses.
38. Miller, *Second Only to Grant*, 96, 120.
39. Richard D. Goff, *Confederate Supply* (Durham, N.C.: Duke University Press, 1960), 247.
40. Charles W. Ramsdell, "General Robert E. Lee's Horse Supply," *American Historical Review* 35 (July 1930): 758, 776, 777; *OR* I, vol. 46, pt. 1, 1267; McPherson, *Battle Cry of Freedom*; Gerleman, "Unchronicled Heroes," 111.
41. Edmund Russell, *War and Nature: Fighting Humans and Insects with Chemicals from World War I to Silent Spring* (Cambridge, U.K.: Cambridge University Press, 2001), 10; Judith McGaw, *Most Wondrous Machine* (Princeton, N.J.: Princeton University Press, 1987), 3; *Report of the Commissioner of Agriculture, 1862* (Washington, D.C.: Government Printing Office, 1863), 336; Gabrielle Hecht and Michael Thad Allen, "Authority, Political Machines and Technology's History," in *Technologies of Power: Essays in Honor of Thomas Parke Hughes, and Agatha Chipley Hughes*, ed. Michael Thad Allen and Gabrielle Hecht (Cambridge, Mass.: MIT Press, 2001), 13. Richard White's phrase "the organic machine" is useful here. See Richard White, *The Organic Machine* (New York: Hill and Wang, 1995).
42. Hughes, *Networks of Power*, 2–3.
43. Alfred D. Chandler, *The Invisible Hand* (Cambridge, Mass.: Harvard University Press, 1977), 8, 490. An overview of recent Civil War literature contains no references to Meigs, Quartermasters, or logistics. James M. McPherson and William J. Cooper, Jr., eds., *Writing the Civil War: The Quest to Understand* (Columbia: University of South Carolina Press, 1998).
44. Robert E. Lee began this debate when, in the wake of his surrender to Grant, he told his troops "After four years of arduous service, marked by unsurpassed courage and fortitude,

the Army of Northern Virginia has been compelled to yield to overwhelming numbers and resources." *OR* I, vol. 46, pt. 1, 1267. A nuanced argument about Union resources often misinterpreted as deterministic is Richard Current, "God and the Strongest Battalions," in *Why the North Won the Civil War,* ed. David Donald (Baton Rouge: Louisiana State University Press, 1960), 3–22; Weigley, *Quartermaster General,* 255.

Turbo-Cows
Producing a Competitive Animal in the Nineteenth and Early Twentieth Centuries

BARBARA ORLAND

According to available statistical data, the history of the modern cow is a story of ever-increasing amounts of milk produced by one animal. Whether one consults national or international, long- or short-term statistics—all of which correlate the number of cows with milk output in a region—the message remains the same: the yield per dairy farming cow has been increasing constantly since the second half of the nineteenth century. Over a period of about 150 years, milk production became one of the most efficient sectors of agribusiness, with statistics showing that this was due in large part to the seemingly unlimited capacity of the individual animal to increase productivity.

Estimates by nineteenth-century agronomists indicate that the milk production per cow in that period averaged not more than one-quarter of today's cows. In 1812 agronomist Albrecht Thaer noted that a cow "in a well organized farm" produced about four quarts (one quart = $1^1/_7$ litre) of milk a day. Within a lactation period of 280 days, he said, a yield of 2,560 pounds a year would be an excellent result.[1] In 1868 Georg von Viebahn, director of the Prussian-German Bureau of Statistics, reported that a poor quality cow would give at best 2,592 pounds per annum, whereas a high quality animal yielded about 3,456 pounds. Viebahn averaged the milk performance of the Prussian cow at 3,000 pounds, although he conceded significant problems with data acquisition.[2] In comparison, modern cows serve as perfect milk machines. While the numbers of dairy farmers and cows per farm are steadily decreasing, the average performance of a cow per annum is on the rise. In Switzerland, production per cow grew from 8,360 pounds in 1980 to 10,700 pounds in 1998. In Germany production rose from 11,150 pounds in 1997 to 11,500 pounds in 1998 and to 12,224 pounds in 2000.[3] Apparently, increasing milk production per cow has not yet reached its limit.[4]

The intention of this essay is neither to confirm the statistical evidence for high-yielding milks cow nor to discuss whether we can "trust in numbers," Theodore Porter's concern.[5] For the purpose of this essay, I will accept the remarkable statistical increase of milk production per cow as a matter of fact.

167

As a friend of mine, a dairy farmer, put it, today's farmers who are not able to produce more than six thousand litres of milk (or twelve thousand pounds) per cow are perceived as losers.

If one takes the statistical increase in milk yield per cow as given, then the framework for further study becomes apparent. Apart from the fact that collecting data about milk performance is itself part of the history of improving the cow's natural capacity to lactate, the thorny question of what these data mean also arises. Can we explain the phenomenal increase in milk production as a successful technical manipulation of the biological productivity of an animal's body? Or must the high-yielding cow be seen as part of a larger process of agro-industrialization that not only transformed the practices of animal husbandry in the nineteenth and twentieth centuries but also facilitated a profound restructuring of the relationship between man and animal, nature and technology?

Historians of agriculture argue that the tremendous success of milk production is evidence of the dissemination and application of scientific knowledge to farming practices. The most influential sciences in changing animal husbandry, we are told, have been animal nutrition research, on the one hand, and scientific breeding and population genetics, on the other.[6] As if practical value and efficiency are inherent properties of scientific knowledge, statistical data on meat and milk production are read as a confirmation of scientific success. From the perspective of the practitioner, however, the transformation that resulted in the modern dairy farming system depended on much more than the application of clearly defined knowledge.[7] In support of the practitioner's opinion, high-yield cows are not entirely a new phenomenon. As early as the nineteenth century, there were reports of milk performance that were nearly as high as that of today's cows. According to observations made by August Meitzen, a well-known economist and historian at Humboldt University in Berlin, even ordinary cows sometimes gave surprisingly high yields.[8] At the same time, the above-mentioned director of the Prussian Bureau of Statistics, Viebahn, observed a quasi-natural rise of milk production in areas near big cities. Referring to von Thünen's "Isolated State" model from 1829,[9] Viebahn suggested how trade between city and country influenced the efficiency of animal husbandry. Increasing milk production depended on how much people were willing to pay and on the cost of transporting milk products to urban markets, in his view.

As Viebahn noted, Prussian agriculture centered on grain production. Commercial dairy farming was found only in areas where intensive agriculture directly bordered on towns. In other cases, when dairy farming was not organized around market trade, output depended on the natural resources available for agriculture, in particular, on the relative amount of greenland in a region. Thus, Viebahn acknowledged the risk of misinterpreting average values. Data from various German states (in this period, Germany was a confederation of about forty states, not a unified nation-state) registered differences between one thousand and ten thousand pounds per cow.[10]

To contemporaries, large local and regional differences in milk performances were evident. The productivity of the cow, as farmers knew, was a question of nature and nurture, of innateness and labor. Soil fertility, the specific breed, lactation stage, seasonal and day-to-day variations, and many other factors influenced the quality and quantity of a cow's milk. Most notably, it was a question of determining goals and labor input. One had to make a choice. Dairy farming required specialized skills and working conditions. It was elaborate and risky. The cow had to be treated carefully at all times. Skillful management and humane care were fundamental attributes of a cow's caretaker. Although the mechanistic worldview of nineteenth-century life sciences and economics led agronomists to define animals in terms of machines,[11] an individual farmer would never treat a cow that way.[12]

From this point of view, "high-yield" becomes a very relative term. This label reminds us that the improvement of an animal's body presumably is caused by a complex of different practices for using and managing bodily capacities. Therefore, a deterministic view of progress is not helpful when investigating the story of the high-yielding cow. In my judgment, the high-yielding cow is a result of various contested representations, practices, and economic problems, rather than a product of scientific research. Instead of succumbing to a simple model of "improving nature," we should examine the new forms of communication and new intellectual configurations that emerged in the nineteenth century and shaped agriculture as well as science. The high-yielding cow tells the story of a new culture of competition, measurement, selection, and predictability. It is a figure that provides evidence of the shift from local practices to networks of impersonal information, from local relations between landscape, animal, and people to large-scale institutions and control mechanisms.

This essay focuses only on the early phase of fundamental cultural and intellectual changes that transformed the cow into a competitive animal devoted exclusively to the production of milk. I will begin by reviewing strategies showing agricultural improvements in the late eighteenth century and trace some of the changes in the cattle trade, breeding, and dairy practices in the nineteenth century. Drawing upon source material from Germany and Switzerland, I hope to demonstrate that, as early as the beginning of the twentieth century, all relevant categories that define today's cows had already been established.

Soil, Animal, and Fodder as a Unit

The production of dairy goods, in the sense of butter- and cheese-making, has a long commercial tradition in Europe, but the cow was employed for this purpose only under specific regional conditions.[13] Making milk, butter, and cheese was part of the way people understood and were engaged with their local landscapes, both practically and symbolically. The saying "Dairy farming is land use"[14] was just as self-evident as the idea of a "dairy zone."[15] Agricultural boundaries were drawn in different ways, one of which was the dairy zone. It was conceptually well-defined in the sense that it was perceived as having clear-cut natural limits. As one agrarian writer put it in 1853, most plants and animals have their natural habitat, outside of which they do not exist.[16] Depending on the perception of the landscape and its potentialities, people created certain modalities, skills, and techniques for appropriating local environments. Thus, the meaning of an animal's capacities and assigned tasks was rooted in the perspective of commitment to one's surroundings. "Locality" and "identity" were aligned for both humans and animals. Dairy farming was a situated knowledge and activity.

But a given landscape is not only a product of natural resources and of human activities (together producing the environment), it is also a place full of memories. Even after an environment has effectively changed, farming practices can continue to be a repository for traditions. In fact, specific ways of thinking about relations between animals and land reflected the long-standing rule that animal production for human food depended heavily on the soil. Achieving a balance between agriculture for grain production and animal husbandry was one of the fundamental rules of all premodern farming systems.[17] Unlike today, there was no market for animal feed. Quality and quantity of available fodder were instead dependent on the way the land was being used, that is, the relationship between all types of land management, such as arable farming, permanent meadows, forests, pastures, orchards, and vineyards. Today, subsidies and market access have resolved this issue. In earlier times, farmers were forced to a much greater extent to adjust farm management to the specific nature of available land resulting from existing geological formations, soil conditions, elevation, climate, and vegetation.

As a result, cattle husbandry became dominant only in those regions where farmers faced problems with grain production. High altitude mountain meadows as well as coastal wetlands, river meadows, and highland and lowland moors were such places. Besides such coastal regions in Holland, Denmark, Ireland, and Sweden, the northern Alpine countries, in particular, had already established a pronounced and often highly developed dairy farming system in the sixteenth and seventeenth centuries.[18] Not technical expertise but site-bound meadow farming was the starting point of Swiss cheese's road to fame. In the regions more heavily oriented toward arable farming, milk from cows, goats, and sheep was at best a byproduct of animal husbandry, available in pitifully insufficient quantities and on a seasonal basis only.[19]

Although most premodern farmers in Europe owned cattle, all accepted the rule that milk cows only made sense in meadow farming regions. An old farmer's adage holds that "Cows give milk through their mouths!"[20] According to old beliefs, milk was a foodstuff that had been transformed in the animal body. Since antiquity, scientists, physicians, and ordinary people believed that food was regularly transformed into blood, which then circulated through the body in order to feed it. This steady cycle was only interrupted after giving birth. The Aristotelean (and folk) tradition accepted without qualification the notion that menstrual blood was redirected to the breast or udder and then reappeared in the form of milk.[21] Cartesian physiology assumed that the chyle (the nutritional juice built during digestion) was pressed into the breast. At any rate, it was not until metabolic theories associated with the new field of organic chemistry were well established that these old beliefs were eradicated. As late as 1842, Justus von Liebig polemicized against the old ways of thinking in his *Animal Chemistry*. Whatever the exact mechanisms of transmutation involved, to him and his fellow chemists it was undeniable that "the herbs and roots consumed by the cow contain no butter."[22]

Early Stages of Agricultural Modernization

The grand notion that the essential elements of dairy farming were soil, fodder, and animal inhibited the development of diary farming in regions with diversified agriculture. Farmers in grain production regions as well as in preindustrial dairy zones had to grasp the idea that butter and cheese could be produced year-round in all landscapes.[23] The rise of industrialized milk production depended on two crucial preconditions: separating arable farming and animal husbandry and thinking in terms of different branches of production.

The first challenges to the old agricultural systems occurred in what Paul Bairoch called the "organic phase"[24] of agricultural modernization, beginning in the late eighteenth century and lasting until the 1870s. In this period the traditional relationship between land, fodder, cattle, and dairy production had been torn apart. In the discourses that marked agricultural reforms, natural spaces were reevaluated, but not with the specific purpose of improving animal production or dairy farming. The industrialization of dairy farming in the sense of establishing a complex technical system from barn to storage began not before the last quarter of the nineteenth century.

Around 1760, proponents of the new physiocratic school of economic thought began to argue for reforms that would improve the arable farming system. To the physiocrats, land was deemed the key to economic progress.[25] The idea was to accomplish a fundamental land reform, with a view to using the natural nitrogen cycle more effectively. To reach this goal, more dung had to be produced. Consequently, all of the physiocrats' proposals and measures for agricultural reform were based on four innovations: (1) abolishing the old style of meadow farming in all arable farming regions, that is, no longer allowing land to lie fallow and parceling out communal pastureland; (2) planting fodder plants, such as clover and sainfoin; (3) summer stall feeding, to make it easier to collect cattle dung and liquid manure. The added quantities of cattle dung could be used to (4) ensure that fields and meadows were fertilized more intensively. Implementation of these ideas was very controversial and took several decades. Nevertheless, the rising interest in livestock farming also meant new opportunities. As agricultural reform began to take effect, enterprising farmers learned to benefit from the "animalization" of field crops, as Curt Lehmann, a professor of breeding science, put it in 1874.[26]

Since the 1840s, livestock husbandry, long regarded by many central European farmers as useless and burdensome, suddenly seemed profitable. Developments in international grain markets created the harsh framework for this change of heart. Europe's last major famine occurred in 1846–47; in the years that followed, the liberalization of the grain trade and the effects on distribution of an increasingly wider and more closely linked network of railroad lines offset an under supply of grain, and not just at the local level. National and international markets established themselves for the long haul and were soon followed by price wars. Agriculture based traditionally on grain production became unprofitable and needed to be replaced with other sources of income.[27]

Suddenly many farmers in the grain-growing regions hoped to increase their incomes through artificial meadows or artificial feed and expanded cattle husbandry.[28] Formerly neglected fallow land attracted increasing interest;[29] schools for land improvement were founded; and the profession of the so-called *Wiesenbaumeister* (Master of greenland) developed.[30] Regions where, despite all the natural advantages of abundant vegetation, farmers had stubbornly adhered to the practice of planting grain, also reacted. One example was the Allgäu region in the area at the foot of the Alps, where, around the turn of the nineteenth century, "farmers, having received foreign ideas, thought of the natural possibilities of . . . making the 'yellow' and 'blue' Allgäu (flax growing) into the blossoming 'green' Allgäu."[31] This reorientation was so radical that today, in the Bavarian part of the Allgäu, the farm landscape is said to have been completely "greened," and local officials advertise with an image as Germany's traditional dairy zone.

When the feeding situation improved, the next step, logically speaking, was to bring dairy farming into the former grain production areas. Although the intention was not to produce a dairy cow, for the first time the notion of a milk-producing cow as an isolated natural entity became a real possibility. The separation of land and animal permitted the construction of a self-evident form of animal production. A system of animal breeding, feeding, and husbandry was emancipated from its cultural roots and imagined as a separate branch of production. Swiss author Jeremias Gotthelf, in his 1850 novel *Die Käserei in der Vehfreude*, described with stunning acuity how cows made land more arable:

> clover, sainfoin, and alfalfa came into the land, and stall feeding became possible, the forests were opened-up, meadows made arable and potatoes planted "en masse," not just as a sort of dessert. Where cattle were in stalls, there was dung,

large and small dung, and it was used copiously and sensibly. As more dung was available, arable land grew, as did the herds of cattle and specifically the cows that could be used. . . . As the number of cows grew, so did the milk, for everything is interlinked, and one grows from another in an uncommon manner, and often in such a fine line that man does not even see the thread, a much finer thread than between cows and milk.[32]

A First Cattle Boom

One result of this reorientation was a growing interest in the cattle trade in Europe's former grain production areas. In general, livestock import and export were not new phenomena. While trade concentrated on oxen, sometimes even cows were traded when they were descendants of cows from one of the European dairy zones. By the seventeenth and eighteenth centuries, cows from the North Sea coast and from Switzerland were already quite famous because of their milk. It was said that the landscape, climate, and living conditions had induced natural selection and produced the ability to give high-quality milk.[33] Most of those who imported these cows hoped to acquire the art of cheese- and/or butter-making. Several estates did so in the eighteenth century, but they made little profit if they did not import the herdsmen as well. Even today the names of villages and landscapes in the region around Berlin tell a tale of where the *Holländer* (the Dutch) settled.[34]

While this early trade in cows was limited, the new interest in livestock farming popularized ideas about the economic benefits of dairy farming cows. In the decades following the defeat of Napoleon and the creation of the German Confederation, Prussia's landowners with large estates indiscriminately imported every type of European cattle purported to be good milk and/or meat producers.[35] The imports, especially of well-known stocks from Netherlands and Switzerland, increased considerably. Meitzen reported for Silesia that the cattle boom started with imports from Switzerland. Later, cattle from Oldenburg and Frisia followed, then from Tyrol, Danzig, and Mürztal, followed by imports from the Netherlands, Holstein, and England.[36] Between 1820 and 1870, farmers in the western part of Prussia imported bulls from most regions of Europe solely to improve the milk and meat output of native cattle.[37] By the 1820s, new breeds had been developed. In Rosenstein, Württemberg, the so-called *Rosensteiner Rindviehviehstamm* was said to be a cross between the *Holländer*, *Schwyzer*, and native cattle.[38]

Before the nineteenth century, high-quality butter and especially fatty cheese were expensive lifestyle products, so much so that German landowning nobilities imported cows for their personal use. The goal of investigating Swiss or Dutch cows and recruiting dairy farmers was not, for the most part, the establishment of commercial dairy farming. Most people did not expect to become successful dairy farmers. As noted above, dairy farming was considered a situated knowledge. The animal and its products were symbols of specific landscapes. Because the Swiss cow represented the Alps and the vigor and health of Alpine herdsmen peoples as well, landowners hoped to import a taste of the Alpine terrain. To the educated classes in Europe the fame of mountainous milk products was such a cliché that travelers were surprised not to see cows *"walking up to their bellies in grass."* Ludwig Wallrath Medicus, a German author who in 1795 wrote a report on Alpine farming, was certainly disappointed to find *"just short and rather low grass."*[39]

As European farming regions began realizing the economic potential of dairy farming and investing in milk cow breeding to make use of this potential, the question of criteria for assessing quality in breeding became more urgent. What constituted a "good" cow and how could one evaluate cows on the livestock market? With the increasing transportation of cattle from one place to another, agricultural reformers expressed their fears that cattle imports confronted farmers with too many technical and financial difficulties. On the one hand, the farmers often scorned native varieties, regarding them as useless mongrels that had been mixed indiscriminately and did not contain "improved blood." On the other hand, they criticized the uncontrolled purchase of new stock for breeding purposes.[40]

As early as 1780, an agrarian writer reported that those farmers in graingrowing regions of northern Germany who had bought big Dutch swampland cows were turning no profit on this investment. In the author's opinion, farmers were not able to deal with these cows for two reasons: first, poor fodder, and second, the farmers' incompetence in dairy cattle husbandry.[41] Another agrarian writer commented laconically on imports made by estates owned by the gentry: "The gentlemen's follies are instructive."[42] While skepticism about famous breeds with remarkable performance records among farmers increased and random breeding with local cattle was criticized, old views about the interrelation between soil, form, and performance were remembered more and more often. "Where soil is meager, the animal can't be fat" was an oftenrepeated insight.[43]

Altogether, practical experience with imports was disappointing. Rather than enjoying flourishing milk production, farmers in former grain production regions and agronomists were confused and dissatisfied. Uncertainty about the respective merits of famous dairy cattle breeds made it clear that breeding was a mystery even to the successful. Judging the capacities of an animal solely on the basis of ancestry seemed inadequate for establishing new breeds and a flourishing dairy farm. The sobering estimate of Charles Flint, an American author of manuals on dairy farming, might also have been published in Prussian Germany. Flint wrote in 1858:

> To work successfully with our common cattle would require great experience, a quick eye for stock, a mind free from prejudice, and a patience, and perseverance quite indefatigable. . . . This mode would require a long series of years to arrive

at any fixed and satisfactory results, owing to the fact that our "native" cattle, made up as they are of so infinite a variety of incongruous elements, do not produce their like, that the defects of an ill-bred ancestry will be continually "cropping out" for several generations.[44]

The Cattle Exhibition—Institution of Competition

For the old livestock farming regions, the situation was far better. Since the 1780s, when livestock farming became more fashionable in arable farming regions, the livestock breeding business underwent economic crises as most farmers were unable to serve the changing interests of the market. The old cattle trade was organized around oxen as working animals and dung producers; the renewed market exchange called for meat and milk. While famous British breeders like Richard Bakewell and Charles Colling since the late eighteenth century set about increasing their profits through the "improvement" of their beef-producing livestock, early breeding of dairy cows took place in Switzerland.[45] Livestock export had influenced the economies of several Swiss cantons since the Middle Ages.[46] Thus, Switzerland had a well-established cattle trade in the beginning of the nineteenth century. However, it was not common to focus breeding on cows or milk performance. Livestock farming for export, in general, did not follow a dairy tradition because milk was almost completely used to feed calves. It was not until local authorities acquired a sense of dairy cows' sales value that the promotion of dairy cattle breeding began.

Within the physiocrats' debates on agricultural reform, it was not farmers but politicians, priests, schoolteachers, and veterinarians who became aware of the differences in animal husbandry in different Swiss cantons. They decried the poor condition of cattle husbandry and the animals' poor health and began debating ways out of the crisis. The physiocratic societies founded commissions to improve cattle, sent authorized agents to visit local formers, and distributed questionnaires to local authorities.[47] Several livestock farming cantons like Uri, Schwyz, or Unterwalden were investigated to determine whether the good condition of their livestock farming was the result of specific skills or favorable natural conditions. The commissions were specifically interested in ascertaining what role the cattle trade played in a community, inquiring as to whether communities held enough cattle to turn a profit, which stocks were doing well or poorly, or whether farmers had enough fodder for the winter.

In general, the authorities were not satisfied with the findings of their observations. Two results—subsequently debated by the state economic commission of the canton of Berne from 1803 to 1806—are especially relevant here.[48] Farmers in wealthy livestock farming regions distrusted imported cattle. Furthermore, they did not like to rent out their mountainous meadows to feed foreign cattle.[49] For them the interrelation between soil, animal, and performance was a traditional one that they did not want to change. Members of the commission interpreted these results to mean that a prohibition of cattle imports in such regions might be in order. A second observation taught them to organize competitions instead of using sanctions or coercive measures. Farmers' reactions were cool and bull-headed if agricultural reformers gave instructions. It proved more effective for local authorities to offer cash premiums paid periodically as a disincentive to prevent the interbreeding of various cattle strains.

These proposals were quickly accepted by the local authorities. On April 23, 1806, the small council of Berne decided to create local stud books, to grant prizes, and to organize cattle exhibitions. A few weeks later, the state economic commission sent procedural rules to all communities in the canton of Berne. One rule fixed the prizes to be awarded and determined which proportion of profit should be given for the best stock bull and the best cow. Age requirements for participating animals were set as well as a time period in which the sale of prize-winning animals was prohibited.[50] Above all, authorities developed an evaluation philosophy. The respective juries were to judge nothing more than the form, the physique, and the beauty of an animal. Fat, height, and pregnancy were not to be taken into account. Height, for example, was to be ignored because the mountainous stocks in general were smaller than animals living in valleys.

In autumn 1807 the first cattle exhibition took place in the canton of Berne; the event immediately served its purpose. Farmers from the different villages, communities, and regions competed with one another. The rivalry here was different from that which characterized the old cattle markets. The aim was not to achieve a good price then and there. Awards were not short-dated bills but rather prospects for future profits. Above this, cattle exhibitions were educational measures that intensified local and regional competition. In fact, only two years later, the state economic commission found evidence of positive effects. Exhibition reports revealed that some villages and communes had already improved their cattle. The animals sent to the farm shows were clearly of higher quality than those on view two years earlier.

Before long, other cantons followed suit. In 1811 the city fathers of Luzern passed regulations governing the scheduling of cattle exhibitions at regular intervals. In 1818 the local government of the canton of Luzern awarded prizes for sires for the first time, and in 1837 the law on cattle exhibitions ("Gesetz über die Schau von Zuchtvieh") was passed.[51] When the first cattle exhibition in the canton of Appenzell was planned in 1846, the organizers hoped to profit from the long-standing experience of other cantons.[52] The goal of these measures, to direct farmers' attention to cattle breeding, was reached. By the 1860s, cattle exhibitions in combination with annual fairs were well established in many Swiss cantons. Swiss cattle were sent to world exhibitions in Paris in 1855 and London in 1862 for the first time. To convey a consistent image of the Swiss cow, delegates took with them only two stocks—the *Freiburger* and the *Schwyzer*. A report from the London exhibition outlined tasks to follow these competitions. The author linked the establishment of the Swiss confederation in 1848 not only to citizenship but also to standardized Swiss cattle.[53]

Pure Bred Cattle and Herdbooks

Those who promoted dairy farming also evaluated the landscape in terms of its meaning for nation and society. Agricultural reformers soon described dairy farming as a kind of political economy. "Milk is the fundamental material of our nation, to build and support its ability to work," wrote Rudolf Schatzmann, a priest who was active on behalf of Alpine dairy farming.[54] A strong relationship between self-image, state of mind, landscape aesthetics, and dietary habits was constructed and became an important element of the Swiss nation's specific symbolic representation. The healthiness of milk and cheese was to become one of the most prominent features of this self-image.[55]

Because the cattle exhibitions became events of national interest, it is no wonder that agricultural reformers began thinking about the universal character of the nation's cattle. Local cattle lacked distinguishing characteristics. In the early nineteenth century, cows in dairy zones as well as in other regions were a motley lot; they varied in size, shape, and color. They were the result of unintended cross-breeding within herds, or because they roamed and bred freely, they reflected geographical characteristics. In any case, they were not uniform. Within the new culture of competition, it was of national interest to find a consistent appearance for local cattle. For Swiss agricultural reformers, the task was to ascertain external traits that marked the already famous dairy cow. The existing agricultural literature was of no help. At best, agricultural writers differentiated between so-called lowland and highland cattle.[56] In a country like Switzerland all breeds counted as highland animals, regardless of other traits.

Because animals are characterized by unique physical and behavioral traits, the notion of breeds exists primarily as a social construct. That is, in order for a breed to exist, enough people have to agree that the animals in question are sufficiently distinct from other breeds. One of the first activities to initiate the social construction of a breed of cattle was the demarcation of geographical boundaries. In other words, reformers tried to stop cattle trading for the purpose of breeding. Second, it seemed necessary to define uniform breeding procedures. However, which prominent features could be identified and should be conserved? What was the true type cow or a dairy animal at all? And besides developing detailed classifications for describing such cows, how could one attain new breeding goals?

For centuries Swiss livestock farming concentrated on breeding bulls. Cows had been part of the reproduction cycle; their milk was needed almost exclusively to suckle young cattle, with little left over for butter and cheese.[57] With the international recognition of the Swiss cow, specialized breeds not only replaced diversity but also concentrated on the milk cow. Although only a small fraction of the national herd consisted of animals that were recognized as belonging to definite breeds, in the 1880s the quasi-official Swiss cow was the *Schweizer Braunviehrasse* (Swiss Brown Race). All other breeds, such as the *Appenzeller, Haslitaler, Prättigauer*, and many more, continued to exist on farms, in villages, and in the minds of the people.[58] But the breed that was sent to exhibitions, which existed on paper and was thought to be the origin of the first Swiss herdbook of 1879, was the Swiss Brown Race. Only one year before, in 1878, the herdbook of eastern Frisia dairy cattle (*Heerdbuch für Ostfriesisches Milchvieh*) had been founded by the association of Frisian farmers (*landwirtschaftlicher Hauptverein*).[59] During the following years, herdbooks sprang up all over Europe.[60]

Documenting breeding results was by no means new.[61] The use of stud books by horse breeders had become quite common with cross-breeding in the eighteenth century. It was in 1791 that "An Introduction to the General Stud Book" appeared for the thoroughbred horse in the United Kingdom, and in 1822 the Shorthorn Cattle herdbook was founded, the world's first such book. Later, the organizers of cattle exhibitions sometimes began to produce herdbooks. However, those early registrations only recorded animals if they won races (horses) or were prize winners at agricultural shows. To normalize herdbooks was not easy. In the beginning, breeders often refused to give pedigree information about the animals they sold, fearing they would be giving away "trade secrets." A first attempt at listing all German breeds (cattle, sheep, and pigs) was launched in 1864. By 1872, when the book was closed, the herdbook of all German breeds (*Stammzuchtbuch Deutscher Zuchtherden*) comprised seven volumes.[62]

The golden age of herdbooks that began in the 1860s resulted in changes in assessment methods. However, a whole range of criteria for selecting milk cows had existed prior to this development. From "feminine appearance" and "soft temperament" to "transparent horns" and good digestive powers, externally indicated by a "large mouth, thick, and strong lips," all descriptions made sense as very subjective judgments. Farmers did not make their decisions about desirable or inferior physical forms and qualities with the help of lists of traits. Long experience with animals taught them to assess the characteristics of every part of the body.[63] Agricultural literature then, very often reflected popular thought or individual experience; one example is Francois Guenon's "milk mirror" (a pattern of hair growth around the udder).[64] With the rise of herdbooks, however, the cultural and intellectual climate changed. To breeders, the classification of dairy cattle became a more objective measurement, independent from local customs and local knowledge.

National standards of appearance, height, weight, and performance, as well as methods of measurement and rating required a process of standardization. Standardization, however, entails new organizations. In fact, beginning in the 1860s, breeding became more clearly separated from farming and animal hus-

Fig. 6.1 "Competing Animals"—On the left, the cow named Queis from an Eastern Prussian breed, on the right a "local" cow, as shown at the jubilee cattle exhibition, Königsberg 1913. *Source:* J. Hansen, *Lehrbuch der Rinderzucht. Des Rindes Körperbau, Schläge, Züchtung, Fütterung und Nutzung* (Berlin 1921), 15.

bandry.[65] And because more and more Swiss farmers tried to take up the dairy business by themselves, they lost interest in the breeding business. The work of breeding was affected by this tendency; after 1850 and especially in the 1870s many cattle dealers were accused of shunning Swiss breeders and importing low-grade animals, as one chronicler claimed.[66] This was the background for the twofold task facing the first Swiss breeding cooperatives, which began developing after 1887. One goal was to bring together like-minded people to support one another. As organizations, however, they were also supposed to represent improvement strategies and the use of well-established breeds. As one author, a veterinarian from the canton of Freiburg, put it in 1892, breeding associations were the best instrument for maintaining the purity of a clean breed because they promoted self-discipline within the breeding community.[67] In particular, control by fellow members was seen as a means of accomplishing the objectives of the organization, namely, ensuring compliance with regulations and providing for the external recruitment of stock bulls. Besides organizing and managing cattle exhibitions, one of the most important tasks was the business of writing herdbooks.

Herdbooks became the central documents of classification. While several descriptive terms distinguishing animals were in common use (e.g., scrub or mongrel, cross-bred or pure-bred), the system of herdbooks formulated scales of points that were designed to aid in acquiring the skills needed to select cows by conformity. A scale of points was utilized to describe the constitution of an animal that, in the opinion of the authors, represented the best manifestation of the characteristics sought. One can imagine that agreeing on a scale of points was a highly contested terrain, because it was difficult to formulate a suitable classification that served to define the so-called pure breeds. As already mentioned, one means of control was to draw boundaries around a geographical terrain. Since the animals might vary greatly within a breed, a second step was to define terms of body description. The first Swiss Brown Cow herdbook of 1879 (*Verzeichnis edler Thiere der Braunviehrasse*) included not less than twenty-three body characteristics and positions that had to be evaluated.[68] Every animal had to be measured and weighed.[69] The results were coded in registers; each item got its own score.

Later, pedigrees were mapped; every animal selected for breeding purposes got its own pedigree, it was "identified." Thus, individuals' characteristics were transposed from spoken to written language. With the use of pedigrees and written documentation, telling a story of origin was in fact an administrative act of authentication and identification. Within a few decades, the use of written documents in the administration of cattle became a matter of course. The bureaucracy of breeding was rounded out with instructions and guidebooks for farmers and breeders. Even breeders' associations received instructions on how registers were to be kept in an adequate and orderly fashion.[70] Animals were "baptized," and the date of birth and the name of the father and mother legally attested to in a kind of "passport." To prevent mix-ups, all animals received an ear tag. Finally, new identification skills, such as the use of the medium of photography, became popular. Photography was said to be a useful teaching aid.[71]

In any case, standardization processes and the administration of body characteristics played an important role and became a major task in the breeding business. Together with the emergence of a centralized state and large-scale economic institutions, the demand for a more rigid and comparable definition of "race" brought not only a new type of breeder but also scientists into the breeding business. With increasing professionalization of the breeding business, a growing number of actors participated in the definition, control, and management of pure-bred animals. In 1881, the newly founded University of Agriculture (*Landwirtschaftliche Hochschule*) in Berlin established the first chair for breeding science.[72] Soon, other universities followed.

The True Type Cow

While the second half of the nineteenth century saw an expanding exchange of technical information about individual cows and bulls, the question of how to "identify" performance remained unsolved. Of course, the simplification and standardization of dairy cattle became a specialty of the old dairy zones in northern Europe (Netherlands, Denmark, England, Schleswig-Holstein in Germany) and in Switzerland.[73] Yet, until the turn of the century no breeder would have bred a cow for just one purpose. As yet, no single, entirely satisfactory way of selecting cows for dairy purposes had been identified. An animal's regional provenance, in combination with the description of body characteristics, the so-called *exterieur* (appearance) seemed to be adequate criteria.

In general, breeding for special purposes was still unusual and the combination of geography and genealogy much more important. The true type cow remained a question of aesthetics: "plain, substantial, and well-proportioned—although rather fleshy—. . . somewhat coarse in the bone and in general make-up . . . large, well-shaped udders with teats of sufficient size to be milked conveniently . . . milk veins and milk wells of medium development"—these descriptive phrases, for example, were used to describe the functional type of the Brown Swiss cow to American students and dairy farmers in 1924.[74] The same author described the general characteristics of the dairy type as follows:

> A person familiar with cattle in general, but not with highly developed dairy cattle, looking for the first time upon a high-class dairy cow in full flow of milk would have his attention especially directed to three points as follows: 1. The extreme angular form, carrying no surplus flesh, but showing evidence of liberal feeding by a vigorous physical condition. 2. The extraordinary development of the udder and milk veins. 3. The marked development of the barrel in proportion to the size of the animal.[75]

Of course, such traits in the development of the true dairy type were very controversial among experts. Especially the question of whether a good cow must have a big udder was debated again and again.[76] For these reasons, it seemed inadequate to define breeds by body characteristics or to present the merits of a regional breed. More and more frequently, aesthetic criteria were criticized. As a German professor of agriculture at the Technical University in Munich wrote in 1899: "Even at the well-organized exhibitions of the German Society of Agriculture (*Deutsche Landwirtschaftsgesellschaft*), it is possible to win a prize with a cow of lesser value, just because the juries award prizes to the most beautiful, pure bred cows but not to the high-performance animal."[77]

But the problem remained: How could criteria other than the aesthetic ones be applied and assessed? Furthermore, how could characteristics be transmitted to the future generations of milk cows? For breeders it was a long-standing fact that when any characteristic or function had been developed to a high degree in a breed of animals, the acquired characteristics might not be transmitted uniformly.[78] There was an ever-present tendency for some of the ancestors' characteristics to reappear. The more highly developed the animal, the more difficult it became to retain desirable acquired characteristics. Moreover, breeders were aware that even a well-marked breed, if exposed to greatly changed life conditions, might produce further variability. At any rate, farmers expected wide variability in the capacity of individual cows to produce milk. It was (and is, until today) not uncommon for one cow to produce four, five, or even more times as much milk as another individual of the same breed held under similar conditions in the same herd.

It was Charles Darwin, who, in his book *The Variations of Animals and Plants Under Domestication*, written in 1868, introduced the term "variability" to describe these well-known rules of inheritance.[79] Darwin demonstrated great respect for the business (he would have said art) of breeding. Because new strains or sub-breeds are formed so slowly that their first appearance passes unnoticed, he was aware that one needed much sensitivity and perhaps even more hands-on experience to become a successful breeder.

Keeping Milk Records

The professionalized dairy cattle breeder looked for methods to master the phenomenon of individual variation in performance. Breeders found a solution in milk records.[80] Since the 1860s, test milking had been undertaken sometimes during cattle exhibitions to aid juries in assessing cows. However, by the end of the nineteenth century, things had changed completely. Not surprisingly, the herdbook societies were among the first organizations to call attention to the importance of milk records as a criterion of performance. At its founding meeting in November 1893, the Herdbook Society of the Allgäu (*Allgäuer Herdbuchgesellschaft*) decided to require that its members provide milk records, since the majority considered them to be a helpful tool in improving yields.[81]

The first incorporated society to test cows through complete milking periods was founded in northern Germany in 1895.[82] A union of thirteen breeders in Vejen, near the border of Denmark, copied a system they had learned about from their Danish counterparts. These farmers employed an inspector for two reasons. First, they wanted to identify the best cows in order to gain information on hereditary potential. Second, they wanted to ascertain the amount of milk and fat produced in direct relation to the cost of feed per cow. Thus, milk records became a method for analyzing production costs, and it is not surprising that this idea first emerged in Denmark. After Denmark had lost Schleswig-Holstein to Prussia in the war of 1864, it switched to a more intensive type of agriculture based on importing grains and growing fodder crops and feeding both to livestock for the production of bacon, butter, cheese, eggs, and meat. In contrast to German farmers, it had become routine for Danish farmers very early on to compare feed costs with the prices they got for milk and butter.[83] These farmers had observed that, as milk yield increases, there is also an increase in the total cost of feed and other items; however, yield and costs do not increase constantly in the same proportion. If one compared the relative cost of producing one hundred pounds of milk, then yield differences observed between a very poor cow and a cow of medium quality proved to be much greater than those that distinguished a good cow from a medium quality cow. The consequence was obvious: the most rapid way to achieve herd improvement was to eliminate cows with the lowest milk yields.

Thus, the Danish farmers exported not only butter to their neighbors but also the insight that the more you know about each cow, the more efficient your herd might become. This idea spread rapidly. Other societies of the same kind were founded, not only in Schleswig-Holstein but all over Germany. In 1900, only four societies for testing cows existed in the German Reich; by World War I, the number of societies increased to 792, controlling about 350,000 cows (3.4% of all German cows). In 1933 the number was increased to 2,897 societies.[84] At this point, the German Society of Agriculture (*Deutsche Landwirtschaftsgesellschaft*), the headquarters of cattle exhibitions, decided to follow the example of Denmark and Sweden. Only tested animals were allowed to participate in cattle shows.[85] Moreover, breeders were asked to publish their milk testing results at exhibitions and markets. The effect was that breeding organizations now were forced to generally test their animals. From then on, the circulation of milk performance data became more and more common. And when milk distribution became a business in itself, cow testing and milk records were used to control contracts between dairy farmers and milk distributors. Soon, there was a call for neutral, state-organized milk-control boards or milk commissions, with the power to negotiate a balance between the interests of producers and those of distributors and to establish milk prices for producers and consumers. When, in 1935, Nazi officials forced every German farmer to be associated with a dairy, it was only a small step from there to keeping milk records for every cow.[86]

Conclusion

Of course, the story told so far has not yet come to an end. Not unexpectedly, as milk records became a standard in dairy farming, other questions became more important for the dairy industry. How can dairy cattle be fed more efficiently? What is the cheapest fodder regime in relation to top milk output? Other stories that might be told are the implementation of statistics in the breeding business since the 1920s, the introduction of artificial insemination since the 1940s, and, last but not least, the normalization of reproductive technologies, hormone research (e.g., the controversial question of bovine growth hormone BST, bovine somatotropine), and, finally, cloning in cattle breeding.

My goal in this essay has not been the production of a general history of the high-yielding cow, nor has it been a general essay on breeding technologies and breeding knowledge. Rather, I have probed different data from agricultural history to understand how knowledge about cows and human relations with cows changed in specific local settings. Since the end of the eighteenth century, three major developments or transformations have been at work to change the perception of the cow. First, the idea of separating soil, landscape, and animal husbandry and assessing individually every part of what was formerly an "organic" farming unit took hold. This step resulted in a substantial and sustained rise of interest in animal husbandry. Farmers in the grain-growing regions hoped to increase their incomes through cattle husbandry; this in turn promoted the cattle trade in Europe. Second, with the new livestock evaluation methods at exhibitions, new forms of competition and new impersonal methods of classification emerged. And third, linked to these developments, we have observed a rising influence of standardization in the agrarian context, leading to changed perspectives on the cow. To sum up, today's high-yielding cow came into being within a new culture of competition, standardization, performance control, selection, and predictability, forcing farmers and new institutions like breeding organizations to search for methods and technologies to improve milk yields.

Notes

I thank Paula Bradish and Susan Schrepfer for their helpful suggestions and assistance in translation.

1. Albrecht Thaer, *Grundsätze der rationellen Landwirthschaft*, vol. IV (Berlin: Reimer, 1812), 344.
2. Viehbahn took great pains to collect data from different sources (official statistics, agricultural literature, calendars, journals, and so forth). He had to take into account that there existed only local milk markets. See Georg von Viebahn, *Statistik des zollvereinten und nördlichen Deutschlands*, 3 vols. (Berlin 1858–1868: I. Landeskunde, 1858; II. Bevölkerung, Bergbau, Bodenkultur, 1862; III. Tierzucht, Gewerbe, Politische Organisation, 1868), vol. III, 509. Quite the same data are published in Hans Wolfram Graf Finck von Finckenstein, *Die Entwicklung der Landwirtschaft in Preussen und Deutschland 1800–1930* (Würzburg: Holzner, 1960), 10. General data on the agricultural production are explored in Eberhard Bittermann, "Die landwirtschaftliche Produktion 1800–1950," *Kühn-Archiv* 70 (1956): 1–149.
3. For the Swiss data see Bundesamt für Statistik (ed.), *Statistik Schweiz, Land- und Forstwirtschaft, Milchproduktion und –verwertung 1980–2001* (Neuchâtel: Bundesamt für Statistik, 2001), retrieved May 5, 2003, from http://www.statistik.admin.ch/. For Germany see Milchindustrie-Verband e.V., Milch & Markt Informationsbüro (ed.), *Zahlen und Daten der deutschen Milchindustrie* (Bonn: MMI, 1998) retrieved September 24, 2002, from http://www.zmp.de/milch/marktkommentar.htm.
4. The world milk production averaged 491 million tons in 2001, with an average increase of 0.7 percent in the past five years. The European Union is still the biggest milk producer (122.2 m.t. in 2000) in the world, with Germany as the biggest milk production country in Europe, retrieved May 21, 2003, from http://www.rentenbank.de/deutsch/bank_d/veroeffentlichungen_d/geschaeftsbericht_d_2001/Seite_14_31_2001.pdf.
5. Theodore Porter, *Trust in Numbers: The Pursuit of Objectivity in Science and Public Life* (Princeton, N.J.: Princeton University Press, 1995).
6. See Alois Seidl, *Deutsche Agrargeschichte*, Schriftenreihe der FH Weihenstephan, Bd. 3 (Freising: Fachhochschule Weihenstephan, 1995), 177; Volker Klemm, *Agrarwissenschaften in Deutschland. Geschichte—Tradition. Von den Anfängen bis 1945* (St. Katharinen: Scripta Mercaturae Verlag, 1992), 272.
7. "The dairy industry owes much of its present economical and nutritional importance to the result of research in many different fields of scientific effort. Animal husbandmen, agronomists, plant and animal geneticists, bacteriologists, chemists, entomologists, physiologists, engineers, nutritionists—and a host of other scientific workers—have all made a contribution to the development of this great industry," wrote the U.S. Department of Agriculture, Bureau of Dairy Industry Report 1951, *Dairy Cattle Feeding and Management*, 4th edition, ed. H. O. Henderson and Paul M. Reaves (New York/London: Chapman & Hall, 1954), 1.
8. He mentioned 3,000 Prussian quarts (about 6,850 pounds). See August Meitzen, *Der Boden und die landwirtschaftlichen Verhältnisse des Preussischen Staates*, vol. II (Berlin: Wiegandt & Hempel, 1869), 501.

9. The von Thünen model is discussed in William Cronon, *Nature´s Metropolis: Chicago and the Great West* (New York and London: Norton, 1992), 42–54.
10. A full discussion of the Viebahn statistics is offered in Hans-Jürgen Teuteberg, *Die deutsche Landwirtschaft beim Eintritt in die Phase der Hochindustrialisierung. Typische Strukturmerkmale ihrer Leistungssteigerung im Spiegel der zeitgenössischen Statistik Georg von Viebahns um 1860*, Kölner Vorträge und Abhandlungen zur Sozial- und Wirtschaftsgeschichte, H. 28 (Köln: Forschungsinstitut für Sozial- und Wirtschaftsgeschichte, 1977).
11. In the middle of the nineteenth century the cow's productive capacity was quite often compared to a machine. As Sally McMurry points out, not only agricultural reformers but also farm journals used this analogy. See Sally McMurry, *Transforming Rural Life: Dairying Families and Agricultural Change, 1820–1885* (Baltimore: Johns Hopkins University Press, 1995), 23.
12. And they still don't today, see for example the complex instructions to manage high-yielding cows, at http://www.inform.umd.edu/EdRes/Topic/AgrEnv/ndd/business/BEHAVIOR_AND_MANAGEMENT_OF_HIGH_YIELDINGCOWS.html retrieved May 21, 2003.
13. The preindustrial dairy near cities is adressed in I. F. C. Dieterichs, *Ueber Milch- und Kuhwirthschaft im nördlichen Deutschland in Nähe grosser Städte* (Berlin: Verlag Karl Wiegandt, 1856).
14. Ida Schneider, *Die schweizerische Milchwirtschaft mit besonderer Berücksichtigung der Emmentaler-Käserei* (Zürich/Leipzig: Rascher, 1916), 7.
15. The dairying families in Oneida County, New York, thought this way. I took the phrase from McMurry, *Transforming*, 12–15. Even today, Italian farmers in a small Alpine village think in terms of dairy zones, see Cristina Grasseni, "Developing Skill, Developing Vision," unpublished Ph.D. dissertation, University of Manchester, 2001.
16. In regions with decidious forest, he wrote, the farmer would be well advised to breed goats, rather than cattle or sheep, in oak and beech tree forests he should prefer pig breeding. Dry and spacious areas should be allocated to sheep, fertile gardens and orchards to bees, lush meadows to cows. See Ferdinand Stamm, *Die Landwirthschafts-Kunst in allen Theilen des Feldbaues und der Viehzucht. Nach den bewährten Lehren der Wissenschaft, der Erfahrung und den neuen Entdeckungen in der Natur, gründlich, faßlich und ermuthigend erläutert* (Prague: n.p., 1853), 372.
17. This problem is explored in detail by Friedrich-Karl Riemann, *Ackerbau und Viehhaltung im vorindustriellen Deutschland* (Kitzingen-Main: Holzner Verlag, 1953).
18. For the history of dairy farming in various European countries, see Patricia Lysaght, ed., *Milk and Milk Products from Medieval to Modern Times*, proceedings of the Ninth International Conference on Ethnological Food Research, Ireland 1992 (Edinburgh: Canongate Academic, 1994).
19. Here, the farmer was involved in a rural economy with primarily three-field or two-field crop rotation. The practice of growing different crops in different years on the same land in order to prevent the soil's nutrients from being exhausted did not leave enough food for large livestock. The same held true for pasture, forests, and allmende land. Thus, most cattle were slaughtered before winter set in, since there was no food to feed them and they provided meat during the coldest months. Cattle were kept for dung, meat, and work.
20. D. Rüger, *Die neue chemisch-praktische Milch-, Butter- und Viehwirtschaft*, vol. 1 (Löbau: Dummler, 1851), 7.
21. Even a late version of Albrecht von Haller's "Grundriß der Physiologie," first published in 1747, addresses this position. See *Albert's von Haller Grundriß der Physiologie für Vorlesungen mit den Verbesserungen von Weisberg, Sömmering, und Meckel*, umgearbeitet von D. Heinrich Maria von Leveling, Theil (Erlangen: Walther, 1796), 755.
22. Justus Liebig, *Animal Chemistry in Its Applications to Physiology and Pathology* (Cambridge, Mass.: J. Owen, 1842), 82.
23. In the Alpine dairy zones, people thought that the best butter and cheese could be produced only on the high mountain meadows. An old farmer's proverb holds that the grass is always better the higher one goes, and at the top it is so good that even farmers might like to eat it. Quoted in Werner Bätzing, *Die Alpen. Entstehung und Gefährdung einer europäischen Kulturlandschaft* (München: Beck, 1991), 29.
24. Following this model the first "organic" phase started in the late eighteenth century and ended during the major depression that hit European agriculture from 1875 to 1890. The

second phase is described as the mechanical phase, lasting until the Great Depression. The third phase is classified as being in the period of the welfare state and thus the post–World War II period. See Paul Bairoch, "Die Landwirtschaft und die Industrielle Revolution 1700–1914," in *Europäische Wirtschaftsgeschichte*, Vol. 3, ed. Carlo M. Cipolla and Knut Borchardt (Stuttgart/New York: Fischer, 1976), 297–332. Also Christian Pfister, *Im Strom der Modernisierung. Bevölkerung, Wirtschaft und Umwelt im Kanton Bern 1700–1914* (Berne, Stuttgart, Vienna: Haupt, 1995), 176.

25. For this whole section, see Pfister, *Im Strom*, 175–202. Also Hans-Jürgen Teuteberg, "Anfänge des modernen Milchzeitalters in Deutschland," in *Unsere tägliche Kost*, Studien zur Geschichte des Alltags, no. 6, 2nd ed., ed. Hans-Jürgen Teuteberg and Günter Wiegelmann (Münster: Aschendorff, 1986), 163–184.

26. Curt Lehmann, *Zur Frage über die Berechtigung einer stärkeren Viehhaltung im landwirtschaftlichen Betriebe*, Diss. Göttingen 1874, in idem., *Gesammelte Schriften*. Im Auftrage des Kuratoriums der Lehmannstiftung herausgegeben von der Deutschen Gesellschaft für Züchtungskunde, 2 vols. (Berlin: Deutsche Gesellschaft für Züchtungskunde, 1920), vol. 1, 2.

27. For the agrarian crisis in the nineteenth century see Bairoch, "Landwirtschaft"; Teuteberg, "Anfänge"; Walter Achilles, *Deutsche Agrargeschichte im Zietalter der Reformen und der Industrialisierung* (Stuttgart: Ulmer, 1993); Toni Pierenkemper, ed., *Landwirtschaft und industrielle Entwicklung. Zur ökonomischen Bedeutung von Bauernbefreiung, Agrarreform und Agrarrevolution* (Stuttgart: Steiner, 1989).

28. Data on the size and composition of cattle herds demonstrated the new priorities. In the canton of Berne, around 1760, the grain-growing areas still had a large number of draft animals, particularly oxen; whereas in the mountainous regions the cows were in the majority, commensurate with their significance for cheese-making. Transitional areas, also called field grass areas, were also already heavily oriented toward dairy farming. After 1790 (until 1911) the cow population began to outgrow that of horses, oxen, and sheep in all parts of the country to an incredible, yet varying extent. Pfister, *Im Strom*, 189.

29. In Prussia in 1840, about 20 percent of agricultural area counted as fallow land, in 1867 it was only 10 percent, by 1913 the percentage of fallow land had decreased to 2.7 percent. See Teuteberg, "Anfange," 166.

30. See on this greenland movement: Franz Häfener, *Der Wiesenbau in seinen ganzem Umfange nebst Anleitung zum Nivellieren, zur Erbauung von Schleussen, Wehren, Brücken etc.* (Reutlingen: n. p., 1847). The term "greenland" in German means all agricultural area used for the production of fodder.

31. Walter Jahn, "Die allgemeinen physischen Faktoren der Landwirtschaft" in *Geschichte der Allgäuer Milchwirtschaft. 100 Jahre Allgäuer Milch im Dienste der Ernährung*, ed. Karl Lindner (Kempten/Allgäu: Milchwirtschaftlicher Verein, 1955), 20.

32. Jeremias Gotthelf, *Die Käserei in der Vehfreude: Eine Geschichte aus der Schweiz*, reprint of the edition of 1850 (Zürich: Rentsch, 1984), 235.

33. See Hans Eugster, "Zur Geschichte des Schweizer Braunviehs und seiner Organisation," in *Appenzeller Viehschauen*, ed. Mäddel Fuchs (St. Gallen: Typotron AG, 1998), 219; Ulrich J. Duerst, *Kulturhistorische Studien zur Schweizerischen Rindviehzucht* (Bern-Bümplitz: Benteli, 1923), 17.

34. Since the second half of the seventeenth century Dutch settlers were recruited to cultivate the swampland in Brandenburg and set up dairies for the personal use of a number of German noblemen. See Jan Peters, Hartmut Harnisch, and Lieselotte Enders, *Märkische Bauerntagebücher des 18. und 19. Jahrhunderts, Selbstzeugnisse von Milchbauern aus Neuholland* (Weimar: Böhlau, 1989).

35. See J. Hansen, *Lehrbuch der Rinderzucht. Des Rindes Körperbau, Schläge, Züchtung, Fütterung und Nutzung* (Berlin: Parey, 1921), 9.

36. Meitzen, *Boden*, vol. II, 484.

37. See Franz Rasch, *Das westpreussische Rind*, Monographien landwirtschaftlicher Nutztiere, vol. II (Leipzig: Richard Carl Schmidt, 1904), 15.

38. See Matthias Weishaupt, " 'Viehveredelung' und 'Rassenzucht'. Die Anfänge der appenzellischen Viehschauen im 19. Jahrhundert," in Fuchs, *Appenzeller*, 15. For the importing of Simmentaler cows into south Germany, see Richard Krzymowski, *Geschichte der deutschen Landwirtschaft unter besonderer Berücksichtigung der technischen Entwicklung der Landwirtschaft bis zum Ausbruch des 2. Weltkrieges*, 3rd ed. (Berlin: Duncker and Humblot, 1961), 340–341.

39. Quoted in Jon Mathieu, "Agrarintensivierung bei beschränktem Umweltpotential: der Alpenraum vom 16. bis 19. Jahrhundert," in *Zeitschrift für Agrargeschichte und Agrarsoziologie*, 44 (1996): 139–140.
40. Meitzen, *Boden*, vol. II, 484.
41. Albrecht Thaer, ed., *Johann Christian Bergen´s Anleitung zur Viehzucht, oder vielmehr zum Futtergewächsbau und zur Stallfütterung des Rindviehes*, mit Anmerkungen, Besichtigungen und Zusätzen neu herausgegeben von Albrecht Thaer (Berlin: Realschulbuchhandlung 1800), 477 (first published in 1780).
42. Johannes Nepomuk von Schwerz, *Beschreibung der Landwirthschaft von Westfalen und Rheinpreussen*, 2 vols. (Stuttgart: Hoffmansche Verlagsbuchhandlung, 1836–37), vol. II, 143.
43. A collection of quotations from 1840 to 1900 can be found in Ulrich Duerst, *Grundlagen der Rinderzucht. Eine Darstellung der wichtigsten für die Entwicklung der Leistungen und der Körperformen des Rindes ursächlichen, physiologisch-anatomischen, zoologisch-paläontologischen, entwicklungsmechanischen und kulturhistorischen Tatsachen und Lehren* (Berlin: Springer, 1931), 123–125. See also Hansen, *Lehrbuch*, 10; McMurry, *Transforming*, 17–18.
44. Quoted in McMurry, *Transforming*, 20.
45. The extensive agricultural reforms of the late eighteenth century, motivated by the desire to make farm production more efficient, took place all over Europe. However, for political reasons, reforms took very different directions in the old dairy zones. Denmark, for example, was helplessly caught in the conflict between Napoleon and the rest of Europe. And the loss of Norway in 1814 meant that the former dual monarchy, which geographically had stretched from the North Cape to the Elbe, was reduced to Denmark itself and the German duchies. Until the Danish-Prussian war in 1864 almost a third of the nation's greenland area was German. Holstein and Lauenburg belonged to the German Confederation, while Schleswig was nationally divided. Shortly after the loss of the German-speaking population and area in the war with Prussia, the Danish parliament passed several regulations to promote a large-scale shift from the cultivation of plants to livestock farming. See Steen Bo Frandsen, *Dänemark—der kleine Nachbar im Norden: Aspekte der deutsch-dänischen Beziehungen im 19. und 20. Jahrhundert* (Darmstadt: Wissenschaftliche Buchgesellschaft, 1994).
46. Livestock farming was embedded in a complex regional and local division of labor. See Rudolf Braun, *Das ausgehende Ancien Régime in der Schweiz. Aufriß einer Sozial- und Wirtschaftsgeschichte des 18. Jahrhunderts* (Göttingen/Zürich: Vandenhoeck und Ruprecht, 1984), 58–69.
47. See Duerst, *Kulturhistorische*, 21–23. For the canton of Glarus, see Jost Hösli, *Glarner Land- und Alpwirtschaft in Vergangenheit und Gegenwart* (Glarus: Kommissionsverlag Tschudi, 1948), 43.
48. See Duerst, *Kulturhistorische*, 22.
49. Within the Swiss livestock farming regions, specializing in fodder production, which occurred as early as the late Middle Ages, was quite popular. This meant that farmers largely did not own cattle but instead rented out and leased farmstead meadows or Alpine meadows and sold hay fodder (particularly in the winter). Examples of feeding contracts from the seventeenth and eighteenth centuries can be found in Rudolf J. Ramseyer, *Das altbernische Küherwesen*, 2nd ed. (Bern/Stuttgart: Haupt, 1991), 46–49.
50. Duerst, *Kulturhistorische*, 22.
51. See Weishaupt, "Viehveredelung," 19–20.
52. Quoted in Weishaupt, "Vielveredelung," 37. In Germany the first cattle exhibitions with nationwide relevance took place in 1863 in Hamburg, 1868 in Mannheim, 1874 in Bremen.
53. For Germany see H. v. Falck, et al., *Die Milchproduktion. Die Milchviehzucht. Fütterung, Haltung und Pflege der Milchtiere. Entstehung, Gewinnung und Behandlung der Milch*, Handbuch der Milchwirtschaft, vol. 1, part 2., ed. Willi Winkler (Wien: Springer, 1930), 73–74.
54. Rudolf Schatzmann, *Die Milchfrage vor der gemeinnützigen Gesellschaft des Kantons Bern*, 1872, quoted in Isabel Kollreuther, *Milchgeschichten. Bedeutungen der Milch in der Schweiz zwischen 1870 und 1930* (unpublished Lic. Phil., Basel, 2001), 22.
55. See Guy P. Marchal and A. Mattioli, eds., *Erfundene Schweiz. Konstruktionen nationaler Identität* (Zürich: Chronos, 1992).
56. A discussion of this question in retrospect can be found in A. Schmid, *Rassenkunde des Rindes*, vol. 1, Rassenbeschreibung (Bern: Benteli, 1942), 22.

57. If the focus was on butter and particularly marketable cheese, then, conversely, farmers decided against extensive cattle farming. All types of production were governed by the dictum that the use of the limited land resources had to be coordinated depending on the season.
58. Likewise in other regions: when the German Society for Agriculture (*Deutsche Landwirtschaftsgesellschaft*) organized its own cattle exhibition in 1887 for the first time, one of the organizers worried about the diversity of the German races. Only Bavaria counted twenty-eight cattle stocks at this time. During the next thirty years they were reduced to twelve, in the year 1903 the society found ten, in the year 1925 eight and in 1948 today's four "Fleckvieh," "Frankenvieh," "Braunvieh" and "Pinzgauer." See Hans Oskar Diener, "Förderung der deutschen Haustierzucht und der tierischen Produktion im 19. und 20. Jahrhundert durch staatliche Maßnahmen," in *Bayerisches landwirtschaftliches Jahrbuch* 57 (1980), 78–120, quote from 79.
59. See Wilhelm Zorn, *Rinderzucht*, 2nd ed. (Stuttgart: Steiner, 1944), 32. In 1901 the Schleswig-Holstein association of breeding organizations for black and white breeds comprised fifteen local breeding associations; the Schleswig-Holstein association of breeding organizations for red-colored Holsteins united in 1898, creating thirty-one local breeding organizations. There were also Shorthorn/Angus and Angler associations. See Gustav Comberg, *Die deutsche Tierzucht im 19. und 20. Jahrhundert* (Stuttgart: Ulmer, 1984), 264–267.
60. Hansen, *Lehrbuch*, 144–402.
61. Comberg, *Tierzucht*, 247.
62. W. Janke, A. Körte, and G. von Schmidt, *Jahrbuch der Deutschen Viehzucht nebst Stammzuchtbuch deutscher Zuchtherden*, vol. 1–7 (Breslau: n.p., 1864–1872).
63. A detailed analysis of old terms of qualification can be found in Hans Ulrich Rübel, *Viehzucht im Oberwallis. Sachkunde Terminologie Sprachgeographie*, Beiträge zur schweizerdeutschen Mundartforschung, Bd. II (Frauenfeld: Huber, 1950).
64. McMurry, *Transforming*, 22.
65. See Comberg, *Tierzucht*.
66. See Eugster, "Zur Geschichte," 220–221.
67. M. Strebel, *Das Freiburger Rindvieh* (Freiburg: Frangniere, 1893), 16.
68. Weishaupt, "Viehveredelung," 38.
69. Several instruments have been developed for the process of measurement. See Max B. Pressler, *Neue Viehmesskunst*, 3rd ed. (Leipzig: n.p., 1886) [first published, 1854].
70. See for example, Oscar Knispel, *Anleitung für Züchtervereinigungen zur ordnungsgemässen Führung der Zuchtregister* (Berlin: Deutsche Landwirtschaft-Ges., 1914).
71. Visual representation in the form of paintings, prints, and photographs provided forms to which other breeders could aspire. Such representations made it obvious to anyone that formal breeds were a reality. Moreover, stories about prize winners could be popularized much longer than the event lasted. Thus, livestock portraits were not only important in the establishment of a breed, but in its maintainence and improvement. For more details, see Hansen, *Lehrbuch*, 471–474.
72. Until then, problems of breeding had been at best part of training for veterinary students, in Munich for example. See http://www.vetmed.uni-muenchen.de/info/geschichte.html.
73. Until 1884 Dutch breeders increased cattle exports throughout northern Europe so dramatically that farmers in the Netherlands faced problems with their offspring. See J. Hansen and A. Hermes, *Die Rindviehzucht im In- und Auslande*, vol. 2 (Leipzig: Richard Carl Schmidt, 1905), 47.
74. Clarence H. Eckles, *Dairy Cattle and Milk Production* (New York: Macmillan, 1924), 98.
75. Eckles, *Dairy Cattle*, 29.
76. Several other controversial questions are outlined in Adolf Kraemer, *Das schoenste Rind. Anleitung zur Beurteilung der Körperbeschaffenheit des Rindviehs nach wissenschaftlichen und praktischen Gesichtspunkten*, 3rd renewed ed. (Berlin: Parey, 1912).
77. Emil Pott, *Der Formalismus in der landwirtschaftlichen Tierzucht* (Stuttgart: Ulmer, 1899), 17.
78. *Tierzüchtungslehre*, Ein Gemeinschaftswerk zusammengestellt und herausgegeben von Wilhelm Zorn (Stuttgart: Ulmer 1958), 63.
79. Charles Darwin, *The Variation of Animals and Plants Under Domestication*, (London: John Murray, 1868), vol. II, Ch. 24–26.
80. I have purposely left out other methods of breeding here because they were used in general.
81. Falck et al., *Milchproduktion*, 76.

82. On this entire paragraph, see Jürgen Hansen, *Zeitfragen auf dem Gebiet des Kontrollvereinswesens* (Berlin: n.p., 1923); Falck et al., *Milchproduktion*, 76–81.
83. For decades Denmark had been one of the leading butter exporters in Europe. See Frandsen, *Dänemark*.
84. Falck et al., *Milchproduktion*, 78–79.
85. Since 1920 the calculation of premiums had been based on performance.
86. The result was a remarkable rise in testing. In the year 1937, 67.4 percent of all German cows were tested, in some districts it was about 90 percent and more. See Comberg, *Tierzucht*, 353.

Canine Technologies, Model Patients
The Historical Production of Hemophiliac Dogs in American Biomedicine

STEPHEN PEMBERTON

Today, we can speak of hemophilia as a manageable disease. In the three decades following World War II, this poorly understood malady was transformed into a well-characterized blood coagulation disorder that could be normalized using clotting factor replacement therapies. A critical turning point in the effort to manage hemophilic bleeding occurred in the early 1970s when clotting factor concentrates became widely available for patient use in the United States and other developed nations. At that time, the American media was portraying the new treatments as a considerable breakthrough for hemophilia patients and their families. In 1971, *Science Digest* went so far as to say that the hemophilia patient had been "freed of the life-threatening, disabling terror of hemorrhage."[1] Or, as one advertising campaign from the era put it, the clotting factor concentrate was the hemophiliac's "passport to freedom."[2] While such descriptions were hyperbolic and failed to capture the complexity of the disease or its management, clotting factor concentrates did help transform the lives of hemophilia patients and their families.[3] Moreover, the availability and subsequent use of clotting factor concentrates represented the culmination of historical efforts by hematologists to translate their laboratory endeavors into techniques for controlling hemophilic bleeding in clinical settings and beyond.[4] For many advocates of hemophilia management, these concentrates affirmed the power of medical science to shape our experience of nature, bodies, people, institutions, technologies, and knowledge. Of course, the freedoms enjoyed by hemophilia patients today were also won through the discipline and sacrifice of previous sufferers. For one largely invisible population of hemophiliacs those sacrifices were quite literal; and here I am speaking of the scores of dogs that researchers have bred, maintained, and used in laboratories since the late 1940s to study hemophilia and its management. This essay describes how dogs with inherited bleeding disorders were brought into laboratory settings in the United States, suggests how this move contributed to the discipline of hemophilia management, and speaks

Hemophilic Dogs as Model Organisms

In 1947, Kenneth Brinkhous, the chair of pathology at the University of North Carolina at Chapel Hill, began breeding dogs with an inherited bleeding disorder. His purpose was to use these dogs to study hemophilia in humans. In the post–World War II era, no one knew whether research with these sickly dogs would actually yield useful knowledge about human hemophilia and its treatment. Yet, on the basis of Brinkhous's initial research proposal, the project won immediate funding from the newly reorganized National Institutes of Health (NIH).[6] Brinkhous argued that his strain of bleeder dogs would allow scientists to answer fundamental questions about the nature of the coagulation defect in hemophilia; his was an argument for "basic science," which the Hematology Study Section of the NIH was keen to support through its extramural funding program.[7] Once granted, the federal funding allowed Brinkhous and his colleagues to mark the "canine hemophiliac" as a valuable resource for post–World War II studies of blood coagulation disorders.

As it also happened, the closed colony of hemophilic dogs established at the University of North Carolina became a model organism for understanding and treating hemophilia in the human. Between 1947 and 1970, Brinkhous and his colleagues made a series of contributions to the fields of hematology and hemophilia research using the dogs, which in turn encouraged other scientists to develop their own colonies of hemophilic dogs or to discover and develop animal models of hematological disorders in other species.[8] The dogs had obvious value for those individuals trying to understand coagulation disorders in canine species. Yet, their value as model organisms derived from the fact that the clinical symptoms and inheritance pattern of disease found in the dogs were analogous to those found among human hemophiliacs. Furthermore, by the early 1950s, scientists demonstrated that the coagulation defect in the animals was functionally identical to that found in the blood of human patients diagnosed with classical hemophilia.[9] The latter finding helped stabilize claims that these bleeder dogs could be deployed as model organisms of human hemophilia.[10] See Figure 7.1.

Throughout the 1950s and 1960s and on the basis of their status as model organisms, scientists used the dogs as a ready source of blood plasma for the biochemical and physiological characterization of the antihemophilic clotting factor (AHF) and as experimental subjects for the production and testing of a series of clotting factor concentrates that eventually culminated in the devel-

1. Rapid loss of antihemophilic factor (AHF) during clotting demonstrated, indicating that coagulation defect in hemophilic dogs is functionally identical to that found in human patients with classical hemophilia.	1949–1951
2. Two bioassay methods developed: one of them (the partial thromboplastin time) became basis for diagnostic scheme employed in clinical laboratories to detect hemophilia and other blood coagulation disorders.	1951–1956
3. Schedule for maintaining hemostasis in hemophilia patients developed on basis of short half-life and disappearance curves of transfused AHF.	1954–1956
4. Classic hemophilia produced in females; completed all major genetic crosses with sex-linked recessive for first time; low level of AHF in homozygous females demonstrated.	1950–1964
5. Biochemical characterization of AHF started; glycine-precipitated concentrates of AHF developed for therapy of hemophilia.	1954–1966
6. AHF shown to be a cryoprotein, forerunner of cryoprecipitate therapy for hemophilia.	1954–1968
7. Found that hemophilic dogs were resistant to disseminated intravascular coagulation (DIC), suggesting treatment of DIC with anticoagulants.	1955–1958
8. Spleen established as site of storage of AHF.	1957–1964
9. Curative role of transplanted liver established for hemophilia, role of kidney explored.	1967–1970
10. Loci of the genes for hemophilia A and B shown to be widely separated, and a new strain of dogs with hemophilia AB established through mating of hemophilia A and B dogs.	1968–1970

Fig. 7.1 Some Contributions Made Using Hemophilia A Dogs at the University of North Carolina, 1947–1970
Sources: Kenneth M. Brinkhous, "Summary of Contribution of Hemophilic Dogs to Knowledge and Human Welfare," in *Research Animals and Medicine*, ed. L. T. Harmison (Washington, D.C.: National Heart and Lung Institute, 1973), 501–503; T. C. Jones, "The Values of Animal Models," *American Journal of Pathology* 101 (December 1980): S3–S9.

opment of the first widely marketed AHF concentrate.[11] As Brinkhous put it in a 1975 talk at the National Academy of Sciences:

> The advent of modern therapy of hemophilia was a direct outgrowth of studies of hemophilic dogs. The animal model has helped us advance from essentially no technology a quarter of a century ago to a present half technology, and it may well be that the same models will be instrumental in helping us arrive at a full technology, where the hemophiliac's body will make its own [clotting factor] and will thus be no different in its hemostatic capacity from the nonhemophiliac.[12]

On the basis of such contributions, the NIH has continued to support laboratory studies of the closed-colony of hemophilic dogs at the University of North Carolina. In fact, in 1998, Brinkhous became the first scientist to receive fifty years of continuous extramural funding from the NIH. Given this history, it is easy to see why the canine hemophilia animal model continues to play a valuable material role in biomedical research, most visibly today in the context of genetic therapy research for hemophilia.[13]

In what follows, I suggest that scientists endowed hemophilic dogs with practical and programmatic qualities that made their use within biomedical science a success story. It remains controversial in many circles to say that organisms were "created" or "fabricated"—particularly, those organisms that occur "in nature." Certainly, Brinkhous and his colleagues created a novel situation when they brought these dogs into the laboratory and framed them as exemplary organisms for the study of human hemophilia. From my perspective, however, a new kind of organism came into being when scientists brought dogs with inherited bleeding disorders into the laboratory for the purpose of understanding and managing hemophilia in human patients. As such, this discussion treats the birth of the canine hemophiliac, by which I mean the emergence of a whole new class of organic life: a way of organic being that was novel and capable of becoming a model of how a whole existing class of beings might experience themselves.[14]

To clarify and amplify my point, we will explore how research using bleeder dogs was born out of the evolving circumstances of Brinkhous's professional life and the changing landscape of American biomedicine and society in the twentieth century. Here, I will describe Brinkhous's own understanding of experimental medicine and link this understanding to the social milieu that engendered his embrace of hemophilic dog studies. Most of the history that I will relate is an unapologetic origin story. As Karen Rader and other historians of science have observed, much historical writing about living research materials fits nicely into an origin story genre that might be called "humans and their organisms."[15] Despite increased sensitivity to the drawbacks of a genre that valorizes scientists and their achievements, scholars have had great difficulty departing from narratives that subordinate experimental organisms and other research materials "to the careers of the individuals who developed them" or to the institutional settings these persons inhabit.[16] The irony, as Rader points out, is that "existing organism histories reinforce the conclusion that the scientific fates of individual organisms are intimately tied . . . to the actions of individual scientists." Such ties are particularly strong during the early stages of a research organism's use. Given that I am also describing the origins of canine hemophiliacs in terms of their "creator's" actions and intentions, I am compelled to follow Karen Rader's lead and ask, "What does it mean to speak about the 'creation' or 'origination' of experimental organisms, both for biology and medicine, and for the history of those fields?"[17]

Blood Coagulation Research in the 1930s

Brinkhous was introduced to laboratory research in 1930, late in his second year of medical school, when pathologist Harry Pratt Smith offered him a research assistantship.[18] Smith had just been recruited to the University of Iowa, appointed its chair in pathology, and provided enough money to establish a limited program of experimental research.[19] Previously, Smith trained and collaborated with pathologist George Whipple at the Hooper Foundation in San Francisco and at the University of Rochester. In 1930, Whipple was already famous for demonstrating that liver extract could ameliorate pernicious anemia in the dog, work for which he shared the Nobel Prize in Physiology and Medicine in 1934. Smith arrived in Iowa City with plans to form a research group based on the example of Whipple's work.[20] He devoted his research program to understanding the mechanisms responsible for blood coagulation (hemostasis), and he recruited Brinkhous among a few others to work with him. Together, Smith's team of researchers became known in the field of hematology as the Iowa Group.

Over the course of the 1930s, the Iowa Group conducted many laboratory studies aimed at characterizing the hemostatic functions of the known proteins in blood plasma. They isolated these complex proteins, purified them as best they could, and tested them in a variety of ways to see if these purified plasma factors could "correct" samples of blood suspected of having hemostatic deficiencies. The Iowa Group's emphasis upon quantification and purification techniques allowed them to test the key components of the blood coagulation system.[21] Specifically, the group's techniques for measuring the hemostatic action of various substances presented hematologists with a meaningful way to evaluate their efforts to correct bleeding attributed to faulty coagulation. The research strategy of the group was itself modeled on two of the most widely celebrated medical success stories of the era: the management of diabetes mellitus by insulin and the correction of pernicious anemia by liver extract.[22] Diabetes and pernicious anemia were each defined as metabolic deficiencies that could be "normalized" by the administration of a substance that was functionally absent in the patient. Moreover, in both cases, dog experimentation played a critical role in understanding and correcting these deficiencies.

The Iowa Group occasionally employed dogs in their blood coagulation studies, both as sources of organ tissue and blood plasma and as objects for testing dietary supplements and plasma components. Brinkhous, in particular, found numerous experimental uses for the laboratory dogs. In one series of experiments, he prepared bile-fistula dogs identical to those used by Whipple in his Rochester studies of bile formation and function. A bile-fistula dog is "fabricated" by tying the common bile duct of these so as to retain all the bile in the gallbladder. The procedure allows the researcher to collect the bile via an opening (or fistula) in the abdominal wall of the dog and thereby subject the substance to laboratory testing.[23] As a side effect, many bile-fistula dogs acquired a tendency to bleeding. Brinkhous was interested in the latter phenomenon and determined that the acquired bleeding tendency in these dogs was due to a prothrombin deficiency. Following Whipple's model of a metabolic deficiency, the Iowa Group subsequently fed the dogs bile and vitamin K (extracted from alfalfa) to normalize their metabolisms. These dog studies were part and parcel to the Iowa Group's successful treatment by vitamin K of adult patients with obstructive jaundice and newborns with hemorrhagic disease.

In summary, Brinkhous's perspective on laboratory-based medicine was largely shaped by the Iowa Group's experimental practices. These activities were guided in principle by their lineage to Whipple's own research, which oftentimes revolved around dog studies from the 1910s through the 1930s. In practice, however, the Iowa Group's research was characterized by laboratory-based problems proper to a quantitative approach to blood chemistry, to research on hemostasis, and to efforts to correct the coagulation defects in patients with bleeding disorders. Over the course of his training with Smith, Brinkhous learned that animals were crucial to the workings of the laboratory since they provided the researcher with the ready-to-hand means for isolating, purifying, and testing the materials essential to the production of clinically useful knowledge. Brinkhous acquired training in both anatomic and clinical pathology by the time he undertook his formal studies of hemophilia in the late 1930s. He therefore knew that his success as an experimental pathologist depended on an effective passage of research materials and knowledge between bedside and bench. Crucially, he saw dog work as a proven means for facilitating that passage.

Working Up Hemophilia Patients

Brinkhous's initial attempts to study hemophilia in the human were makeshift. Between 1935 and 1938, he teamed up with a medical student to obtain fresh samples of hemophilic blood. The medical student would check hospital admissions to find out whether a hemophilic patient was admitted the previous day. If one had been admitted, Brinkhous would go to the patient's bedside to determine whether the hemophiliac would be willing to donate blood to him. These samples of hemophilic blood became the subject of Brinkhous's bench studies. Brinkhous soon found this "hodge podge" approach to be an unsatisfactory way of doing research, involving as it did an unreliable passage of materials between clinic and laboratory. Fresh hemophilic plasma was necessary for his studies, and its supply was always dependent, in his words, upon whether "a patient with hemophilia was unfortunate enough to have to come to the hospital."[24]

In 1938, two events occurred at the University Hospital in Iowa City to help Brinkhous stabilize his laboratory studies of hemophilia. First, a blood bank was established for the hospital's transfusion service, allowing Brinkhous to store his blood samples. Second, Brinkhous obtained a reliable donor of hemophilic blood when Jimmy Laughlin became a patient at the University Hospital.

Laughlin, a twenty-four-year-old taxi driver with severe hemophilia, was admitted for a life-threatening bleed following a fistfight with a passenger who refused to pay his fare. Brinkhous approached Laughlin at his bedside and conveyed to this patient his concern about the dangers of taxi driving for a hemophiliac. He then offered Laughlin a position as a dishwasher and helper in the Iowa Group's laboratory. Laughlin agreed to the career change and was soon spending fifty to sixty hours a week working in the laboratory with Brinkhous, Smith, and the others. See Fig. 7.2. Thus through a strategy of opportunism, Brinkhous was able to

Fig. 7.2 Doctors and Technicians, Department of Pathology, University of Iowa, 1939–1940. The Iowa Coagulation Group included (first row, left to right) Emory Warner, Kenneth Brinkhous, Robert Tiddrick, Harry P. Smith, Walter Seegers, Joseph Flynn, Edwin Mertz, and (second row) Charles Owen. In later years, each of these individuals established their own laboratories for research into hemostasis and thrombosis. Jimmy Laughlin, the laboratory technician with severe hemophilia, appears next to Owen in the second row. Reprinted with permission of Thieme Medical Publishers. Source: Walter H. Seegers, "A Personal Perspective on Hemostasis and Thrombosis (1937–1981)," *Seminars in Thrombosis and Hemostasis* 7 (December 1981): 178.

remedy the shortage of hemophilic blood for study by taking advantage of blood banking and by bringing a hemophiliac into his laboratory.

This practical organization of the laboratory soon grew complicated. As Brinkhous accumulated experience with Laughlin's condition, he became increasingly confident that he could manage Laughlin's bleeds and might even prevent them. Like many hemophiliacs, Laughlin suffered from toothaches since he could not properly maintain his teeth without risk of bleeding. Brinkhous proposed extracting the offending teeth and controlling Laughlin's bleeding using measured transfusions of whole blood given prior to the surgery and at predetermined intervals of time following it. This proposal led to the first transfusion of a hemophiliac in Iowa City, and Brinkhous remembers the traumatic experience that followed as testament to his bravado and overconfidence. "'Lo and behold,'" he recalled, "[Laughlin] continued to bleed the first four days . . . in spite of transfusions" planned in advance.[25] After Laughlin nearly bled to death in the first forty-eight hours, Brinkhous decided to divert from his experimental protocol and shift to transfusions given more frequently and at smaller volumes (see Figure 7.3).[26] Laughlin's condition did stabilize and, over the next few days, the bleeding finally halted. In light of this traumatic experience, Brinkhous tempered his desire to manage hemophilic bleeding clinically. Instead, he returned to his laboratory studies with greater appreciation of the difficulties involved in translating blood work involving test tubes and centrifuges into reliable clinical practice.

Given his goals, Brinkhous was largely ambivalent about keeping the differences between clinical and laboratory life distinct. The point was to bridge the gap, to make laboratory science applicable to the clinical management of hemophilia. Yet, his tooth extraction experiment indicated the dangers of such ambivalence. The near loss of Laughlin forced Brinkhous to consider the uncertainties and risks of his pathologist's perspective, while also reinforcing it. Brinkhous knew that the use of laboratory animals might well present a solution to the problem of managing hemophilic bleeding. The pattern of discoveries in Banting's and Best's diabetic dogs, Whipple's anemic dogs, and the Iowa Group's own bile-fistula dogs all suggested that a therapeutic breakthrough in the laboratory was possible for hemophilia given the right circumstances and organism. However, too little was known about the coagulation defect in hemophilia in the late 1930s to create the condition in a laboratory animal, and no nonhuman hemophiliacs were known to exist in nature. Without such organisms to study, Brinkhous returned to his samples of hemophilic plasma, hoping that the solution lay there.

The Making of the Canine Hemophiliac

In 1946, Brinkhous returned from wartime service and began to address the problems that he had encountered in his earlier prewar investigations of he-

Canine Technologies, Model Patients • 199

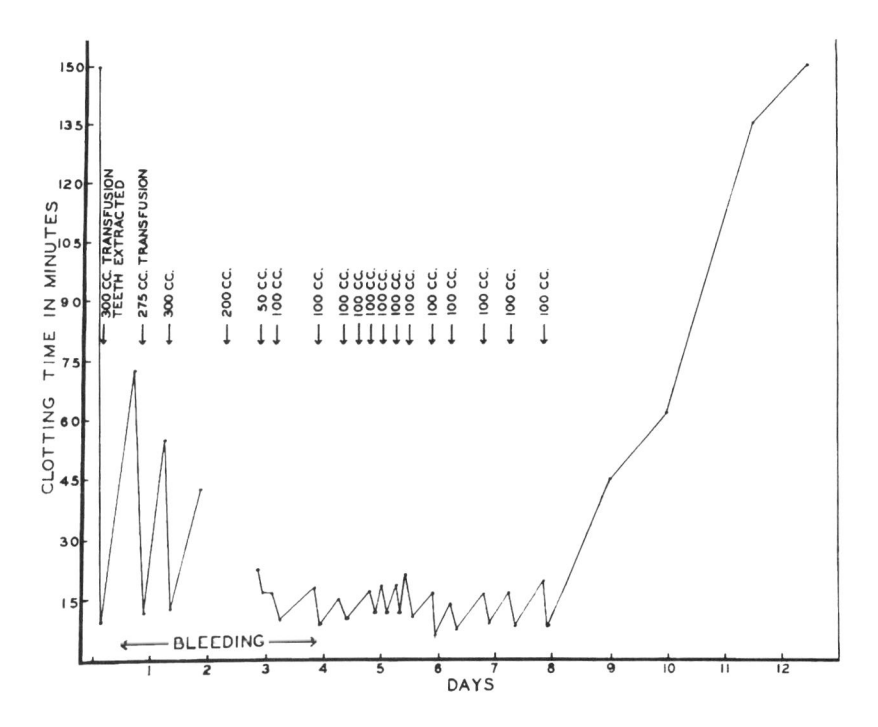

Fig. 7.3 Original experiment dealing with a comparison of widely spaced and closely spaced transfusions of whole blood in a person with a severe degree of hemophilia. This graph depicts Brinkhous's efforts to control the dangerous levels of bleeding following Jimmy Laughlin's 1938 tooth extractions. Brinkhous found that closely spaced, smaller-volume transfusions of whole blood were more effective at controlling Laughlin's bleeding than widely spaced transfusions using larger volumes of blood. Reprinted with permission of American Society of Clinical Pathology. Source: K. M. Brinkhous, "Hemophilia: Pathophysiologic Studies and the Evolution of Transfusion Therapy," *American Journal of Clinical Pathology* 41 (April 1964): 343.

mophilia. The University of North Carolina appointed him its new chair of the department of pathology, thus enrolling Brinkhous in the medical school's quest to establish itself as a leading institution in biomedical research and education. The post–World War II era was a time of expansive growth in biomedical research, when smaller teaching institutions like the University of North Carolina could aspire, on the basis of newly available federal funding, to the status of a major research university.[27] Immediately then, Brinkhous concerned himself with achieving these institutional goals in conjunction with his personal endeavors.

Early in 1947, Brinkhous heard from H. P. Smith (then chair of pathology at Columbia University) that a pedigree Irish setter living in Courtland, New York, named Terry Bay, had given birth to some very unhealthy pups. Terry

Bay's owners were distressed to find that their new puppies had developed severely swollen joints and appeared to be dying. Their veterinarian was also concerned when the joint swelling in the puppies did not respond to doses of vitamin D. Since Terry Bay was a prized show dog, her owners decided that she and her pups should be examined by the veterinary experts at Cornell University in nearby Ithaca.[28] These experts performed an autopsy on one dead puppy and discovered that internal bleeding was responsible for its death. They then performed bleeding and whole blood clotting time tests on all of the remaining puppies and found that while the bleeding times in all of them were normal, the clotting times were quite prolonged in those afflicted. The Cornell veterinarians then pointed out to Terry Bay's owners that only the male pups were affected and learned that two of Terry Bay's apparently "normal" daughters had subsequently given birth to bleeder male pups as well. This canine malady appeared identical to the clinical symptoms and inheritance pattern found in human patients afflicted by hemophilia. These animal doctors therefore had a ready-made diagnosis for what they saw. Terry Bay was judged to be a hemophilic carrier, and her male pups were diagnosed as full-blown hemophiliacs.[29] While this diagnosis was distressing to Terry Bay's owners, it absolutely delighted Brinkhous.

The Cornell veterinarians were the first scientists ever to confirm a case of hemophilia in a dog. So upon hearing about these dogs, Brinkhous phoned Terry Bay's owners and made arrangements to purchase some of Terry Bay's daughters, who had already transmitted the bleeding disorder to their offspring. Soon thereafter, two hemophilic carriers, named Nora and Lynne, were delivered to Brinkhous in North Carolina. Nora was already pregnant upon her arrival and soon delivered some bleeder pups. A few years later, Terry Bay joined her offspring at Chapel Hill after her owners decided to forgo their dog-breeding activities. By this time, Brinkhous had already established a colony of bleeder dogs for the explicit purpose of conducting research on hemophilia and blood coagulation in the human.

When Brinkhous purchased Nora and Lynne, he was making an investment in the future of his research program. It must be recognized, however, that this investment had clear risks in 1947. The Cornell veterinarians had already established that these dogs had a bleeding disorder similar to a well-known human disease. Both the clinical symptoms and inheritance patterns in these dogs were identical to those in the human hemophiliac. But for the experimental purposes that Brinkhous had in mind, the similarities between the bleeder dogs and human hemophiliacs would have to be demonstrated at a much more fundamental level than the Cornell experts' diagnosis provided. Brinkhous needed to show that the defect being transmitted and exhibited in these dogs was functionally identical at the biochemical level to the coagulation defect in human hemophiliacs. This was no small order given that the defect in humans still remained poorly understood in 1947. In recognition of this fact, the title of

Brinkhous's grant over these formative years was "Clotting Defect in Hemophilia, Study of a Strain of 'Hemophilic Dogs.'" Brinkhous's use of quotes around the words hemophilic dogs indicates his sensitivity to the fact that his newly purchased dogs might not have a disorder that was fully analogous to the coagulation defect in the human hemophiliac. Yet, in his own mind at least, he was aware of the complications involved in fabricating these laboratory animals as "models" for the study of human hemophilia.[30] Thus, when canine hemophilia and human hemophilia were later shown to be indistinguishable on the functional level, Brinkhous and his colleagues could look back on the uncertain early years of their study with an authentic sense of accomplishment.[31]

The uncertain and open-ended character of the study became very apparent to Brinkhous only a few months into their effort.[32] In his first NIH progress report, Brinkhous highlighted the two main problems his group faced in 1947. The first related to keeping the dogs healthy and reproducing. Not only had the whole line of dogs proved "unusually susceptible to upper respiratory infection," but the "basic stock at the kennel . . . was practically wiped out during the summer by a distemper epizootic." Moreover, breeding was being "complicated by the low sexual receptivity" of the strain, and artificial insemination was therefore required for certain females to reproduce. The second, more severe problem was the task of raising the male bleeder pups into adults.

> Rearing of hemophilic males requires constant attention. Heretofore, affected males practically always succumbed during the first two months of life. We have succeeded in rearing two bleeders, now 9 months of age, by administering blood or plasma transfusions regularly every 3–4 days. In spite of this regimen, one animal is badly crippled due to hemarthroses and resulting joint deformities. The other animal is in good condition. Both of these animals are at sexual maturity, and we plan to use them in the breeding program.[33]

Brinkhous makes reference here to the crude prophylactic strategy he devised in the late 1940s to ensure the health of his bleeder dog population. In subsequent years, this form of disease management gradually grew more sophisticated and expert.

Central to Brinkhous's prophylactic strategy was his standing order to the animal technicians to give blood transfusions to the dogs at any hint of possible hemorrhage. "Quick and early" was the motto he related to the laboratory technicians. To this end, Brinkhous maintained a small group of normal dogs in the laboratory to serve the bleeder dogs as blood donors and as healthy mates for breeding. At the time, these developments were not recognized as innovative, as they were judged to be a part of the routine management of laboratory animals, deserving little to no comment. By the early 1960s, however, Brinkhous and his colleagues were able to prevent hemarthroses and crippling that would otherwise afflict their hemophilic dogs.[34] These were substantial

Fig. 7.4 This "normal-looking" Irish setter is actually a severe hemophiliac, descended from the original strain of hemophilic dogs brought to the University of North Carolina in 1947. The photograph was taken in the 1960s to document how intensive regimens of plasma transfusion could be used to prevent the joint bleeding and crippling characteristically found in severe hemophiliacs. Photo courtesy of Kenneth Brinkhous, Francis Owen Blood Research Laboratory, Department of Pathology and Laboratory Medicine, University of North Carolina at Chapel Hill.

feats not only in the management of laboratory animals, but also in the management of disease. These dogs were also among the first hemophiliacs to experience the benefits of routine clotting factor replacement and prophylactic therapy.[35] So what does this tell us about the ontological status of these dogs—both as model organisms and creations of the laboratory?

Brinkhous's way of organizing the laboratory was innovative in that it modeled these dogs as both research technologies and clinical subjects.[36] In other words, like any other laboratory animal, these hemophilic dogs were tools whose significance was supposedly secondary to understanding the phenomenon under investigation. In practice, however, the dogs became primary. They themselves became the object of scrutiny because the task of managing their "clinical" condition often took practical precedence over the explicit objects of investigation (e.g., characterizing the clotting defect in the dogs). It was only possible to frame the dogs as technologies insofar as they were already patients. So, in addition to being exemplary tools for research, these dogs were model patients as well.

This institutional organization of the laboratory animals was, therefore, a crucial scientific innovation in that it situated a clinical microcosm within the laboratory. By bringing the canine hemophiliacs into the laboratory, Brinkhous was forced to "treat" the dogs as clinical patients. Indeed, the fact that his patients were dogs (and not humans) allowed him to manage their condition more closely and aggressively. And as Brinkhous was well aware, these nonhuman clinical patients were subject to the rigorous controls of the laboratory without being subject to the moral risks that were implicit in ordinary clinical research (involving humans).

Brinkhous's laboratory management of hemophilia not only positioned the canine hemophiliac as an integral feature of post–World War II research on bleeding disorders. His maintenance and use of hemophilic organisms also elevated his experimental program at the University of North Carolina into one of the world's foremost research centers on hemophilia and other blood-related matters.[37] This effect was seen most clearly in the research that took place between 1962 and 1968. During this period, the dogs along with other laboratory-based technologies allowed Brinkhous and his colleagues to develop numerous lines of hemophilia research, including tests of the antihemophlic preparations that culminated in the first widely marketed AHF concentrates.[38]

Passage Between Laboratory and Clinic

This essay has focused upon an unusual transition in what historians have come to regard as "the laboratory revolution in medicine."[39] This revolution relied not only upon a specific, disciplined way of organizing tools and materials in the laboratory, but also upon a manner of relating clinical phenomena to a laboratory setting. Thus far, I have suggested that Brinkhous's means of managing hemophilia in the laboratory has had largely unexamined origins and implications in the context of biomedical research in twentieth-century America. Certainly, this scientist's use of bleeder dogs was not only a way of managing the clinical presentation of hemophilia, but also a means of amplifying his personal and political influence—as an experimental pathologist and hematologist—within the medical community as well as beyond it. Yet this origin story also reminds us that biological and biomedical research materials achieve their transformative effects through acts of representation. As the philosopher Maurice Merleau-Ponty suggested in the early 1960s, the body is no mere object of perception but rather a project.[40] The origins of the canine hemophiliac are a concrete reminder of this difference; this story indicates that we might better understand the lives and uses of laboratory organisms if we recognize that such organisms are both born and made.

Today, Brinkhous is remembered as one of the most important scientists to have ever worked on the problem of hemophilia, and much of his reputation derives from his organizational abilities both inside and outside the laboratory. His efforts to maintain bleeder dogs as part of his laboratory enterprise and his efforts to represent these dogs as useful "animal models" for hematology and beyond, document his own practical and programmatic skills and the processes through which laboratory organisms are created by scientists. Significantly, the canine animal model of hemophilia was not given by nature. Rather, hemophilic dogs were a natural resource that Brinkhous and his colleagues cultivated after recognizing their potential value as a technology and as an object of study. Moreover, since the bleeder dogs' entry into the laboratory was framed by Brinkhous's desire to bring clinical concerns and subject matter into his laboratory research on hemophilia, it is not entirely accurate to say that these bleeder dogs were model organisms of hemophilia before the early 1950s. These hemophilic dogs only became "animal models," as we understand this concept today, after Brinkhous and his colleagues framed these dogs as clinical subjects within a laboratory organized for the purpose of studying the clotting defect in the hemophiliac and the possible means for its correction. Indeed, this claim gets to the heart of the issue. What historians of biology and the biomedical sciences mean when they say that laboratory organisms are "creations" or "productions" is that the being of these organisms is constituted through the representational acts of scientists — that is, scientists whose aim is to engender truthful claims about the scientific value of a certain organism as well as engineer agreement among multiple actors and social worlds about the validity of those claims.[41]

Of course, the canine hemophilia animal model cannot be judged apart from the circumstances of its making. The narrative presented here suggests that organisms and other research materials also achieve their effects in relation to their environments. Since the magnitude of this power also hinges on how the relation between organism and environment is represented by humans, I am assuming that scientists can exert large measures of control over their research (however much their objects of analysis or their research tools might impinge on or propel the knowledge-making process). Moreover, given the scientist's expert capacities for representation, it is not inconsequential that scientists also adjudicate the power of life and death with respect to their research organisms. The sacrifices that laboratory animals make in the name of science return us to the fact that the scientists and technicians who care for them also, in a certain sense, are responsible for creating them. The lives these organisms experience are conditioned upon the scientist's capacity both to create the means for sustaining the organism and (by extension) to exercise the creator's "right" to sacrifice the organism. After all, were it not for Brinkhous's staff providing careful attention and treatment, none of the bleeder puppies would have survived long enough to become viable research subjects.

I return now to Karen Rader's question: "What does it mean to speak about the 'creation' or 'origination' of experimental organisms, both for biology and medicine, and for the history of those fields?"[42] In light of the origins of the canine hemophiliac, it seems clear that the "creation" of an experimental organism can be framed as a proper object of historical analysis and interpreted in a way that accounts for the being of experimental organisms vis-à-vis the being of their human caretakers. For instance, this essay might very well have focused on the ramifications of using dogs as model organisms rather than some other species of animal. Dogs have a particular place, both culturally and socially, in the lives of Americans; and the varied and complex ways that people relate to dogs is most certainly a consideration for the scientists who both care for these organisms and represent them. By extension, historians of biology and biomedical sciences must eventually take the history of human and dog relations into consideration when charting their place within laboratories. Of course, this essay has itself skirted the relevant histories of dog and human relations, focusing instead on the moral dimensions of scientists' creation stories. This approach mirrors my belief that historical investigations of the laboratory should be making more of the changing ethical relations between scientists and their organisms.

Thus, the origin story related in this essay affirms that Brinkhous and his colleagues were only able to demonstrate the material and functional similarities between canine and human hemophilia by caring for and sustaining their canine subjects. Significantly, then, scientists' manipulations of their subjects were preceded by a necessity to care. The hemophilic dogs were not only treated "humanely" but also framed as patients in this setting. Or, to put this differently, the story of Brinkhous and his hemophilic dogs suggests that a moral imperative is operative in the passage between laboratory and clinic. For historians interested in the relations between human and nonhuman organisms, this imperative means that we cannot understand how scientists discipline their experimental organisms without understanding how these organisms also discipline scientists, forcing them to care.[43]

Karen Rader has suggested in relation to her own historical study of the creation of the standard laboratory mouse that by "listening to and studying existing creation stories, historians are made aware that [laboratory] creatures are valued parts of (to borrow Nathan Reingold's phrase) 'living traditions,' which human scientists construct but are not (in principle or in practice) entirely bound by in their knowledge-making." In a reflexive turn, Rader notes that by telling their own creation stories historians can "make others aware of how scientists make organisms and how people make history."[44] What moral, then, might we want to draw from the creation story that I have just told about the canine hemophiliac?

Certainly, the canine hemophilia origin story raises ethical issues that deserve further reflection.[45] As evidenced by the potentially tragic 1938 experiment with Laughlin, cultural and moral concerns clearly impacted Brinkhous's decision to invest so much energy in later years to the maintenance and use of hemophilic dogs. Certainly, the presentation of hemophilia in a strain of dogs afforded Brinkhous and his colleagues opportunities for studying human hemophilia without the risks that attended much biomedical research in the postwar era. The two decades following World War II were years in the United States when concerns about experimentation with normal human subjects drove many medical scientists to conduct their experimentation not only upon nonhuman organisms but upon captive peoples as well—including prisoners, children, the poor, and illiterate.[46] The availability and malleability of bleeder dogs allowed Brinkhous and other researchers to advance their biomedical research on hemophilia in a way that radically minimized their need to exploit the misfortunes of human patients.

Yet, as is inherently the case with all animal models of human physiology and pathology, their usefulness must be demonstrated; in other words, situations always arise where it is necessary to determine if what works in a laboratory organism also proves "true" or "practicable" for human subjects. For this reason, Laughlin's near fatality in 1938 did not dissuade Brinkhous in later years from bringing human patients into his laboratory and making use of their bodies. The imperative to care demanded that his canine studies be fully integrated with human studies.

In the late 1950s, for instance, Brinkhous accepted a graduate student named Murray Thelin into the University of North Carolina to work with the Department of Pathology's main biochemist Robert Wagner. In the course of his studies with Wagner and Brinkhous, Thelin became accustomed to using his own blood as research material in the group's ongoing efforts to develop effective clotting factor concentrates for hemophilia. After receiving his Ph.D. at Chapel Hill, Thelin went to Los Angeles to work with Hyland Laboratories, which collaborated with Brinkhous and Wagner in the development of the first commercial clotting factor concentrates for hemophilia. Shortly before his untimely death by a heart attack, Thelin's penchant for self-experimentation in the laboratory was chronicled in 1967 in *Today's Health* and *Reader's Digest*.[47] According to these chronicles, Murray Thelin suffered a potentially lethal brain hemorrhage in 1964, which led him and Edward Shanbrom, medical research director of Hyland, to test their unproven AHF concentrate.

> Murray, through badly blurred speech, helped give instructions to the lab and laid out a program of tests on himself to observe the drug's effect.
> Then came the first transfusion. Murray's clotting levels soared. There was no untoward reaction. In ten days he walked out of the hospital, smiling. And now he speeded his efforts to learn how to produce the powder[ed concentrate] in quantity.
> Tragedy threatened again. Murray began to bleed from a peptic ulcer, usually a death warrant for a hemophiliac. Once more his concentrate worked, quickly, safely.
> Still the experts raised questions. How would repeated doses affect a bleeder over a period of time? Murray made himself a guinea pig.[48]

Fig. 7.5 Three organisms with hereditary bleeding disorders studied by researchers at the University of North Carolina at Chapel Hill, 1951. Left to right: Missouri hog (porcine von Willebrand's disease), patient with hemophilia, and Irish setter (canine hemophilia A). Photo courtesy of Kenneth Brinkhous, Francis Owen Blood Research Laboratory, Department of Pathology and Laboratory Medicine, University of North Carolina at Chapel Hill.

On the basis of these events, Shanbrom began giving Thelin routine infusions of the concentrate, and "weeks passed without a bleed."[49] Yet, since hemophiliacs often experienced cycles of bleeding and not bleeding, Shanbrom and Thelin could not determine if the concentrate was really working until Thelin miraculously walked away from a bad car crash without so much as a bruise. Subsequent to that auto accident, the two researchers learned that when concentrate was withheld, Thelin bled at the slightest provocation.[50] Thelin's willingness to self-experiment proved decisive in the effort to develop an effective AHF concentrate for commercial use. Yet, the ethic of experimentation exhibited by him was not without precedent in the history of efforts to manage hemophilia. Not surprisingly, like Laughlin before him, Thelin grew accustomed to making his blood and body available for study in the years he worked in Brinkhous's laboratory.

In the years before the institutionalization of stricter controls on human subjects research in the United States, Brinkhous and other biomedical researchers did capitalize on the proximity of patients to their laboratories and clinics. Such stories of experimentation seem both evocative and questionable today given the regulations and controls that now attend the use of both human and animal research subjects. As I have been suggesting, however, Brinkhous had long cultivated a spirit of voluntarism and sacrifice in his working environments. That spirit encouraged laboratory and clinical workers—animal and human, "hemophilic" and "normal"—to donate their talents and bodies to the scientific enterprise whenever practicable. Moreover, the prevailing norms in Brinkhous's laboratory amounted to a distinctive moral economy that allowed workers to recognize that they were subject to the demands of the hemophiliac and scientist alike.[51] In other words, the norms of the laboratory required researchers to conform to the needs of hemophilic organisms as much as it required hemophilic organisms to subject themselves to scientists' demands.

The origins of the canine hemophiliac speak to the fact that moral imperatives are forces in the historical evolution of both human and animal life, and suggest the need for scholars to analyze further the changing relationships between humans and animals in working environments where the production of knowledge is the aim. In bringing bleeder dogs into his laboratory and framing them as model organisms for the study of human hemophilia, Brinkhous cultivated an environment in which dogs were framed not only as technologies, but also as organisms with agency. In short, the hemophilic dog's transformation into model organism entailed that Brinkhous and his colleagues treat them with a level of care (and potential risk) that most human hemophilia patients in the United States did not experience until later eras.

Notes

This essay is based on work and ideas that appeared in Stephen Pemberton, "Bleeder Dogs and the Framing of Hemophilia Management: From Dismal Prospects to Potent Medicine in the Laboratory of Kenneth Brinkhous, M.D., 1930–1960," M.A. thesis in history, University of North Carolina at Chapel Hill, May 1997. This project would not have been possible without the cooperation and encouragement of Kenneth Brinkhous and the guidance and critical insight of my graduate history advisor, Keith Wailoo. Thanks also to Todd Davis, Rosa Haritos, James Marcum, Michael McVaugh, Philip Scranton, Susan Schrepfer, and Karen Rader for their generous readings and comments at various stages in my writing.

1. Arthur J. Snider, "Breakthrough for Bleeders, If They Can Afford It," *Science Digest* 70 (August 1971): 55–56.
2. Advertisement for Courtland Antihemophilic Factor (AHF) Concentrate, 1970. Courtesy of Kenneth Brinkhous and the Francis Owen Blood Research Laboratory, Department of Pathology and Laboratory Medicine, University of North Carolina School of Medicine, Chapel Hill, North Carolina.
3. Susan Resnick, *Blood Saga: Hemophilia, AIDS, and the Survival of a Community* (Berkeley: University of California Press, 1998).
4. Stephen Pemberton, "Normality within Limits: Hemophilia, the Citizen-Patient, and the Risks of Medical Management in the United States, from World War II to the Age of AIDS," Ph.D. dissertation, University of North Carolina at Chapel Hill, 2001.
5. A wide variety of organisms have been domesticated within laboratories for scientific purposes. See, for example, Adele Clark and John Fujimura, eds, *The Right Tools for the Job: At Work in the Twentieth-Century Life Sciences* (Princeton, N.J.: Princeton University Press, 1992); Doris T. Zallen, "The 'Light' Organism for the Job: Green Algae and Photosynthesis

Research," *Journal of the History of Biology* 26 (1993): 269–279; Bonnie Clause, "The Wistar Rat as a Right Choice: Establishing Mammalian Standards and the Ideal of a Standardized Mammal," *Journal of the History of Biology* 26 (1993): 329–349; Robert Kohler, *Lords of the Fly: Drosophila and the Experimental Life* (Chicago: University of Chicago Press, 1994); Karen A. Rader, "Making Mice: C. C. Little, the Jackson Laboratory, and the Standardization of the *Mus musculus* for Research," Ph.D. dissertation, Indiana University, 1995; Karen A. Rader, "The Mouse People? Murine Genetics Work from the Bussey Institution to Cold Spring Harbor," *Journal of the History of Biology* 31 (1998): 327–354; Rachel Ankeny, "The Conqueror Worm: An Historical and Philosophical Examination of the Use of the Nematod *C. Elegans* as a Model Organism," Ph.D. dissertation, University of Pittsburgh, 1997; Angela N. H. Creager, *The Life of a Virus: Tobacco Mosiac Virus as an Experimental Model, 1930–1965* (Chicago: University of Chicago Press, 2002); Daniel Todes, *Pavlov's Physiology Factory: Experiment, Interpretation, Laboratory Enterprise* (Baltimore: Johns Hopkins University Press, 2002). For a recent critical overview of this literature, see Gerald Geison and Manfred Laubichler, "The Varied Lives of Organisms: Variation in the Historiography of the Biological Sciences," *Studies in the History and Philosophy of Biology and the Biological Sciences* 32 (2001): 1–29.

6. See Victoria Harden, *Inventing the NIH: Federal Biomedical Research Policy, 1887–1937* (Baltimore: Johns Hopkins University Press, 1986); Stephen P. Strickland, *Politics, Science, and Dread Disease: A Short History of United States Medical Research Policy* (Cambridge: Harvard University Press, 1972).

7. See, for example, Angela N. H. Creager, "'What Blood Told Dr. Cohn': World War II, Plasma Fractionation, and the Growth of Human Blood Research," *Studies in the History and Philosophy of Biology and Biomedical Sciences* 30 (1999): 377–405; Daniel M. Fox, "The Politics of NIH Extramural Program, 1937–1950," *Journal of the History of Medicine and Allied Sciences* 42 (1987): 447–466.

8. Colonies of hemophilic dogs have been maintained at several locations since the first was founded at the University of North Carolina in the late 1940s. Four other hemophilic dog colonies were established before 1970: the College of Veterinary Medicine, Oklahoma State University, Stillwater, Oklahoma, USA; the Ontario Veterinary College, Guelph, Ontario, Canada; the Medical Research Council, Blood Coagulation Research Unit, Churchill Hospital, Oxford, England; and the Division of Laboratories and Research, New York State Department of Health, Albany, New York, USA. See Derek Edwin Hall, *Blood Coagulation and Its Disorders in the Dog* (Baltimore: Williams and Wilkins Company, 1972); W. Jean Dodds, "First International Registry of Animal Models of Thrombosis and Hemorrhagic Diseases," *ILAR News* 21 (1977): A1-A23.

9. See J. B. Graham, J. A. Buckwalter, L. J. Hartley, and K. M. Brinkhous, "Canine Hemophilia: Observations on the Course, the Clotting Anomaly, and the Effect of Blood Transfusions," *Journal of Experimental Medicine* 90 (August 1, 1949): 97–111.

10. The term "animal model" was not employed by researchers using hemophilic dogs until the late 1960s and early 1970s and reflects the relatively recent history of this term. I use the terms "animal model" and "model organism" to describe the way that bleeder dogs have been utilized in the twentieth century. The term "model" in this context is analogous to the concept of the exemplar (or example). These organisms are not models in the sense that mathematicians, statisticians, engineers, or physical scientists usually employ the term.

11. Antihemophilic factor (AHF) concentrates are commonly known today as factor VIII concentrates. Classic hemophilia or hemophilia A is usually characterized as a functional deficiency of factor VIII (AHF). For reasons of historical consistency, I have used the terms that researchers themselves used in the period.

12. K. M. Brinkhous, "Animal Models of Hemophilia," in *Animal Models of Thrombotic and Hemorrhagic Diseases* (Washington, D.C.: National Institutes of Health, 1975), 3–13. For discussion of term "half-way technology," see Lewis Thomas, "The Technology of Medicine," in *The Lives of a Cell: Notes of a Biology Watcher* (New York: Penguin Books, 1974), 31–36.

13. See Jean Marx, "A First Step Toward Gene Therapy for Hemophilia B," *Science* 262 (October 1, 1993): 29–30; M. A. Kay, S. R. Rothenberg, C. N. Landen, D. A. Bellinger, F. Leland, C. Toman, M. Finegold, A. R. Thompson, M. S. Read, K. M. Brinkhous, and S. L. C. Woo, "In Vivo Gene Therapy of Hemophilia B: Sustained Partial Correction in Factor IX-Deficient Dogs," *Science* 262 (October 1, 1993): 117–119; J. N. Lozier and K. M. Brinkhous, "Gene Therapy and the Hemophilias," *Journal of the American Medical Association* 271 (January 5,

1994): 47–51; M. A. Kay, C. N. Landen, S. R. Rothenberg, L. A. Taylor, F. Leland, S. Wiehle, B. Fang, D. A. Bellinger, M. Finegold, A. R. Thompson, M. S. Read, K. M. Brinkhous, and S. L. C. Woo, "In Vivo Hepatic Gene Therapy: Complete Albeit Transient Correction of Factor IX Deficiency in Hemophilia B Dogs," *Proceedings of the National Academy of Sciences USA* 91 (March 1994): 2353–2357.

14. My thinking here is greatly influenced by Ian Hacking, "Making Up People," in *Reconstructuring Individualism: Autonomy, Individuality, and the Self in Western Thought*, ed. Thomas C. Heller, Morton Sosa, and David E. Wellbery (Stanford, Calif: Stanford University Press, 1986), 222–236.
15. Karen A. Rader, "Of Mice, Medicine, and Genetics: C. C. Little's Creation of the Inbred Laboratory Mouse, 1909–1918," *Studies in the History and Philosophy of Biology and the Biomedical Sciences* 30 (1999): 319–343.
16. Adele Clarke, "Research Materials and Reproductive Sciences in the United States, 1910–1940," in *Ecologies of Knowledge: Work and Politics in Science and Technology*, ed. Susan L. Star (Albany: SUNY Press, 1995), 183–225.
17. Rader, "Of Mice, Medicine, and Genetics," 319–343.
18. Maxwell M. Wintrobe, *Hematology, The Blossoming of a Science: A Story of Inspiration and Effort* (Philadelphia: Lea and Febiger, 1985).
19. Kenneth M. Brinkhous, "Harry P. Smith (1895–1972): Pathologist, Teacher, Investigator, and Administrator," *American Journal of Clinical Pathology* 63 (1975): 605–608.
20. H. P. Smith's own preference for team research derived from his interactions with George Whipple. While the 1930s was still an age in which many scientists worked alone, a trend was rapidly emerging in medical research circles where the great advantages of scientific collaboration were increasingly recognized. See George W. Corner, *George Hoyt Whipple and His Friends: The Life-Story of a Nobel Prize Pathologist* (Philadelphia: J. B. Lippincott Company, 1963).
21. For references to the quantitative approach of the Iowa Group, see E. D. Warner, K. M. Brinkhous, and H. P. Smith, "The Titration of Prothrombin in Certain Plasmas," *Archives of Pathology* 18 (1934): 587; and E. D. Warner, K. M. Brinkhous, and H. P. Smith, "A Quantitative Study of Blood Clotting: Prothrombin Fluctuations Under Experimental Conditions," *American Journal of Physiology* 114 (1936): 667–675. The Iowa Group gained early recognition for its work on the biochemistry and pathophysiology of blood coagulation due to the demonstrated sensitivity of its two-stage assay. Yet, the two-stage assay was only one of two ways known at the time for measuring the functional activity of the coagulation factors. The other was Armand J. Quick's one-stage assay, which gained wider use in clinical laboratories. See Oscar D. Ratnoff, "Why Do People Bleed?" in *Blood, Pure and Eloquent: A Story of Discovery, of People, and of Ideas*, ed. Maxwell M. Wintrobe (New York: McGraw-Hill, 1980).
22. Michael Bliss, *The Discovery of Insulin* (Chicago: University of Chicago Press, 1982); Keith Wailoo, "The Corporate Conquest of Pernicious Anemia: Technology, Blood Researchers, and the Consumer," in *Drawing Blood: Technology and Disease Identity in Twentieth-Century America*, ed. Keith Wailoo (Baltimore: Johns Hopkins University Press, 1997), 99–133
23. Corner, *George Hoyt Whipple and His Friends*, 98.
24. Kenneth M. Brinkhous, "Interview with Kenneth M. Brinkhous, M.D.," conducted by Susan Resnick, November 29, 1991. Use courtesy of Susan Resnick.
25. My narrative is based on Brinkhous's recollections of the 1990s, as they were related to Susan Resnick in her 1991 interview and to me on two different occasions, in 1994 and 1995. Quote from K. M. Brinkhous, "Understanding Factor VIII," Karl Landsteiner Award Lecture, annual meeting of the American Association of Blood Banks, San Diego, California, November 17, 1994.
26. In 1964, Brinkhous wrote: "My interest in this problem began in 1938 while studying the consumption of prothrombin in hemophilic blood and plasma. The first patient studied was a young man, a laboratory technician who assisted with the experiments. Whole blood transfusions were administered for tooth extractions. A preoperative transfusion promptly brought the whole blood clotting time (CT) to near normal, and after the extractions no unusual bleeding occurred immediately. After several hours, severe secondary bleeding (called *Nachblutung* in the older German literature) began, and the CT was again more than an hour. Another transfusion brought the CT back to near normal, but severe bleeding continued and the outcome was in doubt. Only when frequent and closely spaced transfu-

sions were administered were wide swings in CT prevented. Then the bleeding stopped." K. M. Brinkhous, "Hemophilia: Pathophysiologic Studies and the Evolution of Transfusion Therapy," *American Journal of Clinical Pathology* 41 (April 1964): 342–351. Brinkhous's comments, which appeared below the graph (in Figure 7.3), suggest that his experiment on Laughlin was "designed" to compare the relative efficacy of closely spaced transfusions vis-à-vis widely spaced transfusions. This representation from 1964 runs counter to Brinkhous's more recent recollections of the event (and my narrative), insofar as it equates his original experimental intentions with the actual outcome of the experiment. The discrepancy is a telling artifact, indicative of the role of reconstructive narratives in scientific discourse as well as the tendency within scientific writing to obscure both the uncertainties and moral complications that have often accompanied experimentation with human and animal subjects. Also, Brinkhous was interested in hemophilia long before 1938—as early as 1932—but only able to conduct sustained research on hemophilia once Laughlin came into the Iowa laboratory.

27. For overviews of the growth of the University of North Carolina's Medical School, see Dorothy Long, ed., *Medicine in North Carolina: Essays in the History of Medical Science and Medical Service, 1524–1960* (Raleigh: North Carolina Medical Society, 1972), and W. Reece Berryhill, William B. Blythe, and Isaac H. Manning, *Medical Education at Chapel Hill: The First Hundred Years* (Chapel Hill: Medical Alumni Office, University of North Carolina, 1979). For a history of the Department of Pathology, see John B. Graham, *How It Was: Pathology at UNC, 1896–1973* (Chapel Hill: University of North Carolina at Chapel Hill, 1997).

28. This summary of events was composed using the testimony of both Kenneth Brinkhous and John Graham. Brinkhous relayed his recollections of this event in conversation to me, while I have used Graham's account of the event as told in Graham, "The Hemophilic Dogs," in *How It Was*, 110–115.

29. Veterinarians at Cornell University published two papers describing the condition in Terry Bay's progeny, thereby becoming the first scientists to document hemophilia in dogs. See R. A. Field, C. G. Rickard, and F. B. Hunt, "Hemophilia in a Family of Dogs," *Cornell Veterinarian* 36 (1946): 285–300; and F. B. Hunt, C. G. Rickard, and R. A. Field, "Sex-Linked Hemophilia in Dogs," *Journal of Heredity* 39 (1948): 3–9.

30. Daniel Todes provides an extremely instructive account of how Pavlov and his assistants fabricated "normal dogs" for physiological experimentation, demonstrating that variation always plays a critical role in the scientific enterprise and that no animal is ready-made for laboratory research. See Todes, *Pavlov's Physiology Factory*, 80–122; Geison and Laubichler, "The Varied Lives of Organisms," 5–10.

31. Brinkhous's early studies on the dogs characterized their condition as fully as possible, helping "fix" the identity of the bleeder dogs as canine hemophiliacs. The key publications during this definitive period of experiment (1948–1951) follow: J. B. Graham, J. A. Buckwalter, L. J. Hartley, and K. M. Brinkhous, "Canine Hemophilia: Observations on the Course, the Clotting Anomaly and the Effect of Blood Transfusions," *Journal of Experimental Medicine* 90 (1949): 97–111; K. M. Brinkhous and J. B. Graham, "Occurrence of Hemophilia in Females," *Journal of Laboratory and Clinical Medicine* 34 (1949): 587–588; K. M. Brinkhous and J. B. Graham, "Hemophilia in the Female Dog," *Science* 111 (1950): 723–724; R. D. Langdell, J. B. Graham, and K. M. Brinkhous, "Prothrombin Utilization During Clotting: Comparison of Results with the Two-stage and One-stage Methods," *Proceedings of the Society of Experimental Biology and Medicine* 74 (1950): 424–427; K. M. Brinkhous, R. D. Langdell, and J. B. Graham, "The Problem of Prothrombin Determinations in Serum," in *Blood Clotting and Allied Problems: Transactions of the Third Conference*, ed. J. E. Flynn (New York: Josiah Macy, Jr. Foundation, 1950), 208–211; J. B. Graham, D. L. Collins, I. D. Godwin, and K. M. Brinkhous, "Antihemophilic Factor: Levels in Canine and Human Plasmas, and Following Liver Injury and Dicumarol," *Federation Proceedings* 10 (1951): 355a; K. M. Brinkhous, "Canine Hemophilia: Studies on the Inheritance of the Disease and the Clotting Defect," *Proceedings of the International Society of Hematology, 3rd Congress* (New York: Grune and Stratton, 1951), 439–440; J. B. Graham, G. D. Penick, and K. M. Brinkhous, "Utilization of Antihemophilic Factor During Clotting of Canine Blood and Plasma, *American Journal Physiology* 164 (1951): 710–715; K. M. Brinkhous, J. B. Graham, G. D. Penick, and R. D. Langdell, "Studies on Canine Hemophilia," in *Blood Clotting and Allied Problems: Transactions of the Fourth Conference*, ed. J. E. Flynn (New York: Josiah

Macy, Jr. Foundation, 1951), 51–65; J. B. Graham, D. L. Collins, I. D. Godwin, and K. M. Brinkhous, "Assay of Plasma Antihemophilic Activity in Normal, Heterozygous (Hemophilia) and Prothrombinopenic Dogs," *Proceedings of the Society of Experimental Biology and Medicine* 77 (1951): 294–296; K. M. Brinkhous, F. C. Morrison, and M. E. Muhrer, "Comparative Study of Clotting Defects in Human, Canine and Porcine Hemophilia," *Federation Proceedings* 11 (1952): 409.

32. In his memoir, John Graham outlines some of the difficulties that he and Brinkhous faced since neither one of them had ever reared dogs. See "The Hemophilic Dogs," in Graham, *How It Was*, 110–115.
33. Both quotes in this paragraph are from Kenneth M. Brinkhous, NIH Progress Report, a letter addressed to the Research Grants Division of the NIH relaying progress of RG 978 "Clotting Defect in Hemophilia: Study of a Strain of 'Hemophilic Dogs.'" Found in K. M. Brinkhous, "NIH Correspondence File, 1947–1949," Francis Owen Blood Research Laboratory, Department of Pathology and Laboratory Medicine, University of North Carolina at Chapel Hill, North Carolina.
34. In fact, it wasn't Brinkhous who first started looking at the dogs in this fashion. Margaret Swanton first approached him to study hemophilic arthropathy and its management in the bleeder dog. See Margaret C. Swanton, "Hemophilic Arthropathy in Dogs," *Laboratory Investigation* 8 (1959): 1269.
35. K. M. Brinkhous, M. C. Swanton, W. P. Webster, and H. R. Roberts, "Hemophilic Arthropathy: Transfusion Therapy in its Amelioration—Canine and Human Studies," *Proceedings of the World Congress of Hemophilia* (Sydney, Australia: World Congress, of Hemophilia, 1966), 18–21.
36. On organisms as "technologies," see Kohler, *Lords of the Fly*, 6–8.
37. As Brinkhous and his colleagues gained expertise at diagnosing and treating the dogs, they also developed quantitative tools and replacement therapies for application to the canine hemophiliacs. The development of the partial thromboplastin time (PTT) in 1953 by Robert Langdell, Robert Wagner, and Brinkhous was the most important of the quantitative tools. PTT test methodology has evolved since its introduction from a simple clotting test for studying hemophilia to a sophisticated and reliable system for the study of a broad range of procoagulant enzymes, their activation, and inhibition. Retrospectively, it is amazing to see how this simple test system has uncovered new procoagulants and new deficiency states, made possible new and safer therapeutic plasma fractions, and provided diagnostic testing and therapeutic monitoring for both hemorrhagic and thrombotic disorders. Concurrently, the test helped spawn enlarged clinical coagulation laboratories, which in turn stimulated industrial production of reagents and instruments needed to keep the laboratories going. See R. D. Langdell, R. H. Wagner, and K. M. Brinkhous, "Effect of Antihemophilic Factor on One-Stage Clotting Tests," *Journal of Laboratory and Clinical Medicine* 41 (1953): 637; K. M. Brinkhous and F. A. Dombrose, "Partial Thromboplastin Time," in *CRC Handbook Series in Clinical Laboratory Science*, ed. David Seligsohn (Boca Raton, Fla.: CRC Press, 1980), Section I. Hematology, Volume III, 221–246.
38. See H. R. Roberts, G. D. Penick, and K. M. Brinkhous, "Intensive Plasma Therapy in the Hemophilias," *Journal of the American Medical Association* 190 (1964): 546–548; W. P. Webster, H. R. Roberts, G. M. Thelin, R. H. Wagner, and K. M. Brinkhous, "Clinical Use of a New Glycine-Precipitated Antihemophilic Fraction," *American Journal of Medical Science* 250 (1965): 643–651; K. M. Brinkhous, E. Shanbrom, H. R. Roberts, W. P. Webster, L. Fekete, and R. H. Wagner, "A New High-Potency Glycine-Precipitated Antihemophilic Factor (AHF) Concentrate: Treatment of Classical Hemophilia and Hemophilia with Inhibitors," *Journal of the American Medical Association* 205 (1968): 613–617.
39. Andrew Cunningham and Perry Williams, eds., *The Laboratory Revolution in Medicine* (New York: Cambridge University Press, 1992).
40. See Maurice Merleau-Ponty, *Phenomonology of Perception*, Colin Smith, trans. (Atlantic Highlands, N.J.: Humanities Press, 1992).
41. Here, I echo Rader, "Of Mice, Medicine, and Genetics," 338.
42. Rader, "Of Mice, Medicine, and Genetics," 319–343.
43. See Bruno Latour, "The Costly Ghastly Kitchen," in *The Laboratory Revolution in Medicine*, 295–303. See also Susan Lederer, "Political Animals: The Shaping of Biomedical Research Literature in Twentieth-Century America," *Isis* 83 (1992): 61–79.
44. Rader, "Of Mice, Medicine, and Genetics," 338–339.

45. On the historical relations between human and animal experimentation in the United States, see Susan Lederer, *Subjected to Science: Human Experimentation in America Before the Second World War* (Baltimore: Johns Hopkins University Press, 1995).
46. See David J. Rothman, *Strangers at the Bedside: A History of How Law and Bioethics Transformed Medical Decision Making* (New York: Basic Books, 1991).
47. Patricia Deutsch and Ron Deutsch, "One Man's Fight Against Hemophilia," *Today's Health* (August 1967): 40–42; Patricia Deutsch and Ron Deutsch, "Dr. Thelin's Fight Against Hemophilia," *Reader's Digest* (August 1967): 90–94.
48. Deutsch and Deutsch, "Dr. Thelin's Fight," 93.
49. Deutsch and Deutsch, "Dr. Thelin's Fight," 93.
50. Deutsch and Deutsch, "Dr. Thelin's Fight," 93.
51. The use of the concept "moral economy" in this context has been explored by Robert Kohler in his book *Lords of the Fly*, 11–13. See also Angela Creager's examination of this concept in *Life of a Virus*, 322.

Making the Chicken of Tomorrow
Reworking Poultry as Commodities and as Creatures, 1945–1990

ROGER HOROWITZ

In June 1948 an enthusiastic three-mile parade wended its way through the tiny town of Georgetown, Delaware, as the final event in the improbably named (to contemporary ears) "Del-Mar-Va Chicken of Tomorrow Festival." The parade celebrated a remarkable event that had been building for several years—the national "Chicken of Tomorrow" contest. Initiated by the A&P retail grocery chain and the U.S. Department of Agriculture, a national committee of poultry industry organizations promoted the contest to encourage "production of superior meat-type chickens." A series of state and regional contests proved cash prizes to winners and determined qualified entries for the national competition in Georgetown.[1]

Forty entrants from leading hatcheries throughout the United States competed for the lucrative national prize. Not only would the winner receive $5,000 but doubtless also many orders from farmers eager to grow the best birds for the market. The winner, the Vantress Hatchery in California, was able to grow a heavier, meatier chicken faster than any other entrant. Within ten years the Vantress-produced birds would be the standard used by the nation's poultry farmers.[2]

This immediate postwar event marked a fundamental transition in the place of chicken within America's diet, and indeed, the popular conception of chicken as a type of food. For two hundred years Americans considered chicken a luxury meat served only on special occasions. The Republican party's unfortunate 1928 campaign slogan, "A Chicken in Every Pot and a Car in Every Garage," reflected chicken's status as an usual, exceptional, and hard to obtain food. Broilers, which were young chickens tender enough to be fried or cooked in an oven, were a byproduct of the much more important egg industry, and hence relatively expensive compared to other meats.

By the 1980s chicken had displaced beef as America's favorite meat. Rarely has there been such a dramatic change in American foodways. Vastly increased consumption practices took place in several arenas. The fast food industry, led by Kentucky Fried Chicken, established chicken as a meal to be eaten quickly and for lunch. Simultaneously, concerns over red meat's fat content, an outgrowth of the health and consumers movement, resulted in consumers shifting their eating preferences from beef to chicken. (It is, of course, ironic that two contradictory trends in American foodways rebounded to the benefit of the poultry industry.) What had once been a food item eaten, at best, on Sunday was now often the center of three or four meals each week.

The very language changed along with these consumption habits. At least through the 1940s chicken was part of a larger category called poultry or fowl that contained many distinct breeds. Most farmers relied on White Leghorns for egg production, but many other varieties circulated through the nation's farms and meat markets. Broilers referred not to a type of chicken, but to a stage of development that suited the animal to a certain kind of cooking. Aggressive cross-breeding of chicken following the war largely eliminated breeds outside of poultry fanciers, in favor of distinctions by form in which the animal would be used: layers, broilers, roasters, and so forth.

Paralleling the metamorphosis of chicken as an animal category, distinguished internally by function rather than lineage, was an incorporation of chicken into the category of meat. Poultry firms helped transform chicken's place in America's diet by literally changing the form in which consumers encountered it in eating establishments and supermarkets. This entailed a conceptual shift that the physical integrity of the chicken could be violated to create new products; that the meat chicken could be transmogrified into chicken meat.

From Eggs to Broilers

Poultry consumption has a long history as an annex to the rural cycle of harvesting eggs for home use and the market. Chickens are relatively easy to raise and keep, and for two centuries both rural and urban areas had plenty on hand. Data are especially unreliable for chickens, but they were ubiquitous in nineteenth-century America. The 1840 census estimated the value of poultry on farms at $12 million in 1840, and their numbers and popularity rose steadily throughout the century. In 1910, 88 percent of all farmers kept chickens, with an average flock of around eighty.[3]

Eggs, rather than chickens, were the preferred commercial product, for the evident value of having a constantly replenishing supply of high-quality protein. Availability was highly seasonal until the twentieth century, as egg production was strongest in the spring and summer and tapered off through the rest of the year. Nineteenth-century household advice books contained various suggestions on how to keep eggs fresh through the year, such as packing them in barley and keeping them in a cool place. By the early twentieth century the large meat-packing firms entered the egg business aggressively, as their chilled railroad cars and national branch house network were well suited to taking the seasonal products of the farm and making them available year-round for urban consumers.

By the 1880s thousands of farmers in the northeast were beginning to specialize in meat chicken production for the growing urban markets of the Washington-Boston corridor. These operations remained marginal, however, as it was very hard to deliver birds of consistent quality for urban consumers. The best young eating chickens were a seasonal specialty at best; for much of the year consumers never quite knew what would be available when they went to the market.

National meat processors such as Swift and Wilson were powerful forces in the national market for frozen chicken that somewhat reduced the seasonality of chicken supplies. But the widespread availability of chickens and corresponding decentralization of slaughtering and dressing to many urban centers precluded a centralization of the industry paralleling beef and pork. Instead, as demand grew so too did highly regional chicken markets, as adjacent farming regions learned to supply the needs of growing urban centers.

New York City was the largest market for chicken in the nation, in part because it remained America's biggest city, and in part because of its peculiar ethnic composition. Forbidden from eating pork by kosher dietary rules, New York's Jews were eager consumers of poultry in order to add variety to their diet and to have a special meal for Sundays. A 1926 Department of Agriculture study found that Jews accounted for 80 percent of the live poultry sales in New York City; with a Jewish population of two million by the 1930s, this was a substantial market. Yet they could not partake of the Midwestern-slaughtered chickens. Orthodox Jews would only eat chicken killed by licensed shoctim, using the proscribed kosher method of cutting the gullet and windpipe with two quick forward and backward strokes, then piercing the veins on both sides of the neck.[4]

Farmers in the Delmarva peninsula were the principal beneficiaries of the burgeoning demand for chicken (Delmarva refers to the peninsula lying between the Chesapeake and Atlantic that includes Delaware, Maryland's eastern shore, and one Virginia county). While their initial success was largely due to Jewish consumption practices, the development of trucking and a decent road network in the 1920s was a necessary precondition for breaking the hold of the railroad-based system of Swift and the other larger meat packers. Within a few years these new players would utterly change the poultry industry.

Jewish demand for live chicken fundamentally changed Delmarva's agriculture. Discovery of this market was an accident generally attributed to Cecile Steele, a Sussex County, Delaware, farm women who maintained a flock of laying chickens to contribute to her family's income. In 1923 she mistakenly received five hundred chicks from a hatchery, ten times her usual order. She raised the chickens rather then send them back, and eighteen weeks later was able to receive 62 cents a pound for them, a huge profit. Steele invested some of her earnings in an order for a thousand chicks; within three years she and her husband had expanded their capacity to ten thousand. Their success encouraged other struggling Sussex County farmers to explore this market. By the end of the 1920s there were five hundred broiler growers in Sussex County. While husbands and children may have become involved in this business, women remained central to raising the broilers for market. In the mid-1930s virtually all of Delaware's chickens, produced by protestants who had lived for generations in the same area, went to New York City for the Jewish immigrant market.

For the first ten years of this new business most chicken were "live-shipped" to New York commission markets, where brokers bought them in lots and in turn distributed the birds to local stores. Beginning in the late 1930s the first processing plants opened in Delaware, and by 1942 there were ten in operation with the capacity to process thirty-eight million broilers annually. There was an ample gentile market in New York for these birds, along with less strict second-generation Jews who were willing to patronize kosher butchers who sold fresh killed chicken shipped in ice.[5]

These plants were primarily hand operations that echoed the methods of late nineteenth-century hog processing methods. Men first hung the animals by their legs from racks attached to a moving chain, which carried them to successive operations. Women cut chickens' throats with a knife, and once the animal was bled it entered the five-stage feather removal process. Just as firms had figured out how to replace the men who had shaved the hair from a hog's skin, poultry equipment suppliers developed machinery that supplanted the men and women who had pulled out the feathers by hand. The birds were scalded, whirled around in a drum with rubber "fingers," coated in wax and picked over by women to eliminate most of the feathers, then finally singed with flame to clean the last remnants. Female consumers or their butchers performed the final processing stage of evisceration on these "New York dressed" birds prior to cooking. Plants employing these methods could process 100,000 chickens a day in 1945.[6]

Broilers' popularity brought new entrants into the industry. Delaware chickens had traditionally dominated the important New York City market and were the main poultry product in other East Coast and Midwestern cities. The armed forces, which requisitioned Delmarva's entire chicken production during the war, disrupted these relationships and allowed Georgia and Arkansas growers to gain footholds in lucrative urban markets. (Military buying also encouraged a flourishing black market as poultry growers tried to evade price ceilings and get their product to New York City outlets.) Following the war chicken producers in New England and the deep South rapidly expanded their sales in Midwestern and New England states. While Delmarva's broiler production held steady, its market share fell from 25 percent in 1940 to only 6 percent in 1955 as national broiler production topped one billion for the first time.[7]

Remaking the Postwar Chicken

The 1948 "Chicken of Tomorrow" contests brought together major institutions that would, collectively, transform the place of chicken in American cuisine in the last half of the twentieth century. A well-established network of chicken breeders, egg hatcheries, feed producers, processing and distributing firms hoped that the postwar years would bring about vast expansion in the market for their products. Large retail chains, principally the East Coast A&P firm, wanted a better product for consumers now interested in regular chicken purchases. But the fragmentation of actors, and conflicts among them, meant that it took an institution both embedded in yet not "of" the industry to align these objectives in a common strategy—the U.S. Department of Agriculture and its Cooperative Extension Service.

Established by the 1887 Hatch Act, state extension services were attached to the land grant colleges of each state with the charge to assist the farm economy. The network of county agents and poultry specialists encouraged farmers whose fruit and cotton crops were damage by bugs and diseases in the 1920s to raise broilers. Disseminating technical information on matters such as proper food mixtures and chicken house design, Extension Service personnel also acted as mediators between the emerging urban markets and the rural growers.[8] Following the war state Extension Services went into high gear to persuade farmers that they could profit from promised postwar prosperity by improving their chickens' quality and reducing production costs. In the deep South Extension personnel encourage white farmers in declining cotton regions to shift into broiler production.

Georgetown's "Chicken of Tomorrow" festival and contest were in large part due to the efforts of Delaware's Extension Service and its Poultry Specialist J. Frank Gordy, Jr. In their modernizing project extension personnel stressed it was in the interest of all poultrymen, "grower, processor, hatcheryman, feed dealer," to subordinate individual needs to that of the local industry. Farmers had a "duty to use growing practices that tend to produce better market quality . . . . even though it may be to his short-run advantage to lower his standards." Poultry processors had to be constrained against lower-quality production, as it "has a detrimental effect on all prices" and can give Delmarva chickens "an unfavorable reputation." This language indicates that the Extension Service was willing to experiment with various mechanisms (primarily markets and government regulation) that would discipline individual farmers and businessmen who would not, or could not, cooperate with the dominant trends in the industry.[9]

Gordy's role indicates the valuable nonpartisan status of the Extension Service among poultry interests. A central part of the industry yet not tied to a particular firm or sector, the Extension Service could play a unique role in advancing commercial broiler production. Prior to World War II, the Extension Service's most important assistance to chicken producers was in the area of disease control. As it was part of the state's land-grant institution, the University of Delaware, the Extension Service could call on university faculty to assist its scientific studies. Gordy built on these established relationships and trust to expand Extension Service resources that could be devoted to assisting the poultry industry. He was above all concerned with restoring Delaware's place in the national industry.

One of Gordy's first initiatives was to persuade a new university faculty member, Willard McAllister, to investigate how Delaware chicken was faring in the retail market. Beginning with a painstaking canvass of food retailers in Philadelphia, McAllister's careful studies identified a series of challenges for poultry producers in the late 1940s. The main obstacle to increased chicken consumption, he emphasized, was changing the prevalent attitude that chicken was a special weekend meal. To make chicken a more regular and consistent part of the American diet, the price of broilers had to be reduced below that of red meat and the quality improved. This meant producing an inexpensive and "meatier" bird that had a fresh and attractive appearance in the retail store. McAllister's studies also showed that retailers and consumers preferred a bird that was fully processed and ready to cook, rather than the traditional "New York dressed" style.[10]

Implementing McAllister's recommendations to produce a better broiler entailed dramatic alterations in regional farming and processing methods, and indeed to the bird itself. To persuade the fragmented and quarrelsome poultrymen of his program, Gordy diligently worked to engender a greater consciousness of the trouble they were in and the urgent need for cooperation among different sectors.

For an Extension Service rooted in imparting technical advice the efforts of Gordy and his associates were revolutionary. Publications were oriented to the farmer, such as "Mr. Poultryman: Marketing Is Your Business," and junkets to wholesale poultry markets in New York City and Philadelphia were designed to persuade growers that they had to change their methods to respond to consumer demand. These far-reaching efforts even extended to using the 4-H program to influence future farmers (and their parents!) through a "Junior Broiler Program" contest modeled on the "Chicken of Tomorrow" competition.[11]

Gordy and other Extension Service personnel avidly participated in the Chicken of Tomorrow contest as a first step in their campaign to persuade consumers that chicken was an everyday meal. However, acceptance of the new bird by Delmarva growers was another matter. Farmers were reluctant to buy the more expensive chicks produced by Vantress birds, and they also responded sluggishly to other aspects of McAllister's recommendations, such as switching to white birds so that small pin feathers not removed in the processing plant would be less visible to consumers. To change the chicken, Gordy and his associates also needed leverage to persuade farmers to change their chicken-raising practices.

Establishing a chicken auction in 1952, called the Eastern Shore Poultry Exchange, was an integral part of Gordy's modernizing strategy. By creating an open market in broilers, the Extension Service hoped that market mechanisms would induce farmers to improve their chicken's quality and pay closer attention to the interests of chicken processors. Half a million chickens were sold on June 24, its first day of business, to local poultry processing plants; within five years, daily sales would frequently top one million. Very quickly the auction became "part of the social fabric" of life on the Delmarva peninsula. In Selbyville, hundreds of growers, feed dealers, buyers, and other "interested persons" would gather at the exchange building at 1 P.M. to learn the auction results. Those unable to attend would turn on their radios, as popular stations broadcast the latest chicken prices with the same drama as television announces today's winning lottery numbers. The Delmarva exchange was quickly emulated in other poultry growing states such as Arkansas.[12]

The Exchange motivated farmers to comply with the Extension Service's recommendations if they wanted to receive good prices for their flocks. "Selling birds through the auction has emphasized difference in price for birds of different quality far more than the average grower could realize before the auction was formed," Gordy noted approvingly. Chickens produced from Vantress stock (the winners of the Chicken of Tomorrow contest) grew from 12 percent of Eastern Shore Poultry Exchange auction sales in 1953 to 76 percent in 1957, and sales of white-feather chickens increased from 52 percent to over 80 percent in the same period. At the same time, the average age of chickens brought to market fell from twelve to nine weeks, reflecting improved breeding stock and feeding methods. There also were significant reductions in mortality rates and food consumption costs.[13]

These improvements came at a price to farmers—they had to change the method of financing their chicken growing operations. Building better chicken houses, buying more expensive chicks, providing improved feeds all required capital, which many farmers did not have. Between 1945 and 1960, most chicken farmers switched from relying on credit at interest rates ranging from 15 to 25 percent to obtain hatchling chickens and necessary supplies, to contracting with feed suppliers or hatcheries to produce chickens. In these new arrangements, the contracting firm retained title to the birds, and the farmer had to raise the chicks through methods decided on in conjunction with a representative of the contractor (generally called a serviceman). Your companies "are putting in 80% of the risk capital," marketing specialist Willard McAllister admonished the servicemen, "you should have at least this much control of the growing operation." Courses offered by Extension Service and other School of Agriculture personnel sought to instill in the servicemen both their responsibilities to the industry, and the farming methods they should advocate. While contracting protected farmers from natural disasters and ruinous interest charges, they also lost a great deal of independence, and had greater difficulty resisting changes in raising practices required by servicemen.[14]

By the end of the 1950s, the Extension Service began to advocate the creation of firms that oversaw broiler production from the egg to the processed carcass—usually termed "integration"—as the best way to improve the Delmarva industry. Its personnel believed that fragmentation of the industry into complex factions of egg hatcheries, feed producers, growers, and processors impeded necessary restructuring of chicken production. Practices "which may make short term profits for an individual or a separate segment of the total industry are minimized in an integrated firm," McAllister explained in "A Plan of Action for the Delmarva Poultry Industry." He concluded on an optimistic note, "Therefore, any wasteful or costly practice is not tolerated."[15]

Most of the new large chicken firms had their origins on the agricultural side of the industry. Incorporating processing operations usually was the last step to building an integrated company. The formation of the Townsend, Inc., poultry company is a good illustration of this. A wealthy agricultural family that sent a member, John G., to the U.S. Senate in the 1930s, the Townsend's began financing farmers to raise poultry in the mid-1930s. By the end of World War II the company had its own chicken hatchery and feed mill and had expanded contracts with local farmers to produce chickens, using its chicks and feed. It sold chickens through the Eastern Poultry Exchange until the late 1950s, when the firm built its own processing plant. Townsend's example was soon followed by most Delmarva poultry producers and paralleled similar trends among deep South chicken firms.[16]

Ironically, the Extension Service's success would doom its creation, the Eastern Shore Poultry Exchange. As integrated poultry growing operations acquired processing plants and internalized broiler production stages, the Exchange's sales volume precipitously declined. The end came when the last major grower to use the exchange, Frank Perdue, finally acquired a Swift processing plant in 1968. He almost immediately stopped selling flocks through the Exchange because, as Perdue later recalled, "there was more money in processing." The exchange closed soon thereafter, as it was now the integrated firms that enforced chicken quality.[17]

The emergence of the "integrators," firms like Perdue and Townsend on the Delmarva peninsula and Tyson in Arkansas, led to renewed attention to the processing side of the chicken industry. Until the late 1950s most chicken left the processing plant as uneviscerated "New York dressed" poultry. The facilities handling these birds were rudimentary operations with relatively low labor demands, able to rely largely on female workers from nearby rural areas.

The expansion of processing to include evisceration and better cleaning of the carcasses, stimulated by the advent of federal inspection in 1959 and the growing use of chain supermarkets to sell chickens, resulted in the construction of brand new facilities.

Evisceration also added enormously to the industry's labor needs. So-called on-the-line eviscerating operations simply entailed having the chicken travel upside down between rows of butchers (generally women) who each performed small cuts on the bird, similar to the old meat-packing lines. Capital equipment needs were minimal—a longer chain operation, metal tables, knives, and sundry ancillary equipment. One study estimated that a $250,000 plant processing five thousand broilers daily in "New York dressed" form needed to invest just $25,000 more in equipment to add an evisceration department. But the labor needs were correspondingly enormous. The same plant hired ninety-eight more employees to handle "line" evisceration operations—more than doubling its total paid labor force to 166. Most accounts indicate that this is when the labor force decisively shifted to African American women and away from white women who had better job options.[18]

With the widespread adoption of evisceration, the broiler industry consolidated its insurgency in American food consumption practices. Annual per capita chicken consumption doubled between 1940 and 1960 to thirty pounds largely due to increased broiler sales, while chicken's retail price fell more than 30 percent. A meat that cost more than hamburger in 1950 was, a decade later, ten cents per pound cheaper. The "Chicken of Tomorrow" had arrived.[19]

Chicken Becomes a Meat

With chicken consumption clearly on the rise, state Extension Services conducted more than a dozen consumer research surveys in the 1950s and early 1960s to assess how the industry could advance further. The studies showed that chicken had made great strides in consumer's eyes, reflecting the accomplishments of Gordy and his peers. Although consumers remained watchful for signs of bruising and poor bleeding, they made few complaints about chicken quality, testifying to the widespread adoption of the meat-type broiler and improvements in processing technology. The bird had, in essence, been successfully standardized; it was a broiler, not a Plymouth Rock or Rhode Island Red. Declining retail prices, even with the added labor entailed in evisceration, indicated that the corporate integration, improved feeds, and concomitant production volume expansion had reduced costs significantly. Consumption patterns also showed that chicken was now a year-round meal with national appeal, and that the new self-service grocery stores were an asset to chicken sales as broilers could be easily wrapped in clear film like cellophane.

All was not well with the new chicken, however. Consumers still considered it a special meal and were far more likely to serve chicken on Sundays as part of making that day distinct from the rest of the week. Housewives consistently ranked chicken third, after beef and pork, as a main course for conventional meals. Consumption also varied wildly by ethnic background (Jews and African Americans were generally the best consumers), and lower income consumers were more likely to treat chicken as an exceptional, occasional dish than higher income groups. The expansion of chicken consumption had stretched, but not yet altered its traditional place within American foodways.

"Chicken apparently has not fully achieved the status and prestige of a meat item," concluded one especially insightful study. "In fact, many housewives do not consider chicken to be a meat." While chicken was eaten more often than before 1945, it still functioned as a substitute for or alternative to meat. Before further advances in per capita consumption could be achieved, the study warned, it would be necessary to rid chicken of its "'weak sister' image and the 'inferiority complex' chicken has in relation to red meats." And it recommended that industry expand consumption by launching "*An all-out attempt . . . to give chicken full status as a meat product.*"[20]

The studies also provided insights into why chicken was not yet a meat. Consumers explained they did not eat chicken more often as "it gets tiresome if eaten more than once a week." Such "chicken fatigue" reflected the relative monotony of chicken products available in 1960 compared to beef and pork. "Not only must she make a choice between beef, pork, and chicken," noted one study, "but she also must choose between beef steak and beef roast; pork chops and ham; whole fryers and foul." These are, of course, not equivalent sets of choices. Beef and pork came in more varied cuts and flavors than chicken.[21]

A similar problem emerged from comments indicating the influence of family size on chicken consumption. A elderly women explained she didn't buy chicken more often as "there's just my husband and I here now and it lasts too long with only two people." While two might have been too few to conveniently eat a three-pound broiler, large families also could be a problem. One women explained she bought chicken only three or four times annually even though they liked it, as "with fives kids, chicken just goes too fast." Unlike beef and pork, which yielded cuts of different size and cost, the broilers that dominated the early 1960s market were a one-size-fits-all product. Leftovers from a three-pound bird were too much for a couple, and it was not cost-effective for larger families to eat chicken when they could instead obtain inexpensive beef and pork cuts.[22]

For chicken to become a meat, manufacturers would have to do more than redesign the bird and transform processing methods. They needed to reconfigure the form in which consumers encountered chicken in grocery stores and restaurants so that there were a variety and array of choices more similar to beef and pork. Product differentiation and market segmentation were now desperately needed for chicken to move out of third place as Americans' meal of choice. The chicken had to be transformed into meat with different forms and uses.

Making the Chicken of Tomorrow • 225

As poultry firms entered this new phase they could build on some very positive associations with chicken as a food item. Consumers of all types shared a conception that chicken was a low-fat healthy food that was easy to digest, hence good for adults and well-suited for children. "I think of it as a way of pleasing my family," one women commented. The absence of fat also meant that consumers saw chicken as an efficient, less wasteful food compared to pork and beef. "You can get more out of it for the money invested," reflected another housewife. With evisceration, survey respondents also agreed that chicken was easy to prepare. "Chicken fits well into the concept of modern living," shrewdly observed a 1960 study. Capitalizing on these notions over the next two decades would catapult chicken consumption ahead of beef and pork.[23]

Two of the most successful—albeit quite different—product differentiation strategies were developed by Frank Perdue and the father and son duo of John and Don Tyson. Like Gustavus Swift a century before, Frank Perdue was remarkably knowledgeable of the animal that his firm transformed into food. John Tyson (father of Don) bore more similarity to Philip Armour because they shared a savvy knowledge of the market for their product, albeit a century apart. Through quite different paths the Perdue and Tyson firms would effectively promote higher chicken consumption levels and turn what once was a type of poultry into a form of meat.

John Tyson's initial entry into the poultry business gave his firm a commercial orientation from its inception. A Missouri produce buyer and trucker unfamiliar with the chicken industry, Tyson moved to Arkansas in 1931 to ply the hay market for animals in the drought-ravaged area. The growing broiler industry in the state's northwestern corner attracted his interest, and in 1935 he expanded trucking operations to convey chickens to Kansas City and St. Louis commission markets. Not until the early 1940s did Tyson actually enter the broiler business by starting a hatchery and a chicken growing operation, just in time to ride the enormous increase in deep South chicken production following World War II. Anticipating opportunities in the food business distinguished Tyson's firm from its birth.

Don Tyson took over leadership of the firm in the 1950s and made it America's largest poultry company by creating new markets for its products. In 1964 Tyson sold the U.S. Armed Forces Commissary on precooked portion controlled chicken. A few years later he became the principal supplier of McDonald's Chicken McNuggets and persuaded Burger King to sell Tyson chicken patties. For retail store purchases Tyson expanded into chicken hot dogs and precooked chicken products for home use. And as the white-meat dominated products created a surplus of dark meat, Tyson expanded his international export operations to regions of the world, especially Asia, whose residents preferred chicken legs and thighs. His firm's broilers, instead of simply coming to the store whole, might end up in chicken hot dogs, military rations, and Japanese dinners.[24]

Tyson's market differentiation strategy for chicken echoed beef's appeal, especially the way different cuts and products catered to distinct markets. The weakness of this beef-style approach, however, was that the chicken remained, well, just a chicken, with the firm that produced it hidden behind the commodity. Frank Perdue pursued a strategy more akin to early-twentieth-century pork producers who used branding to develop consumer loyalty to a particular firm's products and thus secure a reliable market share. While Tyson's success proved to post-1960 poultry producers that product differentiation promised a path to success, Perdue demonstrated that chickens could be branded similar to bacon and ham and hence removed from the status of a beef-style commodity food.

Branding entails close attention to product quality, a feature of the Perdue firm since its inception. The company began when Frank's father, Arthur Perdue, established a table egg farm in 1920 in the middle of the Delmarva peninsula. For twenty years the family firm remained a hatchery producing high-quality egg-producing birds. Similar to other Delmarva hatcheries Perdue Farms, Inc., entered broiler production by contracting with local farmers to raise their chicks. Frank became company president in 1950 and augmented company operations by building mills to supply feed to its farmers. By 1968 Perdue was the largest broiler producer in the United States, selling 800,000 birds weekly to processors through the Eastern Shore Poultry Exchange.

When Perdue finally entered the processing business in the late 1960s, the father and son team had almost fifty years of chicken raising experience behind them. "I grew up having to know my business in every detail," recalled Frank in 1973. They also were late-comers to a highly competitive market and under severe sales pressures with close to a million birds to sell every week. Perdue set out to distinguish his chickens from the pack, to in essence emulate successful pork producers like Oscar Mayer that had used advertising to develop a following for their branded goods.[25]

The television advertising campaign launched in 1970 by Perdue in the New York City area remains one of the most successful initiatives in marketing history. The innovative advertisements featured Frank Perdue repeating, in many different situations, the company slogan, "It takes a tough man to make tender chicken." The notion that the firm's president personally monitored the quality of its chickens was the theme that permeated the campaign. Humorous advertisements also played on chicken's perceived virtue as a low-fat product. One depicted two overweight customers eyeing the red meat freezer as Perdue cried out, "Come on folks, shape up! Start eating my chickens." The advertisements successfully established Perdue broilers as a distinctive product with highly positive results for the firm. In fifteen years from the commencement of the advertising blitz, Perdue's output increased sixfold as it rose from twelfth to fourth among America's poultry companies.[26]

Much of the branding strategy rested on the bird's allegedly superior yellow hue. "My chickens always have that healthy, golden-yellow color," Perdue declared in one advertisement. The color had little to do with actual taste; instead, it was an indicator of quality, that the chickens "eat better than people do." Yellow had been a preferred color for chicken meat since the turn of the century, and Perdue admitted he got the idea from Maine processors who produced yellow chickens and "got a premium of three cents a pound." He tried for the same advantage by added xanthophyll to chicken feed through natural additives such as marigold petal extract. Whether there was any difference in taste is disputable, but that was not the point or impact of Perdue's approach. He tapped into the way consumers rely on visual and odor signifiers to evaluate meat's quality. The yellow in chickens was for consumers like the red in freshly cut beef or cooked bacon; an ineffable quality testifying, in some intrinsic way, to the product's natural wholesomeness and value.[27]

Perdue also systematized marketing branded chicken parts that appealed to different markets. Plant-packaged cut-up birds, once known as "three-legged fryers" and "double-breasted chickens," accounted for less than 10 percent of the market in 1962. Perdue expanded these types of products in the 1970s and 1980s, priced to appeal to consumers at different economic levels, and packaged so that couples as well as large families could obtain a convenient amount. Consumers were able to buy packs of legs or thighs for low prices or breasts (sometimes boned and skinned) at higher prices. Other firms (including Tyson) emulated Perdue's branding strategy, so by the 1990s a consumer approaching a supermarket's meat department could find almost as much variety among chicken as beef. Cut-up and "value-added chicken" (including boneless parts, nuggets, hot dogs, and patties) comprised 86 percent of chicken sales in 1995. That same year whole birds accounted for only 14 percent of the market, a long drop from 1962, when they accounted for 83 percent. During the same period annual broiler production grew from 1.8 to 7 billion and per capita chicken consumption topped beef and pork. Through product diversification and branding, chicken had finally become a meat, and American's most popular one at that.[28]

Changing the form of chicken meat had drastic consequences on production methods and the chicken processing workforce. Plants became larger, production processes more complex, with industry employment doubling between 1975 and 1990, and then reaching 200,000 in 1995. Unions that had once held a foothold lost legal certification as older plants closed or changed hands. Segmenting the chicken into boneless breasts, patties, and McNuggets entailed adding extensive cutting, boning, and processing operations that remained labor intensive, for despite the best efforts of firms, the chicken remained an irregularly shaped natural product.

Managers of mid-1990s poultry processing plants usually emphasized to visitors the modernity of processing methods. As has been the case in meatpacking since the "modern" production methods of Cincinnati's antebellum pork plants, "modern" is a commentary on innovations in meat-processing equipment, rather than a comparative judgment with other industries. In the plants of the 1990s, processing more than 100,000 chickens daily, workers still stand shoulder to shoulder performing necessary tasks with their hands and small tools.

Major technological change took place after 1970 in the killing and eviscerating operations. Establishing a detailed division of labor in the 1950s and 1960s subsequently facilitated mechanization as inventive equipment suppliers and company engineers developed ways to automate repetitive hand labor. In doing so chicken processors had an advantage over their counterparts in the red meat industry. The smaller size of the animal permitted greater use of machinery than with the far larger and more irregular cattle and pigs.

Chickens entered 1990s plant in bins unloaded from trucks, brought from farms where they were caught during the night. After men hung the chickens by their feet on a moving line they were stunned by traveling through a shallow salt water bath charged with electricity. A deheading machine cut through the neck, allowing the chicken to bleed out while traveling through the blood tunnel. The killing room was saturated in red light, as birds could not see that spectrum and hence were blind to what was coming. Vacuum fans designed to suck up dust and feathers filled the area with a loud din.

The bled carcasses went through a scalding bath and a series of "rubber finger" defeathering machines followed by a flame-scalder that burned off remaining feathers. At this point the chickens' feet were cut off (later packed and shipped to China) and the chickens tumbled down a chute to workers laboring furiously to rehang them by their wings.

Once back on the moving chain, the headless and feetless chickens move into the evisceration room, where dramatic technological improvement had taken place. Chickens were gutted by a machine with twisting, piston-like plungers entered the animal's cavity and extracted intestines and internal organs. After examination by a government inspector, a worker positioned the carcass so that a machine could cut off the intestines and organs. Another worker, using her hands, separated the hearts, livers, and gizzards into separate receptacles, later recombined and inserted (wraped in paper) back into whole chickens. As the chicken continued along the chain a worker used a suction devise resembling a turkey baster to pull out the lungs and any remaining viscera before it went into the chill bath to cool for four hours with thousands of other carcasses. An evisceration department that once required dozens of workers was staffed by only a handful in the mid-1990s.

Once the body temperature reached 34° F in the chill bath the carcass moved into the cutting operation. Inspectors diverted the best chickens directly into the packaging, where they were wrapped and sold whole. For the chickens destined to be marketed in cut-up or further processed form, a great deal more labor was required. As these are the forms of chicken meat that dominate the late-twentieth-century-chicken market, the cutting and subsequent boning operations were generally post-1980 additions to poultry processing.

In the cutting room late-twentieth-century mechanization took the form of automatic conveyer belts moving chicken from one workstation to the next and machines that performed highly specific cutting operations. This process deskilled labor by allowing machines to assume actual cutting tasks. Nonetheless, the unique capacity of the human eye and hand to work in concert to properly control the irregularly shaped chicken pieces rendered abundant labor still necessary for these operations. Machines severed carcasses into fore and aft sections, as well as into quarters and legs. Workers positioned the chicken for these operations by carefully placing it on the conveyer belt that fed into the machine performing the cut. As the parts fell back onto the conveyer they were whisked to more workers, who readied the pieces for the next cutting operation. The spectacles seemed archaic, hardly modern, as workers standing should to shoulder along the production line touched and moved chicken every few feet, and at times trimming the carcass with hand-held knives to ensure that the machines performed their tasks properly.

The truly labor intensive stages followed—boning and packaging chicken parts. Breasts went to the table boning operation, where women used sharp sheers to cut the meat from bone and remove the skin. Chicken thighs moved into a boning room from the cutting line. Thigh boning machinery automatically pushed the bone out of the piece with a slowly moving piston, but to work properly it depended on a worker to insert the thighs by hand into two dozen slots. The boned parts were then dumped onto a table surrounded by workers who used sharp scissors to remove fat and skin. Similar to the breast boning area workers labored shoulder to shoulder, closely observed by supervisors.

The boned and unboned parts met again along the packing line. Brought there by conveyer belts or carried by hand trucks, dozens of workers pack the pieces by hand into trays traveling along moving belts. The containers were weighed, labeled, and sealed by automatic machines, and then packed into large containers for movement by powered hand trucks into the dry chill area in preparation for shipping. Whenever possible the various forms of wrapped chicken were priced in the processing plant, so that work need not be performed by supermarket clerks. Chicken companies that produced "house" brands for supermarkets labeled and priced items right there, if necessary creating separate streams for separate retail outlets even though the chickens were virtually the same.

One hundred thousand chickens per day flooded through this complex production process at line speeds of up to ninety birds per minute. The hanging, evisceration, and cutting jobs were especially relentless as the chickens arrived at workstations at such short intervals that there were few opportunities for breaks. A 1989 study reported that each "drawhand" along the evisceration line "pulled, twisted, and placed viscera of chickens in excess of 10,000 times per shift." The same report noted that deboners could handle up to twelve thousand breasts in an eight-hour period. Handling chickens at such a rate caused many abrasions and cuts, and too often produced repetitive motion injuries such as carpal tunnel syndrome. In 1989 chicken processing ranked second in the nation in repetitive motion injuries—only behind the red meat industry.[29]

While chicken processing labor generally required little skill, the need for hand labor remains staggeringly high in the many positioning and transitional stages of the dismembering, cutting, and packaging operations. "If you see one person processing four pounds on the eviscerating line," explained an industry spokesperson, "you'll see four people processing that same four pounds in further processing." The increasingly important deboning operations remain the province of relatively skilled workers who used sharp hand tools rather than machines. Despite impressive advances in mechanization of cutting and packaging operations, the modern poultry processing plant bears little resemblance to the highly automated "flow" operations in foods like hot dogs.[30]

Complicating accelerating employment needs was the rural locations of chicken processing operations, limiting access to urban labor markets. Consequently, the poultry workforce's composition changed drastically in the late 1980s as immigrants, largely from Mexico and Central America, streamed into the poultry plants, replacing many of the black workers who had dominated in these jobs since the early 1960s.

The workforce's transformation was closely connected to the new production requirements. These onerous jobs, paying about 60 percent of the average wage for American manufacturing since the mid-1960s, drew workers at the bottom rungs of the American labor market. Regardless of the personal inclinations of employers, the highly competitive industry and labor intensive process made paying wages higher than the industry average a prescription for bankruptcy. A high-wage strategy similar to that of the unionized mid-twentieth-century red meat industry was only feasible if all leading firms had to bear roughly similar labor costs. In the absence of unions commanding a majority of poultry workers and able to raise wages across the board, firms followed a low-wage strategy and accepted turnover rates as high as 100 percent annually.[31]

These dynamics meant that the African American workers who had comprised the industry's principal labor force in the 1960s were largely supplanted by the mid-1990s by immigrant labor. While African Americans already working in poultry jobs were likely to remain (especially if they held better paid jobs) younger blacks generally looked elsewhere for employment. In part this reflected the superior options available in service jobs for African Americans who had better language and literacy skills than the immigrants. On the Delmarva peninsula, for example, young African Americans often preferred to enter the thriving coastal tourist industry, just a few miles from the interior chicken plants. But employment of immigrant labor also reflected a widely held judgment among employers that they were better workers. "Our experience is that the Hispanics are very conscientious and grateful for nice jobs," explained one plant manager in the late 1980s. "We have problems with blacks," complained another, "30–40 percent do not care if they work or not." These comments inadvertently admit that what a Guatemalan immigrant might consider a "nice job" simply was not appealing to young African Americans seeking a better future.[32]

High-volume processing posed other problems for poultry operators. Charges of contaminated chicken bedeviled the industry throughout its rise, as the news media and watchdog groups repeatedly found evidence of bacterial contamination on chickens. As early as the late 1960s an unpublished study by the Delaware Extension Service found salmonella contamination rates "as high as 90 to 100% of the dressed broiler carcasses on any given processing day." With almost cyclical frequency in the 1980s and 1990s newspapers, magazines, and television shows ran exposes with titles like "Is Chicken Safe to Eat?" documenting high levels of bacterial contamination. Requirements adopted in the 1990s that firms include safe handling guidelines on packaged chicken reflected an acceptance by government regulators and the industry that contamination was an endemic problem.[33]

Contamination stemmed from the very same production methods that had made chicken an inexpensive and popular meal and vastly expanded consumption. The small size of chicken meant that mass-production processing methods tended to mingle the carcasses together and hence aid and abet contamination. Bacteria could most easily spread in the feather removal operations, which left a great deal of dust and feathers in the air that could move from one bird to the next, and in the chill bath, where one bird could infect the mixture and contaminate other chickens.

Salmonella may well have been widespread in red meat processing, but the different nature of processing operations minimized any impact. Cattle and hogs are not intermingled during processing as much as chicken because of their greater size, and their skin (where contamination begins) is removed and not eaten, quite unlike chicken skin. While processors could turn chicken into a meat in the eyes of consumers, it remained relatively a very small animal, and hence susceptible to a different array of health dangers in processing operations.

Exposes of salmonella and other forms of bacterial contamination such as campylobacter tended to downplay consumers' traditional method for ensuring wholesome food: cooking it thoroughly. Despite the sensational pretenses of these stories (and the well-documented incidence of bacterial contamination of chicken), consumption continued to climb. Between 1990 and 1995, when stories on chicken contamination saturated the news media, per capital consumption increased almost 20 percent to 75 pounds annually. Consumers may have noticed the stories and a few stopped eating chicken, but there was no appreciable statistical impact on the rate of increase of chicken consumption. The vast majority responded by cooking with more care rather than turn to alternative sources of protein.

Conclusion

Georgetown, Delaware, host of the 1948 "Del-Mar-Va Chicken of Tomorrow Festival," was a very different town fifty years later. Almost half its residents were recent immigrants, the largest group men and women of Indian descent who are legal refugees from the Guatemalan civil war. A huge Perdue plant on the town's edge employs over a thousand workers, more than 50 percent immigrant. Since J. Frank Gordy brought the festival to Georgetown the Delmarva industry had grown to employ close to fifteen thousand people, relying on chicken raised on more than two thousand farms to produce over twelve million broilers each week. And even with these impressive numbers the region ranked only fifth nationally in poultry production.

The rise of the postwar poultry industry represents an unparalleled shift of American food consumption practices. For chicken to move from a food for special occasions to an item eaten several times weekly took extraordinary initiatives to radically transform the animal itself and the way it was processed for consumption. In doing so chicken became more of a meat rather than a variant of poultry.

These changes took place all along the chicken-producing axis. Close supervision of farming practices by the integrated firms and intense research by universities into feeding methods utterly transformed the tempo of chicken raising. In 1923 Mrs. Steele's chickens took sixteen weeks to reach 2.2 pounds, and had a feed conversion ratio (the amount of feed to increase weight by one pound) of 4.7. In 1993 broilers took 6.5 weeks to reach 4.4 pounds, with feed conversion ratio of 1.9.[34]

Once this "Chicken of Tomorrow" left the farm it entered an equally transformed business and processing environment. Displacing the congeries of small enterprises from hatchery to farmer to trucker to processor to commission agent were large integrated corporations internalizing these varied transactions. Replacing the crude feather-picking and cleaning of turn-of-the-century operations were factories interwoven with conveyer plants and bristling with machinery that struggled (not entirely successfully) to automate hand labor.

Despite consumers' worries about chicken's wholesomeness (and widespread concern over the industry's labor practices) the popularity of chicken meat rose steadily throughout the postwar era. Chicken's attractive price relative to other meat, already a factor in consumption by 1960s, was even more pronounced thirty years later. At ninety cents per pound in 1990, whole chickens were less expensive than any other type of meat on the market, and fully 50 percent cheaper than ground chuck beef. Chicken breasts, while more expensive at $2.07 per pound, were still cheaper that hot dogs and stewing beef. Encouraged by these advantageous prices and the variety of chicken products available in grocery stores and eating establishments, chicken consumption reached seventy pounds per capita that year, edging past beef for the first time.[35]

Inexorably and almost imperceptibly chicken changed from a poultry to a meat product. Chicken's transition from poultry to meat was so dramatic and successful that pork producers even sought to improve sales, claiming pork was the "other white meat." It is a testimonial to the success of chicken promoters that rather than pork remaining a companion to beef as it had been through most of American history, twenty-first-century pork aspired to be like the chicken.

Notes

1. *The Chicken of Tomorrow*, video, produced by Cooperative Extension Service, University of Delaware, 1948.
2. Karl C. Seeger, A. E. Tomhave, and H. L. Shrader, "The Results of the Chicken-of-Tomorrow 1948 National Contest," *University of Delaware Agricultural Experiment Station Miscellaneous Publication No. 65* (Newark: University of Delaware, July 1948). Gordy notes that within two decades "more than half of all the broilers raised in the United States resulted from Vantress male breeding." J. Frank Gordy, "Broilers," *American Poultry History, 1823–1973* (Lafayette, Ind.: American Poultry Historical Society, 1974), 400 (hereafter for *UDAES*, all citations are Newark: University of Delaware).
3. Gordy, "Broilers," 373. Lewis Wright, *The Practical Poultry Keeper* (New York: Orange, Judd and Company, 1867), 51–54.
4. F. A. Buechel, "Wholesale Marketing of Live Poultry in New York City," *U.S. Department of Agriculture Technical Bulletin No. 107* (Washington, D.C.: USDA, May 1929). On kosher killing methods see Hugh Johnson, "The Broiler Industry in Delaware," *UDAES Bulletin No. 250* (October 1944): 46–47.
5. Temporary National Economic Committee, *Investigation of Concentration of Economic Power*, Part 7, Hearings March 9–11, May 1–2, 1939 (Washington, D.C.: U.S. Government Printing Office, 1939) 76th Congress, First session, 2867–2871; R. O. Bausman, "An Economic Survey of the Broiler Industry in Delaware," *UDAES Bulletin No. 242* (March 1943); 51. Johnson, "Broiler Industry," *UDAES Bulletin No. 250*, 45–46.
6. Barker Poultry Equipment, *The Barker Catalogue No. 44* (Ottumwa, Iowa: BPE, 1946).
7. Kimberly R. Sebold, "The Delmarva Poultry Industry and World War II: A Case in Wartime Economy," *Delaware History* 25 (1993): 200–214.
8. Survey of extension activity in Delaware can be found in M. M. Daugherty, "A Short History of the Broiler Industry in Delaware," *UDAES Pamphlet No. 15* (July 1944): 3–5.
9. W. T. McAllister, "An Appraisal of Marketing Problems in the Delmarva Broiler Area," 1954, Willard McAllister papers, Hagley Museum and Library, Greenville, Delaware.
10. W. T. McAllister and R. O. Bausman, "The Retail Marketing of Frying Chickens in Philadelphia," *UDAES Bulletin 275* (July 1948). W. T. McAllister et al., "Consumer Preference for Frying Chickens Studies," *American Egg and Poultry Review* (December 1950); "Appraisal of Marketing Problems," 1954, McAllister papers.
11. W. T. McAllister, "Mr. Poultryman: Marketing is Your Business," UDAES Extension Service Bulletin 56, (August 1951). For more detailed advice, see for example, Silas McHenry, "Grow Better Broilers," n.d. UDAES pamphlet; R. O. Bausman and R. J. McMillan, "Costs and Returns in Producing Broilers in Delaware," *USAES Pamphlet 27* (January 1947); and Frank D. Hansing, "Broiler Costs and Returns in Lower Delaware, 1952–1955," *UDAES Bulletin 327* (February 1959). The Extension Service also published a periodic "Timely Notes for Broiler Growers." See Sean Quimby, "The Role of D.P.I.A. in the Lives of Delaware Poultrymen During the 1950s," 1993, unpublished paper in author's possession, and "Annual Report, Extension Poultry Specialists," January 1, 1953–December 31, 1953; Edward H.

Schabinger, New Castle County Extension Service Agent, to County Poultrymen, June 17, 1952, both in Cooperative Extension Service Papers, University of Delaware archives.

12. Don Palmer interview by Roger Horowitz, January 19, 1993. W. C. Evans and R. C. Smith, "The Daily Spread in Prices Among Broiler Flocks Sold on the Eastern Shore Poultry Growers' Exchange," *UDAES Bulletin 330* (January 1960).

13. Evans and Smith, "Daily Spread in Prices;" Gordy from "Annual Report, Extension Poultry Specialists," January 1, 1953–December 31, 1953, Cooperative Extension Service papers, University of Delaware archives. See also University of Delaware, School of Agriculture, *A Situation Report on the Delmarva Broiler Industry* (Newark, Del.: n.d.), 15–23.

14. Interest rate estimates from R. O. Bausman, "An Economic Survey of the Broiler Industry in Delaware," *UDAES Bulletin No. 242* (March 1943): 49–50. W. T. McAllister, "Opportunities and Possibilities for Improving the Economic Position of Delmarva's Broiler Firms," unpublished typescript, n.d., 2, McAllister Papers, Hagley; Idem., *Poultry Handbook* (Newark, Del.: UDAES, n.d); Idem., "The Growth of the Broiler Industry," in College of Agricultural Sciences, *1975 Annual Report* (Newark: Univ. of Delaware, 1976), 8; Silas McHenry, "Grow Better Broilers" (Newark, Del.: Cooperative Extension Service, n.d.) Cooperative Extension Service Papers, University of Delaware archives; S. T. Rice, "Short-Term Financing of Commercial Broiler Production," *Journal of Farm Economics* 33 (May 1951): 248–251; Frank D. Hansing, "Financing the Production of Broilers in Lower Delaware," *UDAES Bulletin 322* (October 1957).

15. Quotes from, respectively, Willard T. McAllister, "A Plan of Action for the Delmarva Poultry Industry" (n.d. but c. 1959), and "Opportunities and Possibilities," McAllister Papers, UD.

16. Edward Covell, Jr., interview by Roger Horowitz, February 1, 1995.

17. Perdue's father had operated an egg hatchery since 1925 and had started contracting with growers to raise its chickens in 1950. Acquisition of a feed mill in 1958 completed the first phase of integrated broiler production. See *Mid-Atlantic Poultry Farmer* (August 25, 1992): Bob Hall, "Chicken Empires," *Southern Exposure* 17, no. 2 (Summer 1989): 15–16; Frank Gordy, *A Solid Foundation . . . The Life and Times of Arthur W. Perdue* (Salisbury, Md.: Perdue, 1976).

18. George B. Rogers and Edwin T. Bardwell, "Marketing New England Poultry: 2. Economies of Scale in Chicken Processing," *Station Bulletin 459* (University of New Hampshire: Agricultural Experiment Station, April 1959).

19. U. S. Department of Commerce, *Statistical Abstract of the United States, 1961* (Washington D. C.: GPO, 1961), 339.

20. Richard Saunders, "Socio-Psycho-Economic Differences Between High and Low Level Users of Chicken," Maine Agricultural Experiment Station, A.E. Progress Report 2 (November 1960).

21. R. C. Smith, "Factors Affecting Consumer Purchases of Frying Chickens," *UDAES Bulletin 298* (Technical) (July 1953): 5.

22. Saunders, "Socio-Psycho-Economic Differences," 34.

23. Saunders, "Socio-Psycho-Economic Differences," 32, 8.

24. Stephen F. Strausberg, *From Hills and Hollers: Rise of the Poultry Industry in Arkansas* (Fayetteville: Arkansas Agricultural Experiment Station, 1995); Douglas Frantz, "How Tyson Became the Chicken King," *New York Times*, August 28, 1994, Section 3.

25. Christian McAdams, "Frank Perdue Is Chicken," *Esquire* (April 1973): 116.

26. McAdams, "Frank Perdue," 114.

27. Thomas Whiteside, "C.E.O., TV," *New Yorker* (July 6, 1987): 39, 46.

28. Strausberg, *From Hills and Holers*, 93, data from Delmarva Poultry Industry.

29. David Griffith, *Jones's Minimal: Low-Wage Labor in the United States* (Albany: State University of New York Press, 1993), 177.

30. Griffith, *Jones's Minimal*, 102.

31. U.S. Department of Labor, *Employment, Hours, and Earnings—United States, 1990–1995*, Bureau of Labor Statistics Bulletin 2465 (Washington, D.C.: GPO, September 1995), 128; U.S. Department of Labor, *Employment, Hours, and Earnings—United States, 1981–93*, Bureau of Labor Statistics Bulletin 2429 (Washington, D.C.: GPO, August 1993), 204–205. U.S. Department of Labor, *Employment, Hours, and Earnings—United States, 1909–90*, Vol. I, Bureau of Labor Statistics Bulletin 2370 (Washington, D.C.: GPO, March 1991), 423–424. U.S. Department of Labor, *Employment and Wages, Annual Averages, 1993*, Bureau of Labor Statistics Bulletin 2449 (Washington D.C.: GPO, October 1994), 1–254.

32. Griffth, *Jones's Minimal*, 156.
33. "Relationship of Processing and Marketing Practices to the Incidence of *Salmonella* on Ready to Cook Broiler Chickens," August 24, 1970, Box 104, Cooperative Extension Service Records, University of Delaware Archives; "Ruling the Roost," *Southern Exposure* (Summer 1989). "Risky Business: Arkansas' Poultry Empire," *Arkansas Democrat* (April 21–25, 1991). "Chicken: How Safe?" *Atlanta Constitution* (May 26 and June 2, 1991).
34. William H. Williams, *Delmarva's Chicken Industry: 75 Years of Progress* (Georgetown, Del.: Delamarva Poultry Industry, 1998), 73.
35. U.S. Department of Agriculture, *Food Consumption, Prices, and Expenditures, 1970–93*, Statistical Bulletin 915 (Washington, D.C.: GPO, 1994); American Meat Institute, *Meatfacts 1991* (Washington, D.C.: American Meat Institute, 1992), 41–43.

Hogs, Antibiotics, and the Industrial Environments of Postwar Agriculture

MARK R. FINLAY

In the mid-1950s, pharmaceutical giant Charles Pfizer and Company launched a new advertising campaign that encapsulated the industrial and pharmaceutical revolution it helped bring to American farming. "Science Comes to the Farm in a Feed Bag," Pfizer boasted in a series of full-color advertisements that stood out among the typical fare of the farm press. The main image in this campaign featured a scientist in a white lab coat, with a bag of feed over his shoulder, appearing as an ominous god amid the clouds. In the foreground stands a farmer, small in stature, who faces the giant deity of medicated feed. The text, taken from an address by Pfizer's president John McKeen, promised that "every" American "livestock producer" (the term itself was a new bit of jargon) could look to feed companies for methods that would set new production records "each" and "every" year. See Fig. 9.1[1]

In actuality, huge, urban-based corporations and industries may not have brought opportunity and profitability to every farmer each and every year. However, in a remarkably brief period after the end of World War II, they did contribute to a series of transformations in the infrastructure of American agriculture that some have identified as the "Second Agricultural Revolution."[2] These changes were especially apparent in livestock production, as farmers and the industrialists who participated in their enterprise sought to reshape and redesign organisms in ways that they deemed appropriate for an industrial society. Innovations with medicated feeds, manufactured housing, and redesigned landscapes spurred farmers to increase the size and capital investment of their livestock operations, to manipulate the natural rhythms of animals' breeding, birth, weaning, rebreeding, and slaughter, and to conduct the business in ever more confined, streamlined, and centralized operations. These changes impacted the broiler (chicken) industry most intensely, but also reshaped turkeys into a smaller size and temperament suitable for a life indoors, moved cattle-finishing operations onto industrial-scale feedlots, and moved hogs from pastures to indoor confinement operations.[3]

This essay focuses on the latter case, or the impact that industrializing strategies brought to the realm of swine production. Agricultural and industrial

Fig. 9.1 "Science Comes to the Farm in a Feed Bag," part of the Charles Pfizer & Co., Inc.'s advertising campaign promoting antibiotics and other growth-promoting ingredients in animal feeds in the mid-1950s. From *Farm Journal*, 80 (February 1956): 67. Permission from Pfizer.

leaders created new forms of the pig that grew more uniformly, more consistently, and more predictably. Farmers sought and found methods to reduce their dependence on relatively expensive and inefficient human labor by transferring labor inputs onto the hog itself. By altering the feeding, housing, and management issues associated with the hog, farmers essentially embraced the role of industrial managers who focused less on animal husbandry and more on controlling labor and energy inputs. As Edmund Russell suggests in his introduction to this volume, farmers increasingly treated hogs as "workers" to be brought into an industrial system, with the time and places for their eating, resting, and socializing increasingly under central control. In all, the changes compressed the time, space, labor, and energy associated with hog production along the model of an efficient industry.

The case of Damon V. Catron, a professor of animal husbandry at Iowa State University, fits neatly into this complex transition toward the industrial hog organism. Catron was born in rural Indiana and, according to one newspaper account, had earned state and national honors as a "pig club boy" in 4-H

Club competitions. After earning degrees at Purdue University and the University of Illinois, he spent a year in the private sector working with Ralston-Purina Company at a time when it pushed for confinement and integration in the poultry industry. He came to Iowa State in 1945; by the early 1950s he had become one of the most prominent swine nutritionists in the country. During that decade, Catron was at the center of every significant development in American hog production, including year-round breeding, medicated feeding, and the emergence of the confined, artificial environments that have reshaped the rural landscape. By 1960, Catron's vision of "assembly-line" hog production had become embedded into the infrastructure of America's industrialized agriculture. Significantly, Catron left academia in 1960 for a position with a manufactured feed company.[4] Although many others also played important roles in the history of the postwar pig, Catron's fifteen-year career at Iowa State provides a useful framework for examining this crucial period of change.

Efforts to industrialize American farming had a long history. A 1916 textbook, for instance, explained that the farmer needed to think of himself as a "manufacturer," for he too converted raw materials into valuable finished goods. Through greater attention to the science of nutrition, the technology of producing one's own quality feeds, and the lessons of sound business practice, the author anticipated that livestock farmers, and hog farmers in particular, could adapt particularly well to changing conditions in rural America.[5]

Still, a number of natural and ecological circumstances limited the applicability of these values to early twentieth-century American hog production. Hogs resisted industrialization because the swine industry followed both the natural rhythms of the animal and the seasonal conditions of Corn Belt America. Farmers typically bred their hogs in the fall so that piglets would be "farrowed," or delivered, in the spring. In the summer, they allowed young swine to feed on pastures of legumes and other fodder crops as well as from corncribs left in the field from the previous autumn harvest. In the fall, pigs ate newly harvested corn, "hogged down" the stubble of grain crops, and "hogged up" the remains of sweet potatoes, artichokes, and similar crops. Fattened hogs could not maintain their weight efficiently through the winter months, so virtually all Corn Belt farmers "finished" their market hogs in late fall, causing an annual glut that overwhelmed meat packers and routinely depressed prices each December and January.[6] The converse of this simple economic equation—relatively short supplies and high meat prices during the balance of the year—proved a frustration for both meat packers and consumers.

Throughout this cycle of hog production, farms operated with relatively few connections to the world of industry and manufacturing. Farmers raised virtually all of their own feeds, for textbooks advised them "to be relatively independent of the feed manufacturers" whose products were often either undesirable, uneconomical, or both.[7] Commercially manufactured housing for hogs was virtually nonexistent; farmers typically built their own feed troughs, farrowing sheds, and corncribs. Animals lived most of their lives in the natural environment, with little more than homemade hog houses and portable sunshades to protect them from the elements. And since many farmers approached hog raising as a part of general farm practice connected intimately with pastures and grain crops, the enterprise remained inherently unspecialized and nonindustrial. As long as farmers treated hogs largely as a means for disposing of a portion of their corn crop, farm byproducts, and wastes, few farmers specialized or invested in hog production.[8]

Farmers also struggled with seemingly intractable problems that made hog production one of the most troublesome sectors of the typical agricultural enterprise. A basic predicament was that 25 to 33 percent of the pigs farrowed died before they were weaned. Most litters had one or more runt pigs that did not reach the sow's milk. Not only did runt pigs fail to reach the standard size, which deterred uniform feeding regimens and efficient marketing, they also cost farmers considerable effort in their often unsuccessful attempts to nurse them along individually. Many others died due to the behavior of careless sows that had an alarming tendency to roll over and crush their baby piglets. Disease linked with poor sanitation caused further losses for hog farmers. In response, reformers promoted the "McLean County System," a scheme that recommended the continual rotation of animals through "clean" and parasite-free pastures. Pastures, however, presented a tremendous bottleneck that precluded an industrialized system. Most important, pastures pulled fields suitable for cash crops out of production. This handicap became especially apparent after soybeans became more common in the 1930s and thereafter, since soybeans offered the twin advantages of being both a suitable feed grain crop and a legume that restored nitrogen to the soil. Moreover, pasture maintenance was a tremendous burden upon farm laborers who had to set up moveable fences, devise portable watering and feeding systems, construct temporary shelters, and the like, which all limited the number of animals they could handle.[9]

The natural rhythms that affected the American hog producer were disturbed further by the chaotic market conditions of the Depression and World War II. Policymakers addressed these two problems with opposite policies— the slaughter of some five to six million young pigs in 1934, and the rapid expansion of hog production during World War II. Both cases demonstrated the government's expanding role in an arena once left to independent hog producers. Once policymakers began to administer a system of minimum and maximum prices for corn, hogs, and other commodities, however, it was very difficult for the price of corn and the price of hogs ("the corn-hog price ratio") to stay in balance.[10] During World War II, secretary of Agriculture Claude Wickard aggressively encouraged hog production through price guarantees, and American hog production hit new records in each year from 1941 to 1945. As a consequence, however, supplies of feed grains fell sharply. This compromised plans to use corn and other grains for production of industrial alcohol, synthetic rubber, and other wartime needs.

Shortages of cottonseed meal, fish meal, and other protein supplements further challenged hog producers. In 1943, in the face of record worldwide demand for the American food products, the USDA launched a voluntary Protein Conservation Program that encouraged the use of vegetable proteins as an alternative to animal proteins.[11] This crisis turned out to be the issue that launched Damon Catron's research career in the realm of swine nutrition and evidently sparked his interest in maximizing feed efficiency.[12] In all, the "feed famine" and other lessons of World War II taught hog farmers that they were increasingly dependent upon commercial feed supplies and international events beyond their control, and that pressures of industrialization were sure to impact the typical hog farmer. As other historians have noted, the war removed the "conceptual blocks" that had limited the application of scientific and industrial innovations to agriculture, which encouraged the adaptation of "scientific" values as appropriate for nearly all social problems.[13]

Circumstances in the agricultural labor market also pushed the redesign of the American hog enterprise along industrial lines. World War II brought intense farm labor shortages; the situation did not improve significantly following the demobilization of the armed forces.[14] In 1943 researchers at Purdue University founded the Work Simplification Laboratory, an endeavor that aimed to reduce human physical labor on the farm. Specialists in hog management used stopwatches, video cameras, and balls of string tied to the farmer to evaluate the time and effort "wasted" in climbing over fences, opening and closing doors, and hauling feed and water. In reports that purported to apply the science of human dynamics and the practices of industrial management to farm operations, researchers claimed farmers could easily save one-fourth to one-half of their labor. Significantly, the work simplification specialists strongly promoted the relatively innovative notion of multiple farrowing—their tables demonstrated that a second litter annually was worth the investment in time and energy.[15]

Stagnant demand for pork products and byproducts created yet another incentive for hog producers to streamline their industry and view their resources as commodities. Traditionally, American meat packers and butchers utilized hogs as much for their byproducts as for their meat. Most pork meat products were not sold fresh but processed through salting or curing. Farmers took hogs to market on rather indeterminate schedules and at a wide range of weights. At times when the corn-hog price ratio was favorable, as in World War II, price supports encouraged farmers to fatten their animals indefinitely, and they often sent a three-hundred-pound porker to market. This approach faced challenges in the postwar period, when new vegetable oils and detergents weakened consumer demand for animal lard, and postwar consumers found the consistent supplies, prices, and quality of chicken, turkey, and frozen fish preferable to pork products. Invented not by "the producer or processor but rather by the housewife," the "meat-type" hog emerged as the new standard, a leaner animal that could be marketed routinely at about 220 pounds. With commercial feeds replacing homegrown grains and scraps in hog rations, it became easier for farmers to raise their hogs consistently and reliably to the ideal of this standardized and industrial organism.[16]

Domestic and global politics provided yet another incentive to industrialize hog production methods. Erratic swings in meat supplies and meat prices proved a continual thorn for the Truman administration, particularly as a "deep . . . meat famine" and threats of a producers' strike appeared on the eve of the 1946 elections. American policymakers also embraced the use of farm products as a component in their suddenly internationalist agenda in foreign affairs. Improving agricultural production became part of what one historian has labeled the "technocratic front in the Cold War" as political and industrial leaders urged farmers to take stock of their role in the global crises. These geopolitical considerations reached the local level when farm conventions and journals began delivering the messages that Soviet spending on industrialization could outpace Americans', and that farmers needed to be vigilant of the possibility of enemies striking the nation's food supplies through bioterrorism.[17]

Journalists also reported that the demographic pressures of the baby boom generation, combined with America's suddenly expanding role in world affairs, were placing new demands on farm production. A 1946 article in *American Magazine*, for example, described the techniques of "genetic engineering," "redesigning nature," and "custom-built livestock" being developed at the USDA research facilities in Beltsville, Maryland. Its author, Frederick Brownell, also recognized the political implications of these new foods. It was time for Americans to repay other nations that helped win the war, he argued, and to fulfill Franklin Roosevelt's promise to provide "Freedom from Want."[18] Similar rhetoric appeared in Maxwell Reid Grant's 1949 essay "Engineering Better Meat" published in *Popular Mechanics*. Employing the industrialists' and engineers' language of "blueprints," "specifications," and "model years," Grant explained that "nature needs help" if Americans were going to rebuild Germany and "feed a hungry world."[19] The sum of these numerous background issues led to a fairly self-evident conclusion for the farmers and industrialists involved in swine production: they had a mandate to reshape both the hog industry and the hog animal in ways appropriate for the postwar era.

It was in this context that Damon Catron of Iowa State began to articulate his interest in industrializing agriculture. In an address delivered in the fall of 1946, Catron explained that Iowa's hog farmers needed to buy "a system of feeding" rather than just a bag of feed. He linked that to changes in the hog organism, for one could not have "poor machinery to put the raw product [or feed] through." Yet his speeches that autumn did not reveal an easy way to reach that goal. In this era, animal husbandry experts typically focused their research on three branches of their science: sanitation, breeding, and nutrition. But influenced by his University of Illinois colleagues who called for a "new fundamental philosophy" in the interpretation of swine feeding, Catron's lectures urged further research on ideal ratios of nutrients, such as fats, proteins, carbohydrates, minerals, and vitamins, a departure from the previous emphasis on searching for an ideal ratio of ingredients, such as corn, alfalfa, milk, and the like.[20] Yet Catron also admitted that his scientific colleagues did not have answers for many of the fundamental problems that Corn Belt hog farmers faced. The solution, Catron suggested, fell under the final theme of his speech, "The Need for College-Industry Cooperation."

Although swine industry observers readily recognized the many incentives to alter hog production practices, few imagined that pharmaceuticals would be at the center of dramatic change. Then antibiotics burst suddenly upon the scene. Led by Thomas Hughes Jukes and E. L. R. Stokstad of the American Cyanamid Company's Lederle Laboratories, researchers discovered in the late 1940s what came to be known as the "antibiotic growth effect"—that the feeding of antibiotics at low levels to agricultural animals resulted in enhanced growth.[21] This research had its origin in a twenty-year search for the animal protein factor (APF), something that seemed to be a combination of vitamin-like substances that could raise the efficacy of feeding animals cheaper plant proteins to the level of commonly used animal proteins. Indeed, Catron and others already had worked with the Lederle laboratory in establishing that farm animals did not gain weight at satisfactory rates when fed vegetable proteins alone. There was something important in animal byproducts, yet whatever it was remained undetected. For a time, vitamin B_{12} seemed to be the mysterious APF factor, and manufacturers at Lederle, Pfizer, and other major pharmaceutical firms produced feeds fortified with Vitamin B_{12} for the 1949 season. Searching for a method that would permit production of the new vitamin on an industrial scale, manufacturers found a ready supply in residues from the fermentation processes used in the production of the antibiotic wonder drugs aureomycin and streptomycin.[22]

Meanwhile, researchers began to notice connections between these common antibiotics and animal growth.[23] Stokstad and Jukes found that vitamin B_{12} produced from the antibiotic organism yielded more rapid growth in chicks than a pure vitamin extracted from liver. They first presumed that the vitamin was the principal cause of the weight gain, but then deduced that the antibiotic alone might be beneficial to the chicks' growth. A second test with antibiotic feeds on piglets produced even more spectacular results. Stokstad and Jukes quickly published two studies in early 1949 announcing the antibiotic growth effect.[24]

That year, American Cyanamid sent samples of antibiotic fermentation materials to prominent animal scientists at several land grant colleges and experiment stations. According to surviving documents, Lederle's scientists did not allude to the emerging significance of the antibiotic growth effect. For example, the company's $1,500 contract with Tony Cunha, a researcher at the University of Florida, suggested that his research would be on folic acid and vitamin supplements, even though the publications that emerged from this study were perhaps the first independent studies confirming the antibiotic growth effect.[25] Lederle's agreements with Catron at Iowa State were similarly vague. After taking Catron and his wife to see the Rockettes in New York, Lederle representatives offered Catron a $1,500 contract to study the animal protein factor in various hog feeds. By December 1949 Catron could see that the pigs thrived on rations fortified with the antibiotic aureomycin, but he did not know the formula and thus could not be sure why growth had accelerated. "I would certainly like to know more about this product with which I am working," Catron told his corporate sponsors.[26] In any case, the studies conducted by Catron and Cunha in 1949 provided the independent confirmation necessary for Lederle to announce its breakthrough discovery.[27]

The public impact of this discovery became evident in April 1950, when Stokstad and Jukes presented their results at the American Chemical Association meeting in Philadelphia. The *New York Times* placed its story on page one under the headline "'Wonder Drug' Aureomycin Found to Spur Growth 50%." The *Times* reported that the new drug was better than any vitamin; it could speed growth of hogs by as much as 50 percent, with similar results for young chickens and turkeys. Early studies indicated an ever more encouraging picture of the antibiotic growth effect. Even small amounts of antibiotics in feeds improved both meat and egg production in poultry. Antibiotics stimulated growth of hogs on both dry lots and in pasture; and they accelerated growth whether administered intravenously or through feed and water. Catron also claimed that medicated feeds reduced the loss of runt or weak animals and produced piglets "four times more uniform in size" than those raised under natural conditions. Early studies also indicated that antibiotic feeding reduced disease among animals, whether administered at therapeutic or subtherapeutic levels, and with no undesirable side effects. Apparently, an agricultural panacea had entered the public arena.[28]

Unsurprisingly, this issue quickly swept the agricultural press in 1950 and 1951.[29] Dozens of researchers published evidence corroborating Stokstad and Jukes's claims. Experiment station and university scientists produced reports that demonstrated the action of antibiotics on the growth of chicks, turkey poults, and hogs; less conclusive experiments on lambs, dairy calves, and steers were also published in that year. Signaling the greater importance of disseminating their findings directly to producers rather than reaching researchers through academic journals, Catron, Cunha, and Jukes regularly spread the gospel of antibiotic feeding before industry groups like the Kansas Formula Feed Conference, the International Baby Chick Association Meeting, and scores of similar state and regional meetings attended by feed dealers, extension agents, agricultural journalists, and other representatives of the emerging agribusiness complex.[30] The Food and Drug Administration (FDA) was called in to investigate the claims. Within a few months, and with only moderate scrutiny, the FDA ruled that antibiotics could be advertised on animal feeds beginning in January 1951.[31]

Damon Catron and Iowa State College seized upon these findings and other postwar opportunities to build the academic-industrial links that he had envisioned. In 1949, for example, Catron could boast that the state legislature had been "good to us" since it had appropriated funds for new swine research facilities and test pastures. In 1951, Catron secured grants of $3,000 from Pfizer, $2,500 from Merck, $5,000 from American Cyanamid, and $20,000 from a number of other pharmaceutical, chemical, and agribusiness firms to fund the new Center for Swine Nutrition Research.[32] Catron welcomed these changes as appropriate and necessary. Through more systematic hog-raising schemes, he also believed the enterprise could become more specialized, and hoped that professional swine producers would take over the industry, thereby removing the "in and outers" who raised hogs only on a sporadic basis.[33]

Catron also vigorously promoted antibiotic feeding in the context of cold war politics. In a report intended to speed federal regulators' approval of antibiotic production methods, Catron recalled the "critical shortages" of feed that hampered the war effort in 1944. In response to the "national emergency" of 1951, Catron cited the power of antibiotic feeds to reduce human labor and to utilize relatively inexpensive vegetable proteins as justification to accelerate production.[34] Although crowding animals together naturally increased their susceptibility to disease, scientists understood that antibiotics and other pharmaceuticals could overcome that natural burden, or at least make it possible for fewer farmers to handle more animals as part of the process.[35]

The major pharmaceutical and feed firms seized upon the success of antibiotic feeding to promote the American farmer as a participant in a modern industrial system. The Lederle firm, for instance, issued a booklet that used simple cartoons to contrast the ragged clothes, limp cigarette, and 1920s-era jalopy of the farmer who eschewed commercial feeds with the starched clothes, fat cigar, and full-sized automobile of the farmer who embraced the products. Throughout the booklet, the text emphasized that commodities reduced the labor required of the farmer, for much of the work had been transferred to the animal.[36]

Meanwhile, Catron collaborated with Pfizer and the American Farm Institute to produce the "Open Door to Greater Hog Profits," a film that offered an informative illustration of the emerging links between antibiotics and interventionist production methods. The film began with Catron's ten-minute lecture on feed rations. Among other lessons, Catron reported that runt pigs fed on antibiotic feeds yield a 54 percent improvement in growth rates. Then, the film portrayed an old, haggard, disinterested female farmer, bedecked in cheap

Fig. 9.2 Depiction of haggard female farmer feeding hogs in an inefficient and imprecise manner. Clip from the 1952 film "Open Door to Greater Hog Profits." Permission from Pfizer.

clothes and ragged scarf, shoveling corncobs and other crude rations into a hog pen. See Figure 9.2. The producers contrasted her inefficiencies with the capabilities of a cleanly dressed male farmer, who was depicted reading a farm journal in front of a real symbol of modernity—a television. Ideal farmers were aware of the latest research and determined to apply the rather self-evident lessons of using scientifically balanced rations, tending to hygiene and sanitation, and trying to save every piglet upon farrowing. In essence, this film captures the moment of transition from traditional to modernist methods of hog production, highlighting the potential of the still experimental confinement hog operations. The film also repeatedly showed a sight that has been quite rare in rural America for decades—isolated hogs feeding themselves in lush pastures as if they were on the open range.[37]

Catron next sought to link improved feeding efficiencies with the goal of year-round farrowing in order to create a more steady supply of market hogs. The discovery that baby pigs could be weaned from the sow within just a few days of birth if given feeds fortified with antibiotics and vitamins to replace their mother's milk further expanded the producers' control over their pigs. The resultant physiological changes meant that the sow could be bred again within just nine weeks of giving birth, thus drastically altering both the natural reproductive patterns and breaking the traditional animal cycle linked to the corn harvest. As meat packer Jay C. Hormel explained, the "sow might immediately be put back to work producing another litter instead of performing no other service than milking her litter."[38] "Multiple farrowing" combined with

Hogs, Antibiotics, and the Industrial Environments of Postwar Agriculture • 247

LIFE CYCLE SWINE FEEDING
BUILD FEEDS INTO AN ECONOMICAL FEEDING PROGRAM BASED ON NUTRITIONAL NEEDS

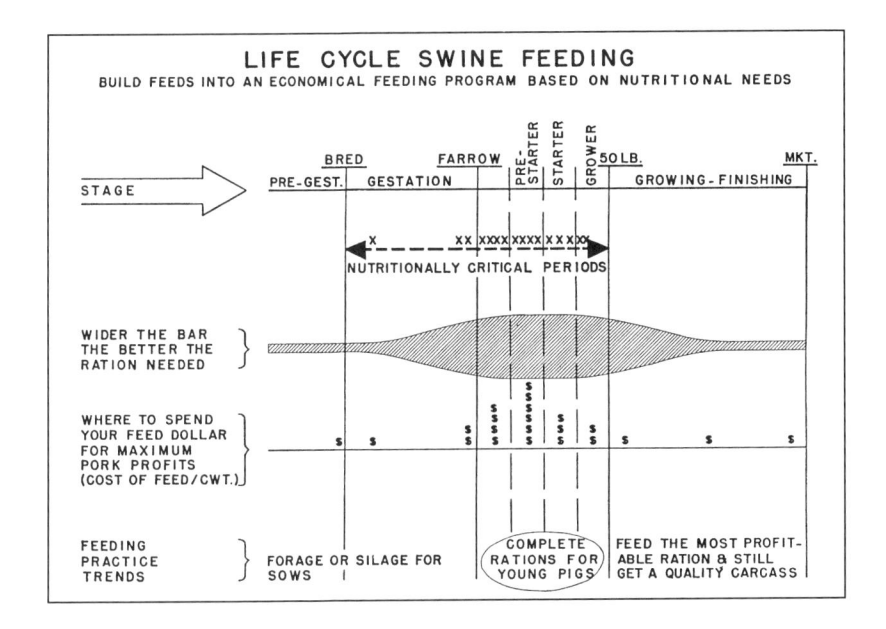

Fig. 9.3 Depiction of "Life Cycle Swine Feeding" system promoted by Damon Catron of Iowa State University, 1955. Note the emphasis on "pre-starter" feeds. This version is from *Feed Age*, 34 (May 1955): 3.

more efficient feeds meant that hogs could reach market weight within five or six months of birth, thus turning Midwestern farmers into producers of a virtually continuous flow of finished products.[39]

To further regulate and standardize animal feeding practices, Catron's next area of research was the notion of "life cycle feeding." According to this principle, pregnant sows, lactating sows, piglets, growing pigs, and fattening hogs each required different and sophisticated feed formulas. In March 1954 Catron introduced the "I. S. C. Pig Pre-Starter '75' Formula," a complex feed mixture intended for the crucial stage in young pigs' lives after their unnaturally early weaning from the sow. For Catron, the formula provided a crucial tool that enabled him "to study the nutritional requirements of baby pigs without interference from the sow." The formula included seventeen basic ingredients, one of which was a mixture of vitamins and antibiotics that contained another sixteen ingredients. Nevertheless, pre-starter feeds struck a chord with Midwestern farmers; in one four-day period, Catron received 1,500 requests for information about the formula and how it could be obtained. Catron made arrangements with several feed companies to sell his formula to the public; he even urged his brother, a farmer in rural Indiana, to jump on the opportunity to invest in this branch of the business. See Fig. 9.3.

By insisting that local farmers could not "do a good job of mixing their own pig starters" Catron again contributed to the commodification of the industry. Yet it is also significant that the pre-starter formula defined the point of resistance between the microlevel research of the scientists and the macrolevel production goals of commercial feed producers. Since pre-starter feeds were especially complex in their formulation but were intended only for a brief period of the animal's lifespan, feed companies balked at pressures to produce such feeds. Meanwhile, Catron introduced a simple tool that contributed to the goal of manipulating the life cycle of the mature hog approaching market weight. The "Pork Costulator," a small, circular slide rule, enabled farmers to abandon imprecise feeding methods and calculate ideal ratios of corn and other feeds. The device had calibrations on two scales, one for "least cost" for farmers who wanted to bring their hogs to market weight at minimal cost, and another for "least time" for those who wanted to get their hogs to market quickly in order to hit anticipated favorable prices. In all, the response to pre-starter feeds and the Pork Costulator indicated that many farmers were willing to accept Catron's boast that "swine nutritionists . . . know more about what the pig needs than the pig himself."[40]

Catron and other land grant university researchers continued to fit additional pieces into the puzzle of industrializing the hog organism and the hog production system. For Catron, an extensive tour in 1954 of the leading animal science research centers in Europe left him with the impression that Europeans were far behind Americans in terms of antibiotic feeding, but that they had much to teach about improving livestock housing and management practices.[41] By the mid-1950s, compressing time and space became the central tenets of this experimental agenda, and "environment," "confinement," and "integration" replaced antibiotics and feeding as the buzzwords of swine research. In a remark that may allude to the Pfizer advertisement used to introduce this essay, Catron explained "in my opinion the farmer has waited too long for the feed man to put the miracle drug in the feed bag."[42] As Louisiana hog farmer Homer Harris put it, it was time to "modernize our entire hog farrowing assembly line" by lessening the time between farrowings and restricting animal mobility.[43] In response, researchers examined countless designs for centralized hog houses, many outfitted with artificial heaters, air conditioners, automatic waterers, electric lights, and even offices. Signaling the opportunities that these trends offered to firms beyond the traditional agribusiness sector, the Reynolds Aluminum Company funded Iowa State's agricultural engineers to investigate connections between swine growth, feed efficiency, the temperature of the animal's environment, and materials used in construction of hog pen buildings.[44]

A shift in Damon Catron's grantsmanship activity during this era offers another illustration of the trend toward the industrialization of the hogs' environment. Explaining "we are reaching the limit in substituting nutrition for good management," Catron presented a new seventeen-point program of efficient hog management that called for manipulations of the swine diet, environment, breeding, veterinary care, marketing, and more.[45] In an appeal for funds from farm building and equipment manufacturers, Catron outlined the need for research on automated feedlot equipment and facilities designed "for comfort of the pig—and of the operator."[46] These studies comprised Catron's program of "life cycle housing," the notion that farmers should coordinate their "life cycle feeding" with investments in separate housing units for the distinct stages of the hog's life.[47] Life cycle housing called for farmers to invest in air conditioning, heated floors, regulated water temperatures, germicidal lamps, and other aspects of a manipulated environment. Manufacturers and hog producers responded first with designs for various systems intended to bring sows in from the pasture during farrowing time. Indoor farrowing and early weaning demanded new housing complexes that separated sows and piglets, and structures built with guardrails, sloped floors, and other precautions to "remove the sow hazard," or the possibility of death by crushing. Also, early experiments with hormone treatments pointed to the possibility that one could precisely schedule the birth process to ensure that sows would give birth at times convenient to the farmer.[48]

The emergence of confined housing systems that eliminated the need for pasture altogether proved an especially significant development. According to one analysis, multiple farrowing was the main incentive for farmers to move toward the "assembly line production" on concrete feedlots, since continual crops of young pigs would simply require too much labor to continually move fences, feed troughs, and waterers through the traditional system of using pastures. The advantages of artificial environments challenged the pasture system in other ways as well. Researchers at Purdue, for example, reported "startling" losses of hogs on hot summer pastures, and issued reports promising that "manmade weather can improve livestock efficiency."[49] In 1959, *Successful Farming* magazine highlighted the operations of J. Herbert Doak, an Ohio farmer who also had been featured in 1952 for his innovative use of pasture and continuous farrowing. By the late 1950s, however, Doak had shifted his operation entirely to concrete, convinced that he could manage disease threats and lessen his labor requirements in more confined spaces.[50]

Antibiotics also played a role, challenging the notion that pastures were naturally more healthful environments than confined hog lots. Although crowding animals together naturally increased their susceptibility to disease, antibiotics, other pharmaceuticals, and regularly dousing pens with lye and creosote were cheap insurance policies that helped reduce disease risks.[51] By the late-1950s, advertisements of the Portland Cement Association, with their slogan "The Mark of a Modern Farm . . . Concrete," appeared alongside the advertisements for artificial housing, manufactured feeds, feeder pig companies, and "high-dosage" antibiotics.[52] Although converting to a system of confinement and concrete required a capital investment too great for tenants and many small farmers, many analysts argued that for others, the benefits outweighed the costs, especially the opportunity to convert pasture into productive cropland.[53]

As powerful as the juggernaut of pressures to adopt the factory farming paradigm appeared to be, it is significant that many farmers questioned the trends. Some blamed "integration" as the root of their problems, for it seemed that packers, grocery stores, feed merchants, and pharmaceutical firms were prospering despite low profits on the farm.[54] Understandably, farmers objected most passionately to the impact of industrialized agriculture on commodity prices. In 1955 hog prices fell from nineteen cents per pound in May to a thirteen-year low of eleven cents per pound in December. Rural citizens from many trades and occupations wrote to President Dwight Eisenhower, Secretary of Agriculture Ezra Taft Benson, and others in government with grievances about prices, pleading to slow the pace of agricultural change. Dozens of Midwestern citizens protested the cost/price squeeze that many attributed to the expenses of pharmaceuticals, commercial feed, and new buildings and equipment. California hog farmer Warren Jaycox sent Secretary Benson a sarcastic Christmas card depicting hog farmers begging with a tin cup. See Fig. 9.4. Thirteen-year old Charles Borstad of Minnesota was among the many who appealed for help, explaining that he faced losses of at least $10 per head on the hogs he had raised through his 4-H Club.[55] Agribusiness concerns also faced pressure from the National Farmers Organization, an aggressive grassroots organization founded in 1955 to fight for higher commodity prices in the corn-hog belt.[56]

There is also ample evidence that farmers did not readily adopt the Fordist, assembly line model that antibiotic feeding promised and Catron promoted. Farm journals routinely featured farmers who creatively improved their operations with inexpensive and homemade equipment. Examples include feed bins and water troughs placed on sleds that could be hauled by tractor from one pasture to another; cheap automatic feeders built from old tires and scrap metal; and walk-through corncribs that enabled hogs to feed and find shade at the same time.[57] Other agricultural experts also encountered signs of resistance. Farm extension agents in Florida, for example, complained in 1954 that local farmers seemed ignorant about the latest recommendations in swine nutrition. Their 1956 report promised to teach more aggressively the doctrine of centralized swine management, but it noted that fatty carcasses, market gluts, and batch production schedules remained common in Florida.[58] In a speech at the 1956 Iowa Swine Producers Day, poultry producer Leon Johnson complained mightily of the "whining" hog producers who refused to see the wisdom of large-scale capitalism.[59] Although limited in their impact, such cases corroborate what historian Ronald Kline describes as the "contested processes" and "individual modernities" with which rural Americans often encountered social and technical change.[60]

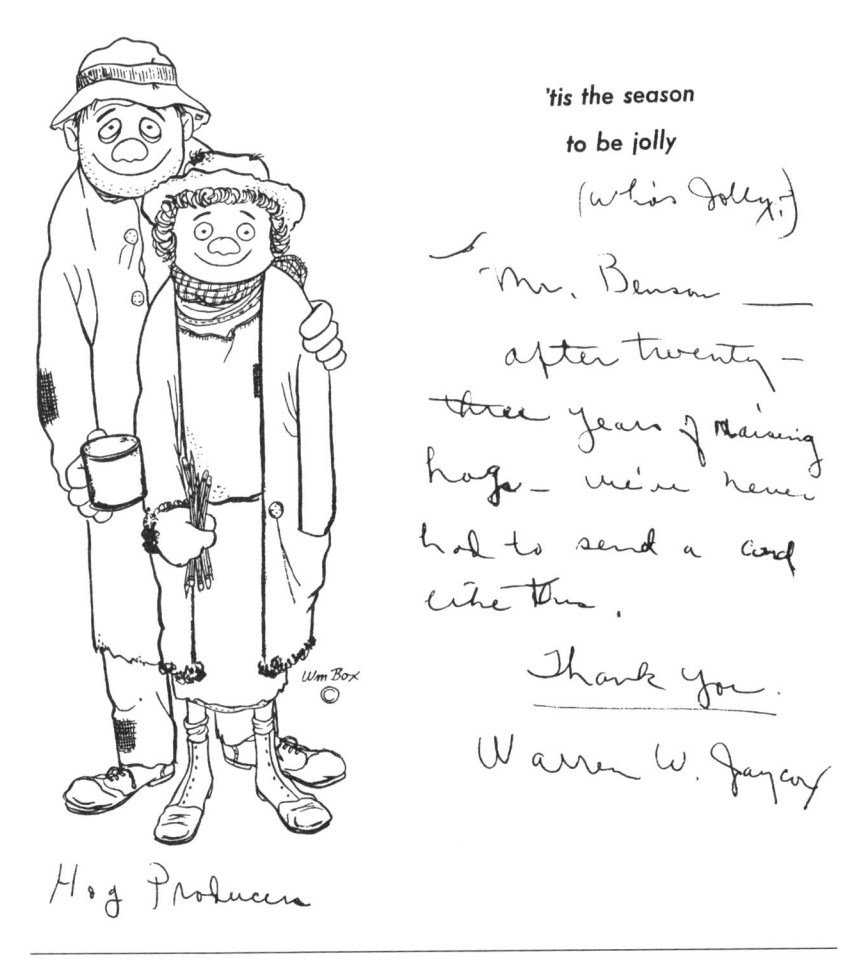

Fig. 9.4 Christmas card sent from Warren W. Jaycox to secretary of Agriculture Ezra Taft Benson in 1955, a time of unusually low hog prices. Note reads: "Mr. Benson—After twenty years of raising hogs—we've never had to send a card like this. Thank you." Warren Jaycox to Ezra Taft Benson [December 1955], in RG 16, Records of the Office of the Secretary of Agriculture, E17 General Correspondence 1906–1976. Box 2717, National Archives, College Park, MD.

Yet such cases could slow development of the industrialized pig only temporarily. By the end of the 1950s, future trends in hog production were already visible, even if their final implementation remained incomplete. Agribusiness concerns, farm journal editors, and academics from the land grant universities continued to push for the further manipulation of nature into artificial agro-industrial systems. Antibiotics and artificial environments continued to bring consolidation and rising productivity to American agriculture in the decades after 1960. Confined operations centered on concrete feedlots defined the new-era Midwestern farms, as hogs on pasture became a distant memory. In all, changes in the hog industry exposed growing interconnections between agriculture, technology, ecology, and political economy.[61]

Most of the changes that occurred between 1945 and 1960 occurred at the level of what Edmund Russell has labeled "macrobiotechnologies," or the interventions that impacted the lifestyle and physiology of the whole organism. But in a speech delivered to the Midwest Feed Manufacturers Association meeting in 1960, Catron hinted that many future changes would be approaching the realm of Russell's "microbiotechnological" interventions, or those that occur at the level of cell and molecules.[62] Catron predicted that hog producers of the future would develop ever more precise and complex formulas for their feeds, aided in part by the "electronic brains" that some feed companies were beginning to utilize. He expected researchers to develop more specialized feed additives that would affect the animal in precisely targeted ways. For instance, predigested foods and certain enzymes might alter the hog's natural digestive function and extend feed efficiency. Other feed additives could alter and improve natural expectations for controlling estrus, ensuring embryonic survival, developing leanness of the animal carcass, and "other biological functions in the animal over which the farmer would like control." Indeed, Catron forecast that farmers of the future could generate performance "exceeding" the animal's actual "genetic potential." Catron's speech also accurately predicted the growing attention to the genetic "plasm" of the boar and sow, to artificial insemination, and to a future of pigs being delivered by caesarian section and reared in isolated, disease-free environments. In all, in this speech and elsewhere, Catron called for farmers, like managers of a factory, to pay greater attention to strategies that would induce the animals, like laborers, to do even more work. With "work" in this context defined as having hogs consume the feeds that maximized growth efficiency, Catron suggested that hog producers learn to pay greater attention to environmental conditions that caused animal stress, such as uncomfortable temperatures, bothersome noise from airplanes and equipment, and bitter or unpalatable feeds. Amid all of his interest in science and technology, Catron also urged attention to the "tender loving care factor" as a stimulus to get hogs to work more eagerly.[63]

Damon Catron issued these predictions at a time when he was taking stock of his own situation. In December 1959, Iowa State University asked him to submit a list of his activities over his fourteen-year career at the school—a list that included increasing the staff involved in swine research from two to forty, expanding the swine herd from thirteen head to 347, extending the value of the physical plant from $14,000 to nearly a million dollars, and acquiring over $539,000 in grant funds. In January 1960, Catron nonetheless announced his resignation from Iowa State and accepted a position with an Iowa feed manufacturing company.[64] Overall, this trajectory in Catron's career suggests that swine research had reached a level of stability and that one principal researcher could now reap in industrial employment what he had sowed during his period of academic research work.

During the fifteen years that encompassed Catron's career in Ames, tremendous and rapid changes took place in the American swine industry. The sights, sounds, and smells of the Midwestern landscape changed drastically as hogs moved from pasture and to increasingly confined operations in artificial environments. Increasing the pace of breeding, farrowing, weaning, rebreeding, and finishing had altered the pig's life cycle. The animal's physiology changed through the development of "meat-type" hogs engineered to reduce backfat measurements and feeds designed to get animals to market weight more quickly and efficiently.[65] Animals also became more dependent on pharmaceutical interventions, as farmers used antibiotics and other veterinary medicines as routine insurance against disease threats.

Each of these changes altered the relationship between hog and the hog producer. The latter increasingly gained the ability to transfer responsibilities of labor and energy conversion onto the former, thus streamlining their operations along industrial lines. Hog producers also saw their occupation grow increasingly connected to the emerging agribusiness complex that included commercial feed manufacturers, pharmaceutical firms, housing and equipment makers, government policymakers, and consumers. Hog existence became increasingly regulated through manipulations of their physiology, social activities, spatial environment, and connections with the natural environment. A common thread impacted both hog and hog producer as both witnessed the disappearance of their independence and autonomy, a process similar to what had occurred in so many other industries.

In addition to the calculus of costs, profits, and efficiencies, many other factors provided impetus for the changes in the hog's organism and environment. After years of false starts, farmers and agribusiness leaders finally brought an industrial program to a farm animal that seemed firmly bound to natural ecological rhythms and traditional market circumstances. On the farm, putting soybeans into rotation with corn made land too valuable for pasturing hogs, while manufactured nitrogenous fertilizers further reduced the incentive to graze livestock for their manure. Increasing mechanization made it possible for a single farmer to handle larger and more profitable acreages and proved an especially pertinent aid for the efficient handling of feed and manure.[66] Off the farm, various external pressures also induced farmers to reshape the hog to play a more vital role in the postwar economy and society. Increasing government manipulation of corn and hog prices, memories of the feed and agricultural labor shortages of World War II, popular pressure to apply wartime science to civilian ends, and changes in the behavior of grocery store consumers who demanded leaner pork products all contributed to the redesign of these animals.

Such changes also had geopolitical implications that went beyond Corn Belt hog lots and farrowing stalls. In the words of one historian of American foodways, "the Cold War ... was waged and perhaps even 'won' through kitchen debates, grain deals, breadlines, and the Green Revolution."[67] The cold war intensified Americans' interest in technological successes and, in the words of another scholar, "created the sense of an automation race" with Soviet bloc counterparts. Americans justified technological innovations for their own sake and because increased production and bountiful supermarkets could be highlighted as symbolic of the free enterprise system.[68] Along these lines, President Eisenhower's Agricultural Trade Development and Assistance Act and President John F. Kennedy's Food for Peace Program both included policies that encouraged consumption of American-grown protein products and helping "friendly" nations with the meat and grain surpluses that remained. It was no coincidence either that Soviet premier Nikita Khrushchev's 1959 tour of the United States included visits to both an American kitchen and to Iowa's efficient corn and livestock operations.[69] For all of these reasons and more, the versions of the pig that had been adequate for centuries were no longer suited for the demands of the postwar world.

Many actors participated in the process of creating the industrialized pig. By the 1950s, most of the players in the drama were already on the stage—pharmaceutical firms, feed manufacturers, animal husbandry experts, regulators, lobbyists, and more. Some farmers made an effort to slow the pace of these changes, and even in the late 1960s some 65 percent of the hogs in Illinois were raised on some pasture. In all, though, little could stop trends toward antibiotic feeding, confinement, and other aspects of industrialized hog production. The number of farms in Iowa that produced pigs fell from 181,400 in 1940 to 134,500 in 1960 and to 84,900 in 1970. The pace of consolidation intensified thereafter; by 2000, the number of pork producing farms in Iowa had dwindled to 10,500.[70] Nowadays, empty and abandoned hog lots are a common sight in the rural Midwest. Even so, the story is not over. Recent reports show that a handful of farmers have carved out niche markets based on the raising of "free-range" and antibiotic-free hogs, and that new cooperative marketing agreements are challenging the major packers. Most strikingly, pharmaceutical firms are quietly reducing their dependence on antibiotics for their growth-promoting effect.[71] The postwar pig, created through manipulation and redesign of the animal's organism and its environment, is still an unfinished product.

Notes

> Portions of this essay had roots in one that I co-wrote last spring with Barbara Kimmelman of Philadelphia University. Mark Finlay and Barbara Kimmelman, "The Origins of 'Farmaceuticals': How Antibiotics Came to Be Used in Agriculture," contribution to conference on Bacterial Resistance to Antibiotics, Johns Hopkins University, April 28, 2001. I also would like to acknowledge Becky Jordan of the Special Collections Department at the Iowa State University Library for her repeated help with the Catron papers; Richard Willham, Palmer Holden, and Vaughan Speer, Department of Animal Science, Iowa State University; Thamon Hazen, Department of Agricultural Engineering, Iowa State University; the family

of A. F. Grashuis; research assistants Thomas Shipley, Grace Fleming, and Sandy Brown; Vernon Kisling of the University of Florida Library; and Caroline Hopkinson and Melissa Jackson of the Armstrong Atlantic State University Library.

1. "Science Comes to the Farm in a Feed Bag," *Farm Journal* 80 (February 1956): 67. Pfizer's "Science Comes to the Farm in a Feed Bag," campaign appeared in farm journals in 1956 and 1957, notably in an eight-page advertising spread in *Feedstuffs* 29 (July 13, 1957): 33–40.
2. Wayne D. Rasmussen, "A Postscript: Twenty-Five Years of Change in Farm Productivity," *Agricultural History* 49, no. 1 (1975): 84–86.
3. Roger Horowitz, "Making the 'Chicken of Tomorrow,'" this volume; and William Boyd, "Science, Technology, and American Poultry Production," *Technology and Culture* 42, no. 4 (October 2001): 631–664.
4. Biographical information from "Russellville Native Is Selected for $1000 Award for Research," undated clipping from the *Frankfort Times*, in Damon Catron Papers, Collection RS September 11, 1955, Iowa State University Archives, Ames, Iowa [hereafter cited as Catron Papers], Box 12, Folder 2; Damon Catron [hereafter DC] to L. E. Johnson, January 4, 1960, Catron Papers, Box 41, Folder 5; Vaughn Speer to Mark Finlay, September 19, 2002; and the finding aid to the Catron papers at Iowa State University. It is pertinent that Catron considered his one-year experience at Ralston-Purina more valuable than any of his academic degrees. See DC to E. G. Cherbonnier, December 14, 1954, Catron Papers, Box 15, Folder 4.
5. F. W. Woll, *Productive Feeding of Farm Animals*, 2nd ed. (Philadelphia: Lippincott, 1916), 1–3. This text appeared in the context of the Country Life Movement, an effort to bring the progressive era's interest in social engineering, expertise, and efficiency to farm issues. Reformers believed that mechanization and urban and industrial values offered effective strategies for improving American rural life. For background information see, David Danbom, *The Resisted Revolution: Urban America and the Industrialization of Agriculture, 1900–1930* (Ames: Iowa State University Press, 1979); Deborah Fitzgerald, *Every Farm a Factory: The Industrial Ideal in American Agriculture* (New Haven: Yale University Press, 2003).
6. [Lew P. Reeve], *Multiple Farrowing Pleases . . . The Farmers . . . and the Housewife*, undated pamphlet produced by George. A. Hormel & Company and Iowa State College, Catron Papers.
7. Woll, *Feeding of Farm Animals*, 210.
8. Arthur L. Anderson, *Swine Management, including Feeding and Breeding* (Philadelphia: Lippincott, 1950).
9. A late 1920s study indicated that 11 percent of young piglets were crushed or eaten by the sows, 5 percent were born dead, 2 percent froze to death. In all, 34 percent of pigs farrowed died before weaning time. William Jackson, "Livestock Farmers Can Obtain Better Profits with Better Husbandry," in U.S. Department of Agriculture, *Yearbook of Agriculture, 1928* (Washington, D.C.: GPO, 1929), 423–425. See also Anderson, *Swine Management*; U. G. Houck, "Hog-Cholera Losses Can Be Much Reduced by Sanitary Measures," in USDA, *Yearbook of Agriculture, 1928*, 361–363.
10. The corn-hog price ratio is calculated from the average price per one hundred pounds of live hogs divided by the average price of one bushel of corn. Typically, ratios above 11 or 12 to 1 encouraged hog production. For more on difficulties with the hog program under the New Deal, see Theodore Saloutos, *The American Farmer and the New Deal* (Ames: Iowa State University Press, 1982), 70–75; and John C. Culver and John Hyde, *American Dreamer: The Life and Times of Henry A. Wallace* (New York: Norton, 2000), 124–125.
11. Claude R. Wickard and J. A. McConnell, "The Feed Situation," *Nation's Agriculture* 18 (April 1943): 4, 14; "Pigs in a Pickle," *Business Week*, no. 723 (July 10, 1943): 46–47; "Feed Crisis Near," *Business Week*, no. 735 (October 2, 1943): 14; "Forage Goals Rise," *Business Week*, no. 738 (October 23, 1943): 40; "Corn for War," *Business Week*, no. 738 (October 23, 1943): 78; "Headache in Hogs," *Business Week*, no. 742 (November 20, 1943): 14; "Corn or Hogs—No. 1 Problem," *Business Week*, no. 757 (April 8, 1944): 15–16; "Feed Crisis Near," *Business Week*, no. 761 (May 6, 1944): 50; and "Farmers Face Feed Famine," *Breeder's Gazette* 109 (May 1944): 5, 22. The imbalance of the hog-corn price ratio also encouraged farmers to fatten their hogs to an unusually heavy weight, up to three hundred pounds, before slaughter. This had the twin effects of further depleting feed grain supplies and bringing unusually fatty meat products to the consumer marketplace.

12. Catron and his colleagues observed "very unsatisfactory results" when plant proteins replaced animal proteins in continuous dry lot conditions. See B. W. Fairbanks, J. L. Krider, and Damon Catron, "Some Observations of Plant versus Animal Protein Supplements for Young Pigs," *North American Veterinarian* 25 (June 1944): 351–355. See also W. L. Robison, "Vegetable vs. Animal Concentrates for Feeding of Swine," *Flour & Feed* 43 (December 1942): 10; and H. W. Titus, "Replacing Animal Protein with Vegetable Protein in Feeding Poultry," *Flour & Feed* 43 (April 1943): 12–13.
13. For analysis of connections between World War II and scientific values, see David A. Hollinger, "Science as a Weapon in the *Kulturkämpfe* in the United States During and After World War II," *Isis* 86 (1995): 440–454. For the agricultural sciences in particular, see Nicolas Rasmussen, "Plant Hormones in War and Peace: Science, Industry, and Government in the Development of Herbicides in 1940s America," *Isis* 92 (June 2001): 291–316.
14. Based on a study that compared the average figures of 1935 to 1939 with those of 1944, total farm production increased 36 percent, while the total number of American farm workers declined 8.8 percent. See "Wartime Changes in Agricultural Employment," *Monthly Labor Review* (September 1945): 442–451. See also Kenneth S. Davis, "The Hired Man—A Vanishing American," *New York Times Magazine* (July 23, 1950): 16, 34–35.
15. For the time management studies, see [E. C. Young], "Activities, Results, and Available Materials in Farm Work Simplification in the 12 States Cooperating in the National Farm Work Simplification Project," August 1944, unpublished manuscript bound with *Work Simplification News Letter*, Purdue University Library; J. W. Oberholzter, "Labor Economy on Farrowing Spring Pigs," *Work Simplification News Letter*, no. 3 (July 1943): 3–7; "Some Conclusions on a Work Simplification Study of the Hog Enterprise in Indiana," *Work Simplification News Letter*, no. 9 (October 1944): 8–9; "Issue Summarizes Two Years of Research Work," *Work Simplification News Letter*, no. 13 (October 1945): 7–9;. Lawrence M. Vaughn and Lowell S. Hardin, *Farm Work Simplification* (New York: Wiley, 1949), 17–18, 57; and John Strom, "Are You Wasting Labor?" *Country Gentleman* 121 (May 1951): 21, 86–88. For other complaints of hog industry inefficiencies, see Howard R. Long and W. H. Yaw, "'Swine Courts' Control the Hogs," *Country Gentleman* 120 (June 1950): 33, 109.
16. Compare Anderson, *Swine Management*, a textbook that depicts three-hundred-pound hogs as a viable possibility, with Tony J. Cunha, *Swine Feeding and Nutrition* (New York: Interscience Publishers, 1957), a text that depicts the 220-pound hog as the norm. See also "America's 'Meat on the Hoof'," *National Geographic* 101 (January 1952): 33–74; and "Farms Retooling for New Model Pig," *Nation's Business* 43 (October 1955): 44–46.
17. Allen J. Matusow, *Farm Policies and Politics in the Truman Years* (Cambridge: Harvard University Press, 1967), 7–8, 14–16, 58–61; John H. Perkins, *Geopolitics and the Green Revolution: Wheat, Genes, and the Cold War* (New York: Oxford University Press, 1997). See also Dorothy Schwieder, *75 Years of Service: Cooperative Extension in Iowa* (Ames: Iowa State University Press, 1993), 128–134. For evidence of the cold war rhetoric reaching local hog and feed circles, see Harvey E. Yantis and George L. Gates, "Western Feed Men Hear National Trade Officials, Research Men," *Feedstuffs* 23 (December 1, 1951): 3–8; and Tom Linder, "Swine Diseases, Inoculation, and Quarantine," *Georgia Farmers' Market Bulletin* 35 (August 29, 1951): 1, 4.
18. Frederick G. Brownell, "Super Cows and Chickens," *American Magazine* 141 (June 1946): 108–111.
19. Maxwell Reid Grant, "Engineering Better Meat," *Popular Mechanics* 91 (February 1949): 174–177, 246, 252.
20. Damon Catron, "Serving a 600 Million Dollar Business," notes for speech delivered to the East Des Moines Club, October 23, 1946; and Damon Catron, "Swine Nutrition—Weaning to Market," speech delivered to Iowa Nutrition School, September 13, 1946, Catron Papers, Box 37, Folder 8. See also "Your Hog Business in 1945," broadsheet published by University of Illinois, Extension Service in Agriculture and Home Economics, Catron Papers, Box 25, Folder 4; and B. W. Fairbanks, "Current Problems in Swine Nutrition," *Journal of the American Veterinary Medical Association* 104 (February 1944): 89–92.
21. Jukes was born in Hastings, England, in 1906. He emigrated to Canada in 1924, and then to California. He earned a doctorate in biochemistry at the University of Toronto and then worked in the field of poultry nutrition, mainly at the Davis campus of the University of California. He worked in the private sector at Lederle Laboratories from 1942 to

1959, before embarking on a second career with a specialty in molecular biology. See "Outspoken UC Berkeley Biochemist and Nutritionist Thomas H. Jukes Has Died at Age 93," retrieved March 2, 2001, from http://www.berkeley.edu/news/media/releases/99legacy/11–10–1999.html; John Maddox, "Thomas Hughes Jukes, (1906–99)," *Nature* 402 (December 2, 1999): 478. Stokstad was born in China in 1913, earned his doctorate from the University of California in 1937; worked at Lederle Laboratories from 1941 to 1963, and returned to Berkeley as a professor of nutrition for the latter stages of his career.

22. For a survey of animal nutritional issues on the verge of discovery of the antibiotic growth effect, see Damon Catron, "APF," *Successful Farming* 47 (July 1949): 31, 70–72. A useful introduction to discovery of the antibiotic growth effect appears in "Antibiotics in the Barnyard," *Fortune* 45 (March 1952): 118–140. A detailed and informative treatment of the work at Lederle also appears in Thomas H. Jukes, "Public Health Significance of Feeding Low Levels of Antibiotics to Animals," *Advances in Applied Microbiology* 16 (1973): 1–30. See also William L. Laurence, " 'Wonder Drug' Aureomycin Found to Spur Growth 50%," *New York Times* (April 10, 1950): 1, 17; and Merck & Co., Inc., *Procaine Penicillin in Animal Nutrition* (Rahway, N.J.: Merck & Co., Inc., 1956). A brief history of the discovery, plus interesting summaries of related regulatory action, is provided by Marc Lappé, *Germs that Won't Die: Medical Consequences of the Misuse of Antibiotics* (Garden City, N.Y.: Anchor Press/Doubleday,1982), 126–149.

23. An earlier study along these lines had appeared in 1946, when P. B. Moore and his coworkers at the University of Wisconsin found that streptomycin and sulfasuxidine (a relatively more toxic sulfa drug) had an impact on the growth rate of chicks. The Wisconsin researchers did not follow the antibiotic trail, but Stokstad and Jukes of Lederle Laboratories did.

24. See E. L. R. Stokstad, A. C. Page, J. Pierce, A. L. Franklin, T. H. Jukes, R. W. Heinle, M. Epstein, and A. D. Welch, *Journal of Laboratory and Clinical Medicine* 33 (1948): 860; T. H. Jukes, E. L. R. Stokstad, and K. C. Gilbert, *Poultry Science* 27 (1948): 434; E. L. R. Stokstad, Thomas H. Jukes, J. Pierce, A. C. Page, Jr., and A. L. Franklin, "The Multiple Nature of the Animal Protein Factor," *Journal of Biological Chemistry* 180 (1949): 647–654; "Antibiotics in the Barnyard"; Thomas H. Jukes, *Antibiotics in Nutrition*, Antibiotics Monographs, 4 (New York: Medical Encyclopedia, 1955), 17–18; and Orville Schell, *Modern Meat* (New York: Random House, 1978), 20–21.

25. T. J. Cunha to R. S. Glasscock, April 6, 1949, and "Agreement between the University of Florida and Lederle Laboratories Division, American Cyanamid Company," May 1, 1949, both in Box 1, Series 100a, Institute of Food and Agricultural Sciences, University Archives, Public Records Collection, University of Florida Library, Gainesville.

26. DC to Allan B. Clow, August 27, 1949; DC to Thomas H. Jukes, December 12, 1949; and DC to Allan B. Clow, December 12, 1949, all in Catron Papers, Box 4, Folder 4.

27. T. J. Cunha, H. H. Hopper, J. E. Burnside, A. M. Pearson, R. S. Glasscock, D. V. Catron, A. B. Hoerlein, P. C. Bennett, P. W. Cuff, and P. G. Homeyer, "The Effect of Aureomycin and Other Antibiotics on the Pig," *Journal of Animal Science* 9 (1950): 653ff; and D. V. Catron, et al., "Effect of Vitamin B$_{12}$, APF, and Antibiotics on Enteritis in Swine," *Journal of Animal Science* 9 (1950): 651ff.

28. "Antibiotics in the Barnyard," 118–140; William L. Laurence, " 'Wonder Drug' Aureomycin Found to Spur Growth 50%," *New York Times* (April 10, 1950): 1, 17. Three days later, another report repeated the enthusiasm for this discovery: Waldemar Kaempffert, "How Aureomycin, Antibiotic, Acts to Increase Rate of Growth is Subject of Speculation," *New York Times* (April 16, 1950): iv, 9. For a parallel case on the emergence of growth hormones in animal feeds, see Alan I. Marcus, *Cancer from Beef: DES, Federal Food Regulation, and Consumer Confidence* (Baltimore/London: Johns Hopkins University Press, 1993).

29. For example, "Comment on Aureomycin Report," *Feedstuffs* 22 (April 15, 1950): 1, 65; "Aureomycin Greatly Accelerates Animal Growth, Scientists Report," *Feedstuffs* 22 (April 15, 1950): 1, 69; "Aureomycin Speeds Gain," *Country Gentleman* 120 (June 1950): 52–55; T. J. Cunha, "Aureomycin: It's [sic] Effect on Pigs," *Duroc News* 24 (September 1950): 155; and Elton L. Johnson, Damon Catron, and Dean Wolf, "Latest on Wonder Drugs in Feeds," *Successful Farming* 49 (April 1951): 36–37, 74–75, 155–158.

30. Damon Catron, "New Developments in Swine Nutrition," *Feedstuffs* 22 (November 4, 1950): 34, 36–37; "Western Grain, Feed Men Hear Reports on Antibiotics in Feed," *Feedstuffs* 22

(December 2, 1950); Lawrence Galton, "Meat-Making 'Wonder' Drugs," *Country Gentleman* 121 (April 1951): 21, 109; H. Ernest Bechtel, "How Far Can You Travel on Antibiotics?" *Feedstuffs* 23 (August 11, 1951): 18–22; and Yantis and Gates, "Western Feed Men Hear National Trade Officials," 3–8. On the trend of reporting agricultural research results through the channels of producers, rather than of academics, see Marcus, *Cancer from Beef*, 16.

31. Significant secondary studies included the work on turkey poults by James McGinnis of the State College of Washington, and J. K. Loosli of Cornell University on dairy calves. Other confirming reports include E. L. R. Stokstad and Thomas H. Jukes, "Growth-Promoting Effect of Aureomycin on Turkey Poults," *Poultry Science* 29 (1950): 611–612; A. C. Groschke and R. J. Evans, "Effect of Antibiotics, Synthetic Vitamins, Vitamin B_{12}, and an APF Supplement on Chick Growth," *Poultry Science* 29 (1950): 616–618; J. McGinnis, L. R. Berg, J. R. Stern, R. A. Wilcox, and G. E. Bearse, "The Effect of Aureomycin and Streptomycin on Growth of Chicks and Turkey Poults," *Poultry Science* 29 (1950): 771.

32. DC to H. J. Prebluda, July 27, 1949, Catron Papers, Box 5, Folder 1; and "Cash Grants-In-Aid Received by Swine Nutrition Research, Iowa State University, Fiscal Years 1946–47 to and Including 1958–59," Catron Papers, Box 41, Folder 5. In view of the tremendous payoff that such research brought to agribusiness firms, these studies are significant indications of the value that corporations were discovering in their support of applied agricultural research.

33. [Damon Catron], "Development of the Iowa System of Swine Production Involving Multiple Farrowing," April 16, 1956, Catron Papers, Box 32 Folder 1.

34. Damon Catron, "The Importance of Vitamin B_{12}-Antibiotic Feeding Supplement (Aurofac) to the Swine Industry of the United States," attached to DC to Earl Reeve, March 17, 1951, Catron Papers, Box 7, Folder 2.

35. "Farms Retooling for New Model Pig," *Nation's Business* 43 (October 1955): 44–46.

36. Lederle Laboratories Division, *Feed for Profit . . . Aureomycin*, promotional booklet issued in 1953 (in author's possession).

37. "Open Door to Greater Hog Profits," 1952 film produced by the American Farm Institute. My thanks to Becky Jordan for obtaining for me a copy from the American Archives of Factual Film, Iowa State University Library.

38. "Antibiotics Used on Livestock by Hormel to Clear Bacteria for Full Effect of Fodder," *New York Times* (December 13, 1951): 53; Hal Borland, "This Little Pig Ate Antibiotics," *New York Times Magazine* (August 31, 1952): 10, 17; Merck & Co., Inc., *Procaine Penicillin in Animal Nutrition*.

39. Damon Catron, "Baby Pigs Don't Need Their Mommies Any More!" *Iowa Farm Science* 6 (April 1952): 147–149, 160; Damon Catron, "Wean 'Em at 7 Days," *Country Gentleman* 124 (April 1954): 38–39, 82; and Dean C. Wolf, "New Early Wean Ration Brings More Pig Profits," *Country Gentleman* 124 (September 1954): 38, 80–81.

40. DC to Robert Stokstad, March 25, 1954, Catron Papers, Box 14, Folder 1; DC to Dale Eugene Catron, February 8, 1954, Box 15, Folder 4; DC to Elvin L. Quaife, May 9, 1955, Catron Papers, Box 15, Folder 5; Frank Nelson to DC, February 28, 1955; DC to Frank Nelson, March 11, 1955, both in Box 16, Folder 2, Catron Papers; DC to O. L. Kline, May 7, 1955, RG 88, Records of the Food and Drug Administration, E5 General Subject Files, File 432.97–10, Box 1977, National Archives; Damon V. Catron, *Future of Hog Raising . . . and Your Hog Business*, promotional pamphlet released by the Walnut Grove Products, Catron Papers. Originally published in *Feed Age* (May 1955); and *How You Can Make More Money with the Pork Costulator*, promotional pamphlet (Newton, Iowa: Vernon Company, 1954), Catron Papers, Box 29, Folder 8.

41. DC to Charles Barnhart, September 28, 1954, Catron Papers, Box 12, Folder 5.

42. DC to Dick Hanson, April 4, 1955, Catron Papers, Box 15, Folder 5.

43. Homer H. Harris, Sr., to DC, September 29, 1954, and DC to Homer H. Harris, Sr., October 5, 1954, both in Catron Papers, Box 13, Folder 2.

44. Dr. Thamon Hazen to Mark Finlay, August 30, 2002; and DC to Thamon Hazen, June 10, 1955, Catron Papers, Box 15, Folder 5. Researchers found that buildings constructed with aluminum did not offer a significant advantage over conventional buildings constructed with wood and asphalt shingles.

45. Catron quoted in "Health Institute Committee to Study Feed Additives," *Feedstuffs* 29 (April 13, 1957): 10–16.

46. Damon Catron, "Development of the Iowa State System of Swine Production Involving Multiple Farrowing," unpublished paper April 1956, Catron Papers, Box 32, Folder 2.

47. DC to Larry Michaud, June 30, 1952, Catron Papers, Box 9, Folder 1.
48. Damon V. Catron, "Future of Hog Raising ... and Your Hog Business," *Feed Age* (May 1955): 1–8; Wayne Messerly, "Farrowing Crate Saved Every Pig," *Better Farming* 125 (July 1955): 33; "No More Late-Hour Sow-sitting?" *Farm Journal* 81 (October 1957): 40. See also the architectural drawings that indicate an artificial environment in Iowa State College, Agricultural Extension Service, "Plan for Farrowing Stalls," AEP-11 (Ames, Iowa: Agricultural Extension Service, February 1952).
49. Arnold Nicholson, "Will It Pay You to Air-Condition Hogs?" *Country Gentleman* 124 (May 1954): 38, 74, 76–77. These achievements, too, would not have been possible in the prewar era, before rural electrification became commonplace. Palmer Holden, interview with author, June 20, 2002.
50. Delmer E. Groves and Vernon Schneider, "He Gets a Monthly Hog Check," *Successful Farming* 50 (February 1952): 33–35, 62–63; and Mike Bay, "Why I Shifted from Pasture to Concrete," *Successful Farming* 57 (May 1959): 52–53, 97.
51. "They Raise Pigs Like Broilers," *Farm Journal* 80 (July 1956): 44–46; Palmer Holden, interview with author, June 20, 2002; and "Farms Retooling for New Model Pig."
52. Portland Cement Association advertisements quite common in this era, see, for example, *Breeder's Gazette* 124 (March 1959): 12.
53. Dean C. Wolf, "Pasture vs. Concrete in Hog Feeding," *Country Gentleman* 124 (July 1954): 22–23; "Confinement Cuts Feed Bill, Too," *Breeder's Gazette* 124 (March 1959): 13; J. A. Hoefer, "Constant Change Ever Occurring in Swine Management Plans," *Feed Bag* 35 (February 1959): 77–81. For typical hog confinement operations, efficient handling of manure remained the most significant bottleneck. By the late 1950, however, some mechanical manuring systems had become available. See for instance Dick Braun, "Clean Hog Lots with a Pump," *Farm Journal* 82 (December 1958): 34–35, 83; and "Automation of a Hog Farm," *Farm Quarterly* (Winter 1960): 78–81, 102.
54. Catron was among those who thought that integration went one step too far, and he urged that the trend be slowed in order to avoid its "socialist tendency" and to not hurt the efficient hog producer. DC to Joe O'Connor, December 16, 1957, Catron Papers, Box 26, Folder 9. O'Connor was president of Walnut Grove Products Company, the firm that eventually lured Catron away from Iowa State. In this letter, Catron stated "it seems to me that the hog producer loses, (especially the better ones) and about everyone else gains, especially the consumer from integration.... I personally would not want to be integrated."
55. Warren Jaycox to Ezra Taft Benson, [December 1955], in RG 16, Records of the Office of the Secretary of Agriculture, E17 General Correspondence 1906–1976, Box 2717, National Archives; Charles Borstad to Mr. Benson, December 5, 1955, RG 16, Records of the Office of the Secretary of Agriculture, E17 General Correspondence 1906–1976, Box 2539, National Archives. See also Salem Hendricks to Senator Hubert Humphrey, September 10, 1955; and Paul H. Muller to U.S. Rep. Henry O. Talle, October 24, 1955, also in Box 2539.
56. Edward L. Schapsmeier and Frederick H. Schapsmeier, *Ezra Taft Benson and the Politics of Agriculture: The Eisenhower Years, 1953–1961* (Danville, Ill.: Interstate, 1975), 146–147; and Gilbert C. Fite, *American Farmers: The New Minority* (Bloomington: Indiana University Press, 1981), 158–164. See also Lauren Soth, *Farm Trouble* (Princeton, N.J.: Princeton University Press, 1957).
57. John Rohlf, "Money-Making Hog Ideas by the Dozen," *Farm Journal* 78 (October 1954): 48–49, 68; "Handy Works for the Hog-Raiser" *Better Farming* 125 (March 1955): 90–91; Wayne Messerly, "Farrowing Crate Saved Every Pig," *Better Farming* 125 (July 1955): 33; "Hog Farmers Intensify for the Sixties," *Nation's Agriculture* 35 (April 1960): 12–13, 21–22. Another interesting illustration of an intermediate approach to confinement operations was the "Pigloo" farrowing system, a twelve-sided, igloo-like pen that forced the sow to lie along the outside of the pen while nursing piglets were confined by a metal guard to an inner circle that prevented them from being crushed. "New 'Pigloo' Hog System," *Farm Journal* 82 (May 1958): 65.
58. J. E. Pace, "Animal Husbandry Activities in Florida," *Annual Report of Extension, 1954*, Florida Cooperative Extension Service, Annual Reports, Series 91b, Box 22, University of Florida Archives, Gainesville; J. E. Pace, R. L. Reddish, and K. L. Durrance, "Animal Husbandry Activities in Florida," *Annual Report of Extension, December 1, 1955 to November 30, 1956*, Florida Cooperative Extension Service, Annual Reports, Series 91b, Box 33, University of Florida Archives, Gainesville.

59. Leon Johnson, "A Competitor's Viewpoint," speech delivered at 1956 Iowa Swine Producers Day, February 20, 1956, Catron Papers.
60. Ronald R. Kline, *Consumers in the Country: Technology and Social Change in Rural America* (Baltimore/London: Johns Hopkins University Press, 2000), 214, 278. In contrast to Kline's focus on farmers' resistance to adopt various new technologies, William Boyd's recent study of poultry focuses on resistance to this pressure among the animals themselves. See Boyd, "Science, Technology, and American Poultry Production." For broader examination of alternatives to the mass production model, see Philip Scranton, *Endless Novelty: Specialty Production and American Industrialization, 1865–1925* (Princeton, N.J.: Princeton University Press, 1997).
61. J. A. Hoefer, "What's Ahead in the Hog Business," *Successful Farming* 58 (February 1960): 42–43, 80–82; "Hog Farmers Intensify for the Sixties," *Nation's Agriculture* 35 (April 1960): 12–13, 21–22. See also Perkins, *Geopolitics and the Green Revolution*.
62. See Edmund Russell essay in this volume; comment here is based upon draft document titled "An Anatomy of Organismal Technology."
63. Damon V. Catron, "Tomorrow's Formulas for Swine Feeding" *Feedstuffs* 32 (March 12, 1960): 26–31; "Iowa State Researchers Seek to Perfect Flavor Compounds for Swine Rations," *Feedstuffs* 32 (February 13, 1960): 50; and DC to Dick Hanson, April 4, 1955, Catron Papers, Box 15, Folder 5. Also, Palmer Holden, interview with author, June 20, 2002.
64. "Cash Grants-In-Aid Received by Swine Nutrition Research, Iowa State University, Fiscal Years 1946–47 to and including 1958–59," and "Development of Swine Nutrition Research at Iowa State University," both December 1959, both in Catron Papers, Box 41, Folder 5. Catron's letter of resignation, effective May 1, 1960, is DC to L. E. Johnson, January 4, 1960, Catron Papers, Box 41, Folder 5. One factor in the resignation may have been controversies surrounding Catron's willingness to release information regarding the efficacy of new feed and pharmaceutical products just before they came onto the market. (Vaughan Speer to Mark Finlay, September 19, 2002.)
65. Through a simple process that involved a scalpel and a metal ruler, farmers learned to measure leanness on their own live pigs by making a deep incision and placing the ruler directly into the animal's back muscle tissues. See L. N. Hazel and E. A. Kline, "Mechanical Measurement of Fatness and Carcass Value on Live Hogs," *Journal of Animal Science* 11 (May 1952): 313–318.
66. I received a useful summary of the main impetuses for postwar changes from Thamon Hazen, letter to Mark Finlay, August 30, 2002. For data on the expansion of the nitrogenous fertilizer industry, particularly in the World War II era, see Vaclav Smil, *Enriching the Earth: Fritz Haber, Carl Bosch, and the Transformation of World Food Production* (Cambridge and London: MIT Press, 2001), 116–119.
67. Warren Belasco, "Food Matters: Perspectives on an Emerging Field," in *Food Nations: Selling Taste in Consumer Societies*, ed. W. Belasco and Philip Scranton (New York: Routledge, 2002), 4.
68. Amy Sue Bix, *Inventing Ourselves Out of Jobs? America's Debate over Technological Unemployment, 1929–1981* (Baltimore/London: Johns Hopkins University Press, 2000), 249–250.
69. Harold Lee, *Roswell Garst: A Biography* (Ames: Iowa State University Press, 1984), 175–178. See also Schapsmeier and Schapsmeier, *Ezra Taft Benson and the Politics of Agriculture*, 98–101.
70. Iowa data provided by Palmer Holden, professor of animal science, Iowa State University. Also pertinent, 93 percent of Iowa hog operations in 2000 had over 500 hogs on inventory, in contrast to the 20 percent of Iowa farmers who expected to bring over 250 hogs to market in 1962. Compare "Hog Business is Changing!" *Wallace's Farmer* 87 (April 21, 1962): 60; and U.S. Department of Agriculture, *Agricultural Statistics, 2001*, vii–25. Illinois data from John Wallize, "The Story of a 'Model,'" *Iowa Farm Science* 22 (July 1967): 3–9.
71. "Eat-Well, Eat Antibiotic Free," retrieved August 12, 2002, from http://www.keepantibioticsworking.com. See also http://www.dubreton.com; http://www.edenfarms.org; http://www.caherbegfreerangepork.ie; and Roy A. Schultz, "Antibiotic-Free Pigs: Is This the Future?" *Large Animal Practice* 21 (March 2000): 32, 34. Organic hog production currently accounts for far less than 1 percent of the total volume of 97 million hogs produced annually. Promoters of organic livestock production consider their goal to be a 1.8 percent market share, like that of organic milk. See Jerry Adler, "A Tale of Two Hogs," *Newsweek* 139 (September 30, 2002): 56–58

Afterword

SUSAN R. SCHREPFER

This volume has been the product of a long-term coming together—of colleagues within Rutgers University, of editors and authors, of those who participated in the 2002 conference sponsored by the Rutgers Center for Historical Analysis, and of those who offered their research for inclusion. *Industrializing Organisms* represents, as well, a meeting of environmental history and the history of technology, fields that had previously followed separate trajectories within their own professional societies and with potentially divergent perspectives. The disciplinary origins of the history of technology were embedded in mid-twentieth-century enthusiasm for industrial development; environmental history emerged somewhat later, during the quickening of concern about the untoward consequences of that enthusiasm. In his Introduction, Edmund Russell makes clear what those who have focused on technology bring to discussions of industrializing organisms; they have been thinking long and hard about human creativity. Environmental historians bring no less to the table, including the salutary caution inherent in a field tracing its lineage to societal demands for political and economic accountability in the 1960s and 1970s.

Donald Worster some years ago defined environmental history as clustered in three areas of analysis: that of ecologies, of modes of production, and attention to the political, cultural, social, and economic contexts of environmental change.[1] His definition points out the field's long-standing interest in both technological issues and the evolutionary shifts implicit in ecologic histories. Worster's definition also indicates a wide diversity among those who choose to call themselves (or are called by others) environmental historians. Indeed, it might appear at times as if, collectively, they have no center, but of course they do. That center is a keen sensitivity to the role that the natural environment has played in human history.

As Philip Scranton explained in the Preface, the inception of the 2002 conference "Industrializing Organisms" can be dated to a discussion begun the previous summer on Envirotech, a listserve then operating out of Stanford University. Edmund Russell launched the conversation with the question: "Are animals technology?" The initial responses of those in technology were mixed; one jokingly questioned Russell's sanity. Most of their opposition to the equation of organisms and technologies, however, seemed to lessen as it became clear the ways in which the history of domestication and breeding of plants and animals has echoed that of mechanical engineering. As Russell suggested, studying the garden facilitates understandings of the machine. As an environmental historian, however, I see an important caveat (one which Russell amply acknowledges) to the equating of organic and nonorganic technologies. Organisms have their own agendas, their own genetic options, and their own limitations. They are self-replicating.

Russell urges historians to use the concept of coevolution to describe the convergence of human design and the invisible hand of biological self-interest. Although this convergence has often proved agreeable to humans, the essays in this volume indicate that even those developments we have worked to engineer have been in substantial part the result of the pursuit by plants and animals of their own desire for survival. As the biblical tale of the Garden of Eden warns, gardens are neither as innocent, submissive, nor amenable to men and women as they may at first appear. Our metaphoric Eden warns that gardens tempt us to hubris. As historians Alfred Crosby and William McNeill, and others, remind us, human history has been dramatically shaped by the impact of infectious diseases borne by pathogens pursuing their own ends and playing by rules we only dimly understand.[2]

The pursuit of self-interest by all organisms has generated coevolutionary patterns that are clear in outcomes as diverse as the pests that kill wheat crops and the white camellias Susan Warren Lanman describes blooming in the greenhouses of New Jersey. The cultivated flower represented the convergence of Peter Henderson's business goals, the human sense of smell, and the flower's requirements for life. Henderson met the plant's need for warmth, food, and shelter as surely as the flower met his for a marketable commodity. Mark Finlay's essay tells us of the ways in which the industrial production of hogs capitalized on the natural rhythms of a reproductively active species. Organisms, Stephen Pemberton reminds us, are both "born and made." Some of the essays in this volume show instances in which plants and animals have shaped large sectors of economic production and global markets with, as in the case of sugar, only minimal changes to themselves.

Edmund Russell calls on historians to establish a new field within which to study such patterns of coevolution. He argues for an "evolutionary history" that will carry forward the mid-century theoretical synthesis of genetics and natural selection by integrating models of societal and biological change. Such an approach should advance the protean interaction begun between historians of the environment and those focusing on technology. One might well ask what makes these two groups receptive to a mutual discussion at this particular point in time. Jeffrey K. Stine and Joel A. Tarr date evidence of a general receptivity within the Society for the History of Technology to Technology and Culture's July 1997 issue, devoted to the theme of technology and the environment.[3] The subsequent establishment of the two listservs, or discussion forums on the Internet, Envirotech and H-Environment, similarly signaled a potential hybridization of these fields. It is clear that historians of technology began some time ago to complicate accepted theoretical assumptions separating man from nature, urban from rural, and industrial from organic. This volume demonstrates that the agricultural and biological revolutions of the nineteenth and twentieth centuries can most productively be analyzed as industrial revolutions, farms as factories, and organisms as technologies.

Environmental historians also began some time ago to question the Cartesian dualism of man and nature. They have reexamined one of the discipline's own icons—the American wilderness. A growing recognition of the long-term and active role played by humans in modifying North American environments has prompted awareness of the artificial and constructed qualities of legally designated wilderness areas. Not coincidentally, environmental historians are increasingly exploring diverse landscapes, expanding their gaze from the landed issues that preoccupied them earlier to those of the city and the industrial workplace. They have become keenly aware of how the human body functions as a technology, especially in its interaction with the machine and other organisms. Richard White further complicated the theoretical line between the natural and the mechanical by describing the Pacific Northwest's Columbia River as an "organic machine," a system of energy flows situated in the intersection of the mechanical action of turbines, the natural course of the river, and the human elements of labor. If rivers can be machines, animals can certainly be technologies.[4]

The present confluence of the interests of environmental historians and historians of technology grows also out of a shared desire to contribute to current debates arising out of microcellular research and genetic engineering. For example, what can historians bring, Russell asks, to the problems posed by present opposition within the global market to bioengineered commodities that the United States proposes to export? One of the ways in which the past informs the present is to reassure; there is, after all, nothing new under the sun. Present-day decisions can be made on the basis of criteria that have guided the domestication and breeding of organisms for thousands of years. The difference is not between natural organisms and biotechnologies but between, Russell tells us, older macrotechnologies as represented by the hybrid flower and the milk cow and modern microtechnologies.

Given the strong arguments for continuities, what are some of the generalizations that one can draw from the macrobiotechnologies described in this book? For one, their histories prompt reflection on the relative roles played by the public and private sectors. These histories tell of "creators" who acted on their own initiative and often for their own profit, from enterprising laboratory scientists to the entrepreneurial Perdue and Tyson families who designed the modern chicken. One cannot overlook, however, the vital part played by the state. Susan Warren Lanman explains how essential U.S. government tariffs and the construction of public infrastructures were to the development of commercial horticulture. The public sector was particularly active in the

United States during World War II and the cold war, as evidenced in Mark Finlay's history of hog production. These essays foreground a broad spectrum of publicly funded institutions, from the University of North Carolina to Iowa State, from the Cooperative Extensions that helped industrialize the chicken to the U.S. Department of Agriculture, from federal and state experiment stations to the National Institutes of Health, which funded fifty years of research on canine hemophilia. Future microbiotechnologies might well require a similar balance between public and private sector initiatives, just as future research will need to take both into account. A pertinent example of this balance of public and private roles is the recent project to map the human genome.

The essays in this volume also indicate a tension to which environmental historians might be particularly sensitive. Industrialization brings standardization, while biotechnologies, macro and micro, require diversity. Scientific production and modern marketing tend to require a simplification of the plants and animals on which they rely. Breeding for desirable traits, such as the meatier chicken, may result if only indirectly in a loss of commercially "unwanted" genetic traits and perhaps even of subspecies and species. The early twentieth-century neo-Darwinian synthesis of genetics and natural selection tells us that species are mutable (new ones are created out of old), but it also shows that nature is remarkably stingy—preserving even seemingly useless genetic traits, favoring diverse populations even within species, and zealously guarding the lines between species.[5] The genetic information thus provided encourages natural selection to continue; it has also proved to be central in the industrialization of organisms. The Olmstead-Rhode essay on wheat reminds us of the extent to which old pests are rarely eradicated and new ones constantly arise. As well, the industrial process itself presents a dilemma, as it tends to dictate a narrowing of the physical environments of productivity, such as replacing the barnyard with the hog pen. Crowding and genetic simplification have invited plagues and pestilence and have required genetic fixes. From where are future fixes to come if the genetic pools have themselves been narrowed and the walls of species breached? One key message of "evolutionary history" may be a warning: the future of organic technologies, as of natural selection, will demand diverse and separate populations.

Perhaps most sobering in these essays is the suggestion that standardization affects not only plants and animals but industrializing labor forces as well, a reality echoed chillingly in the words of those who, Finlay tells us, referred to the hog "assembly line." The parallel between the slaughter of war horses in the Civil War (some five hundred a day) and the loss of soldiers on those same days reminds us of our own stake in stories of industrializing organisms, as does the parallel between the standardization of the chicken and the experiences of the workers who eviscerate that chicken. As Stephen Pemberton's essay on hemophiliac dogs emphasizes, the line between man as scientist and man as experimental subject is fine indeed.

Environmentalists may well ask not only what rights do animals have in biotechnologies but what of human rights? The essays in this volume make clear the need for all organisms to have what Finlay calls "loving care," attention to their needs for warmth, shelter, and food. Occasionally biotechnologies prompt difficult ethical dilemmas, as when the suffering of animals must be weighed against the well being of humans. Pemberton demonstrates how scientists have balanced the two and how humans have at times been spared the requirement of being experimental subjects only by our biological affinities with other animals. The essays in this book tell us much about how organisms, including humans, make demands on industrial systems, among them the demand for discerning ethical sensitivities.

A third caution raised by the macrobiotechnologies discussed in this volume is the sheer complexities of the changes they invoked. None of these are stories of the simple engineering of an organism; each shows how industrialization not only transformed an organism's immediate physical environment—whether farm, garden, greenhouse, or laboratory—but also brought sweeping changes to the societies that produced and consumed them. Each essay emphasizes the importance of systems approaches. It was less the cow that was altered to produce milk year-round than it was changes in the face of European landscapes and global trade patterns that provoked this new pattern. Microbiotechnologies, if they follow the pattern of macrobiotechnologies, will reshape our lives, our economies, and our landscapes as thoroughly as the industrialized production of hogs, chickens, sugar, and wheat altered nineteenth- and twentieth-century life. These changes touched people's everyday lives in a positive fashion, producing goods that met their needs for mobility, food, and beauty.

Historians are well equipped, however, to recognize not only the occasionally dubious reassurances inherent in continuities but also the rending power of discontinuities—real and perceived. When Leo Marx addressed the "machine in the garden" he spoke not primarily of the changing realities of modern life but of how Americans experienced the "new" machine—its noise, its smoke, its fire, its power. From those who spoke of the sweet smell of the locomotive to the cowboys who raced the iron horse out of town or authors like Hawthorne who described the fire of the steam engine consuming workers, contemporaries experienced a revolution. These changes were qualitative and quantitative, alike real and perceived. It is difficult to measure the discontinuities represented by the industrial and the biological revolutions. The future, like the past, as David Lowenthal reminds us, is a foreign country not only because of evolutionary changes but because humans construct the narratives that explain them.[6] The perception of Africans and Europeans as to what constitutes appropriate foodstuffs is as important as the actual corn that the United States offers to export. In the postmodern world, perceptions have become important grist for the historian's mill.

The last caveat here might well be a question raised by several discussants in the Envirotech debate: If animals are technologies, are not humans as well? It seems possible that future historians will be called upon to look very close to home for the meaning of industrializing organisms. The collapse of the Cartesian dualism derives from recognition of how much man has remade his natural environment. Now, however, humans command the microbiotechnologies to remake their own bodies. This volume tells of important success stories engineered by man, from the industrial production of cows to hogs, from sugar to wheat. Each of these stories, however, reminds this environmental historian that the species shaped to meet mankind's needs remain ultimately self-directing organisms. As my cat might say of me, "Finally, she's got it right!"

Notes

1. Donald Worster, "Appendix: Doing Environmental History," in *The Ends of the Earth, Perspectives on Modern Environmental History*, ed. Donald Worster (Cambridge: Cambridge University Press, 1988), 290–307.
2. William C. McNeill, *Plagues and Peoples* (New York: Anchor Books, 1976); Alfred W. Crosby, *Ecological Imperialism: The Biological Expansion of Europe, 900–1900* (Cambridge: Cambridge University Press, 1986); for discussion of gardens, see Michael Pollen, *The Botany of Desire: A Plant's-Eye View of the World* (New York: Random House, 2002).
3. Jeffrey K. Stine and Joel A. Tarr, "Technology and the Environment," H-NET Urban History Discussion List H-Urban@H-NETMSU.EDU
4. Richard White, *The Organic Machine* (New York: Hill and Wang, 1995).
5. Susan R. Schrepfer, *The Fight to Save the Redwoods: A History of Environmental Reform, 1917–1978* (Madison: University of Wisconsin Press, 1983), 92–98.
6. Leo Marx, *The Machine in the Garden: Technology and the Pastoral Ideal in America* (New York: Oxford University Press, 1964); David Lowenthal, *The Past Is a Foreign Country* (New York: Cambridge University Press).

Notes on Contributors

William Boyd completed his doctorate at the Energy and Resources Group, University of California at Berkeley and his JD at Stanford Law School. In 2002–03 he clerked for a federal judge in Baltimore and is now revising his dissertation on southern forestry for publication.

Mark R. Finlay is an associate professor of history at Armstrong Atlantic State University. He earned his Ph.D. from Iowa State University in the History of Technology and Science and has published several articles related to the history of the agricultural sciences in Germany, Britain, and the United States.

Ann N. Greene is completing her dissertation, "Harnessing Power: Horses and Technology in 19th Century America" at the University of Pennsylvania.

Roger Horowitz is associate director of the Center for the History of Business, Technology, and Society at the Hagley Museum and Library. He has published widely on the American meat-packing industry. Currently he is at work on a book titled *Meat in America: Taste, Technology, Transformation*.

Susan Warren Lanman is an assistant professor of history and academic administrator at Metropolitan State College of Denver. Her articles focus on British and American garden history as a reflection of society, technology, and culture. Her current project is a book on American nineteenth-century commercial horticulture.

Alan L. Olmstead is professor of economics and director of the Institute of Governmental Affairs at the University of California, Davis. He is also a member of the Giannini Foundation of Agricultural Economics and director of the All-University of California Group in Economic History.

Barbara Orland is "Oberassistentin" at the Institute of History, History of Technology, at the Federal Institute of Technology (ETH) in Zurich, Switzerland. http://www.tg.ethz.ch/ She studied political science and history, completing her Ph.D. at the Free University Berlin (1991) with a thesis on the history of laundry work and technology. She has published several books and articles on the history of household technology, gender and science, medical technologies, and the popularization of science and technology. Her current research interests include the history of biomedical technologies and the history of agriculture, nutrition, and food technologies. She also produces online courses as an introduction to the history of technology.

Stephen Pemberton is a postdoctoral fellow in the department of history and the Institute for Health, Health Care Policy and Aging Research at Rutgers University, New Brunswick. He is a historian of medicine, science, and technology, currently writing a history of efforts in the United States to manage hemophilia.

Scott Prudham is assitant professor in the department of geography, program in planning at the Institute for Environmental Studies, University of Toronto.

Paul W. Rhode is professor of economics at the University of North Carolina, Chapel Hill, and research associate at the National Bureau of Economic Research.

Edmund Russell is associate professor of technology, culture, and communication and history at the University of Virginia. He is the author of *War and Nature: Fighting Humans and Insects with Chemicals from World War I to Silent Spring* (Cambridge, 2001).

Susan R. Schrepfer, associate professor of history at Rutgers–New Brunswick, focuses her research on environmental history. Her new book on mountaineering in America is forthcoming.

Philip Scranton is board of governors professor of history, Rutgers University Camden and director of the Center for the History of Business Technology and Society at the Hagley Museum and Library. With Susan Schrepfer, he codirected the Rutgers Center for Historical Analysis, 2001–03, and its industrial environments project, which sponsored the conference that led to this volume.

Mark J. Smith is assistant professor of history at the University of Central Oklahoma. His work focuses on agricultural technology, property rights, and ecological change in Latin America and the United States.

Index

Index note: pages references with an *f* or *t* indicate a figure or a table.

Abattoir, 24
Adams, Charles, Jr., 153
Afghanistan, war horses in, 161
African-American population
  consumption of chicken by, 224
  as labor for chicken production, 230–31
Agricultural Trade Development and Assistance Act, 254
agriculture, intentional and unintentional change of species in, 4–5
American Agriculturalist, 32
American Breeders Association (ABA), 111, 112–14
American Chemical Association, 244
American Cyanamid, 243, 245
American Farm Institute, 245
American Florist, 26
American Flower-Garden Directory, The (Buist), 22
American Magazine, 242
American Scientist, 46
Animal Chemistry (Liebig), 171
antibiotic resistance, 4, 5, 13
Arborgen, 128
"artificial selection," 3
Atack, J., 44
Atlantic Monthly, 157
Austin, Lloyd, 107, 115

Bairoch, Paul, 171
Bakewell, Richard, 175
Ballantyne's Nursery, 21
Bateman, F., 44
Benson, Ezra Taft, 250, 251*f*
Bessemer, H., 92
"biopower," 109
"biotechnology," defined, 6–8
Biotechnology Industry Organization (BIO), 5
Borstad, Charles, 250
Bosch, K., 8
Boyd, M., 4
Boyd, W., viii, 4, 7

Breck, Joseph, 32
Brill, Francis, 29
Brinkhous, Kenneth, 10, 192, 194, 195–97, 197*f*, 203, 205–8, 210–11*n*26
Brown, Frederick, 242
Buist, Robert, 22, 32
Buist's Exotic Nursery, 22
Burbank, Luther, 114, 115
Burger King, 225

Cambridge Economic History of the United States, 44
camplyobacter, contamination in chicken meat products, 231
canines, hemophilia research using, 191–92, 193*f*, 194–97, 197*f*, 198–99, 199*f*, 200–202, 202*f*, 203–7, 207*f*, 208–13, 264
Carleton, Mark A., 48, 64, 71, 72
Carson, William, 30
Catron, Damon, 238–39, 241–44, 247, 247*f*, 248–49, 252–53, 256*n*20
cattle/cows
  as biological artifacts, 1
  Freiburger stock in, 176
  Herdbook Society of Allgäu, 182
  history of social construction of a breed in, 167–79, 179*f*, 180, 265
  Holländer stock in, 173
  keeping milk records of, 182–84
  "milk mirror" of, 178
  "organic phase" in production of, 171, 185–86*n*24
  Rosensteiner Rindviehviehstamm stock in, 173
  Schwyzer stock in, 173, 176
  Shorthorn Cattle herdbook of, 178
  studbooks of, 176, 178
  Swiss Brown Race of, 178, 180
  Swiss development of high-yield milk cows, 11, 167–79, 179*f*, 180, 265
  Swiss purebreds and herdbooks in, 177–79, 179*f*, 180, 265
  a true type cow in, 177, 180–82
  viewed as machines, 169, 185*n*11–12
Center for Swine Nutrition Research, 245

269

Chandler, Alfred, 160
"Chicken of Tomorrow," 8, 11, 215, 219, 220, 223, 232, 233*n*1–2
chickens
   African-American consumption of, 224
   contaminated meat products of, 231–32
   history of marketing and consumption of, 216–21
   industrialized production of, 8–9, 11
   Jewish consumption of, 217, 224
   as layers, broilers, or roasters, 216
   meat status of, 215–35
   military consumption of, 218
   postwar development of, 219–23
   processing of, 218, 223–32
   Rhode Island Reds in, 11
   White Leghorns in, 216
Civil War
   Army of the Cumberland in, 151
   Army of the Potomac in, 152, 153, 157
   army use of war horses in, 7, 9–10, 143–46, 146*t*, 147–65, 264
   equine casualties in, 155–57
   horticultural production during, 27–28, 29
   wheat production in period of, 46–47
Clark, J., 67
Clawson, Garrett, 63
Cochrane, Willard, 43
coevolution, 12–13
Cold War, 242, 254
Colling, Charles, 175
Committee on Breeding Nut and Other Forest Trees, 113, 133*n*25
Conners, I., 49
Cooperative Extension Service, impact on chicken production, 219–22
corn, genetically engineered form of, 5
Country Gentleman, 32
Coxe, Tench, 143
Crosby, Alfred, 262
Crown Zellerbach Corporation, 122
Crozier, William, 32
Cuba
   batey factory site in, 99
   colonos system in, 93–94, 102
   development of sugar biotechnology in, 85–88, 88*f*, 89–98, 98*f*, 99–106
   frio cane in, 99
   frio-primavera cane in, 99
   primavera cane in, 99
   sugar production in Central Manatí, 86–87, 88*f*, 97–98, 98*f*, 99–105
   zafra cane, 98

Cuban Reciprocity Treaty of 1902, 104
Cunha, Tony, 244

Danhof, R., 62
Darwin, Charles, 2, 3, 182
Davidson, William, 30
De Bary, Anton, 49
Diamond, J., 4
Die Käserei in der Vehfreude (Gotthelf), 172
Doak, J. Herbert, 249
Dobzhansky, T., 3, 4
dogs
   in development of colonies of hemophilic dogs, 192, 193*f*, 194–95, 209*n*8
   hemophilia research with, 10, 191–92, 193*f*, 194–97, 197*f*, 198–99, 199*f*, 200–202, 202*f*, 203–7, 207*f*, 208–13, 264
   Terry Bay, Nora and Lynne as bleeder dogs, 199–200, 211*n*29
Dondlinger, T., 49, 53
Dreer, Henry, 38
Duffield, Jack, 122
Dye, A., 105

Earle, F., 87, 96
Eastern Shore Poultry Exchange, 221, 222, 226
Eddy, James, 115
Eddy Tree Breeding Station, 115
Edinburgh Botanic Gardens, 22
Eisenhower, Dwight D., 250, 254
Envirotech, vii, 4, 5, 262, 266
eugenics, 12
evolutionary biology, 2–13
evolutionary history, 2–16, 261
   future focus on human social problems, 12–13
   organisms as workers in, 9–10
   plant-animal dichotomy in, 12
   research methods of, 6–13
   standardization of organisms in, 11

"factory farming," 8
famine, in Zimbabwe, 5
Fife, David, 62, 72
Finlay, M., viii, 7–8, 10, 262, 264, 265
fish, impact of narrowed selection of, 5
Fitzgerald, D., 4
Fitzgerald, G., 11
Fleming, James, 30
Flint, Charles, 174

Food for Peace Program, 254
Fordist model, 250
forests, industrial improvement programs of, 107–39
Forrest, Nathan Bedford, 147
4-H, role in chicken production, 220
Fultz, Abraham, 63

Garden and Farm Topics (Henderson), 32
Gardeners' Chronicle, 26
Gardeners' Monthly, 32
Gardening for Pleasure (Henderson), 32, 33
Gardening for Profit (Henderson), 29, 32
gardens
  "machine in the garden" of, 1–2, 265
  market gardening of, 19
  See also horticulture
General Motors, 9
genetic engineering (GE), 13
  of corn, 5
  genetic diseases in, 11
  leak of engineered genes in, 128
  political implications of new food, 242
  quantitative trait loci (QTL) techniques in, 126
  of trees and forests, 125–29
Georgia-Pacific, 127
German Confederation, 173
Gilchrist, Peter, 21
Glass Manufacturers Association, 35
Gordy, J. Frank, Jr., 219–21, 223
Gotthelf, Jeremias, 172
Grant, Maxwell R., 242
Grant, Ulysses, 147, 154, 157
Greeley, William, 121
'green' Allgäu, 172
Greene, A., viii, 2, 6
"Green Revolution," 71, 254
Griliches, Z., 64
Guenon, Francois, 178
Gussow, J., 49
Gustafson, R., 45, 46

Haber, F., 8
Haber-Bosch method, 8
Halleck, Henry, 148, 151
Hamilton, L., 49
Harris, Homer, 248
Hatch Act, 219
Havemeyer Sugar Company, 23
Hayami, Y., 43
Haynes, L., 63
Hays, W., 63, 113

Hee, Stephen, 126
hemophilia
  antihemophilic clotting factor (AHF) concentrates in, 191, 192, 193*f*, 194, 203
  canine research at University of North Carolina, 192, 193*f*, 194, 203*n*37, 264
  Iowa Group research of, 195–96, 197*f*, 210*n*21
  issues of using dogs in research of, 205–8
  research with canines, 10, 191–92, 193*f*, 194–97, 197*f*, 198–99, 199*f*, 200–202, 202*f*, 203–7, 207*f*, 208–13, 264
  Terry Bay, Nora and Lynne as bleeder dogs, 199–200, 211*n*29
Henderson, Alfred, 30
Henderson, James, 21, 23
Henderson, Peter, 11, 19–42, 262
Henderson's Handbook of Plants (Henderson), 32
H-Environment, 262
"High Yield Forestry Program," 124
Hill, James, 125
Hispanic population, as labor for chicken production, 231
hogs
  "antibiotic growth effect" in, 243–49, 251, 257*n*22–28
  assembly line production of, 8, 10
  "free range" and organic production of, 254, 260*n*71
  hog prices of, 250–51, 251*f*
  industrialized strategies for production of, 237–38, 238*f*, 239–46, 246*f*, 247, 247*f*, 248–51, 251*f*, 252–60
  "life cycle feeding" and "life cycle housing" for, 246–47, 247*f*, 249
  "McLean County System" for, 240
  production during World War II, 240–41
  research of porcine von Willebrand's disease of, 207*f*
  time management in production of, 241
Honolulu Iron Works, 99
hops, waste as fertilizer, 23
Hormel, Jay, 10, 246
Horowitz, R., viii, 8, 11
horses in the Civil War
  antebellum uses of, 143–46, 146*t*
  army veterinary corps with, 158
  care and maintenance of, 153–54, 158
  equine casualties in, 155–57
  glanders outbreak in, 153, 163*n*26
  "jaded" horses in, 154, 158

horses in the Civil War *(cont.)*
    specifications and selection of, 5, 7, 9–10, 143–62, 163*n*20, 164–65
    Union army demands for, 143–65
    vulnerability and death of, 154–57
    as war technology in the Civil War, 147–65
horticulture
    consumer culture and, 34–40
    fertilizer sources for, 23–24
    garden writing and, 32–34
    glass prices and, 23, 27, 35
    impact on the urban diet, 39
    mass production and commercialization of, 19–42
    pot sizes for, 26–27
    standardization of seeds in, 11, 20–42
Horton, J., 54
Hounshell, D., 39
Hovey, C. M., 22, 26, 32
How the Farm Pays (Crozier), 32, 33
"Humbugs in Horticulture" (Henderson), 33
Hyslop, J., 52

inbreeding, 11
Industrial Forestry Association (IFA), 121, 122
Industrial Revolution, 104
infant diarrhea, 12
insecticides
    impact on survival/reproduction of insects, 3
    nicotine liquid as, 24
insects, infestations of wheat in the U.S., 44, 46–57
Institute of Forest Genetics (IFG), 115, 127
integrated pest management (IPM), role in wheat production in U.S., 44
International Baby Chick Association Meeting, 244
International Paper, 122, 127
International Union of Forest Research Organizations (IUFRO), 111
Internet, 262
Iowa Group, 195–96, 197*f*, 210*n*21
Israel, P., vii

James, General, 149
Jaycox, Warren, 250, 251*f*
Johnson, G., 45, 46
Johnson, Leon, 250
Jones, R., 150

Judd, Orange, 32
Jukes, Thomas H., 243, 244

Kansas Formula Feed Conference, 244
Kaufman, Clemens, 118
Kennedy, John F., 254
Kentucky Fried Chicken, 215
Khrushchev, Nikita, 254
"killing dry," 21, 27
Kline, Ronald, 250
Klippart, John, 71
Kloppenburg, J., 4
Klotzsch, Johann, 114
Knights of Labor, 35

Lanman, S. W., viii, 11, 262, 263
Large, E., 48
Larsen, Carl S., 116, 129
Larson, Syrach, 122
Laughlin, Jimmy, 10, 197, 197*f*, 198, 199*f*, 205, 206
Lawton, Alexander, 158, 159
Lederle, 243, 244, 245
Lee, Robert E., 157, 159
legumes, 8
Lehmann, Curt, 172
Leopold, Aldo, 107
Liebig, Justus, 171
Lincoln, Abraham, 148, 158
Longview Fiber, 122
Lord & Burnham, 36
Louisiana Sugar Planters' Association, 96
Lowenthal, D., 265

"machine in the garden," 6, 265
"macrobiotechnology," 6–8, 252, 265
Magazine of Horticulture, 22, 26, 32
malaria, 12
Malin, J., 62
market gardening, 23–28
Marlatt, C., 56
Martin, J., 67
Marvey, James, 25
Marx, L., 1, 2, 4, 6, 265
McAllister, Willard, 220, 221, 222
McClellan, George, 151, 154, 158
McCormick, Cyrus, 43, 72
McDonald's Chicken McNuggets, 225, 227
McKeen, John, 237
McMahon, Bernard, 32
McNeill, William, 262
McShane, C., 4
meat production. *See* chickens; hogs

Medicus, Ludwig, 174
Meigs, Montgomery, 7, 148–54, 156–60
Meitzen, August, 168, 173
Mellvain & Orr, 30
Melville Castle, 21, 22
Mendel's law/genetics, 48, 112, 132*n*23
Merck, 245
Merleau-Ponty, Maurice, 203
"microbiotechnology," 6–8, 252, 265
Monsanto, 128
Morris Canal, 30–31
Mosley, S., 5
mules
    antebellum uses of, 143–46, 146*t*, 162*n*9
    army use in the Civil War, 150–51, 154
Munger, Thornton, 121
Mycogen, 128
Myers, Abraham, 158, 159

Nathaniel Tooker's Molasses House, 23
National Institutes of Health (NIH), 192, 194, 201, 264
"natural selection," 3
"new biotechnology," defined, 6
New Jersey Sugar Refining Company, 23
New York Botanical Garden, 114
New York Horticultural Society, 34
New York Times, 244
Norrie, K., 64
North Carolina
    canine hemophilia research in, 192, 193*f*, 194, 203*n*37, 264
    State Cooperative, tree improvement program of, 118, 119
Northern Pacific Railway, 125
Northwest Tree Improvement Cooperative (NWTIC), 122, 123, 124

Olmstead, A., viii, 9, 264
O'Mara, Patrick, 37–38
On the Origin of Species (Darwin), 3
"Open Door to Greater Hog Profits," 245–46, 246*f*
Oregon Forest Conservation Act of 1941, 121
Oregon State Department of Forestry, 122, 123, 127
Orland, B., viii, 11
Osband, K., 56
Oxford Paper Company, 114, 115

P. Lorillard's Snuff and Tobacco Factory, 24
Pacific Northwest

Pacific Northwest Tree Improvement Research Cooperative (PNWTIRC) in, 123, 124
    tree improvement programs in, 120–24
Parker, W., 44
Pauley, S., 110
Pemberton, S., viii, 10, 262, 264, 265
Perdue, Arthur, 226
Perdue, Frank, 222, 225, 226–27, 234*n*17, 262
Perkins, J., 4
pesticides, resistance to, 5
Peter Henderson & Company, 30, 31, 37, 40
Pfizer, 243, 245, 248
Pfizer, Charles, 237, 238, 238*f*
pigs, assembly line production of, 8, 10, 237–38, 238*f,* 239–46, 246*f,* 247, 247*f,* 248–51, 251*f,* 252–60
Pinchot, G., 113
plants, as biological artifacts, 2
Pomeroy, Earl, 121
Popular Mechanics, 242
population genetics of, 3
Porter, Theodore, 167
Portland Cement Association, 249
poultry, transformation to a meat product of, 215–35
Power, J., 63
Practical Floriculture (Henderson), 32, 37
Pringle, Cyrus, 72
Protein Conservation Program, 241
Prudham, S., viii, 7
Prussion-German Bureau of Statistics, 167, 168

Rader, Karen, 194–95, 204–5
Ralston-Purina Company, 239
Reader's Digest, 206
Red Queen, The, wheat production and concept of, 46, 50, 54, 55, 73*n*9
Reid, James, 30
reproductive rate, 4
resistance
    to antibiotics, 4, 5, 13
    to herbicides, 129, 138*n*103
    to pesticides, 4–5
Reynolds Aluminum Company, 248
Rhode, P., viii, 9, 264
Rionda, Manuel, 86, 87, 97–98, 99, 100, 101
Ritvo, H., 4
Roelfs, A., 49
Roosevelt, Franklin, 242
Rosecrans, William, 147, 151, 152, 153, 158

Rose Manual, The (Buist), 22
Royal Botanical Society of Edinburgh, 21
Rural New Yorker, 32
Russell, E., vii, viii, 46, 238, 252, 261–63
Russian thistle, 55
Rutgers Center for Historical Analysis (RCHA), vii, 11, 261
Ruttan, V., 43

Salmon, M., 55, 57, 66
salmonella contamination in chicken meat products, 231
Satterthwait, A., 54
Saunders, C., 64, 72
Schatzmann, Rudolf, 177
Schindel, S., 63
Schrepfer, S., vii, viii
Science Digest, 191
Scott, James, 130
"Scottish system" of gardening, 21
Scranton, Philip, 261
"Second Agricultural Revolution," 237
seeds
  commercialization of, 20–42
  commodification of, 11
  seedmen for, 20
  standardization and distribution of, 28–32
  See also horticulture
Shanbrom, Edward, 206–7
sheep, as biological artifacts, 1
Sherbaker, horse agent, 149–50
Sheridan, Philip, 147, 157
Sherman, William, 152, 157
Silen, Roy, 122
Silvics, 113, 133n29
Smith, Harry P., 195–97, 197f, 199, 210n20
Smith, M., viii, 8
social Darwinism, 12
Society for the History of Technology, 4
Society of American Florists, 26, 31
Southern Forest Tree Improvement Committee (SFTIC), 116
Southwide Pine Seed Source, 116
Spillman, W., 70
Sprang, Charles F., 22
Stakman, E., 46
Stanton, Edwin, 148
Steele, Cecile, 217–18, 232
Sterling, George, 21, 27
Stine, J., 262
Stokstad, E. L., 243, 244
Straw, Levi, 149

Stuart, "Jeb," 147, 151
Successful Farming, 249
"sugar house scum," 23–24
sugar production
  "big sugar" of, 105
  cultivation and harvesting of, 98–104
  industrialization in Cuba of, 8, 85–88, 88f, 89–98, 98f, 99–106
  process of, 87–88, 88f, 89–97
  revolution in plant varieties of, 91, 95–96
  saccharum officinarum in, 85, 87, 105
Sugar Research Foundation, 85, 96, 97
survival rate, 4
Swift, 217, 222
Swift, Gustavus, 225

Tarr, J., 4, 262
Taylor, J., 5
technology
  black boxed form of, 2, 6
  evolutionary history of, 1–16
  nature and, 1–16
  social factors influence of, 4
Texas A&M University, tree cooperative at, 118
Thaer, Albrecht, 167
Thelin, Murray, 206–7
Thorburn, George, 22
Thorburn, Grant, 22
Tillet, Mathieu, 49
Today's Health, 206
Townsend, Inc., 222
Tree Genetic Engineering Research Cooperative (TGERC), 127–29
trees
  biotechnologies in improvement of, 125–29
  economic and biological challenges of improvement in, 109–10, 131n9
  forestation projects in Europe of, 111–12
  improvement programs of, 107–39
  macrobiological and microbiological changes in, 7–8
  organizations in research of, 110–14
  transgenic trees and conifers, 125–29
tuberculosis, 12
tularemia, biotechnology of, 11
Tyson, 222, 262
Tyson, Don, 225–26
Tyson, John, 225

University of Florida, tree cooperative at, 118

U.S. Armed Forces Commissary, 225
U.S. Centers for Disease Control and Prevention (CDC), 5
U.S. Department of Agriculture (USDA), 43, 63, 96, 112, 127, 215, 219, 241
U.S. Food and Drug Administration (FDA), 243
U.S. Forest Service, 114, 115, 120–23

Van Sicklen, Abraham, 29
Vantress Hatchery, 215
Variation of Animals and Plants under Domestication, The (Darwin), 3, 182
Viebahn, Georg von, 167, 168
Vietnam War, 160–61
von Thünen's Isolated State model, 168

Wagner, Robert, 206
Wakeley, Philip, 116, 117
Ward, T., 64
Weekly Register, 144
Weigley, R., 161
West Coast Lumbermen's Association, 121
Westvaco, 126, 127
Weyerhauser, 121, 124–25, 126, 127
Weyerhauser, Frederick, 125
wheat production in America, 43–57, 58*t*, 59*t*, 60*f*, 61, 61*t*, 62–65, 65*t*, 66–68, 68*t*, 69–83

achieving and maintaining yields of, 45–57
biological innovations in history of, 44, 73*n*10
diseases and insect infestations of, 46–57
Green Revolution in, 71
introduction of new varieties to, 46, 48, 57, 58*t*, 59*t*, 60*f*, 61, 61*t*, 62–65, 65*t*, 66–68, 68*t*, 69–70
major wheat regions of, 44–45, 47–48, 57, 58*t*, 59*t*, 60*f*, 61, 61*t*, 62–68, 68*t*, 69–70
winter and spring wheats in, 47, 74*n*20
Wheeler, Joseph, 157
Whillden Pottery Company, 26
Whipple, George, 195, 196
White, Richard, 262
Wickard, Claude, 240
Wilkenson, Sidney, 38
Wilson, 217
Wilson, E., 149, 150
Work Simplification Laboratory, 241
World Forestry Conference, 129
World Trade Center, historic site of, 31
Worster, Donald, 261

Zell, J., 104
Zobel, Bruce, 118, 119, 135*n*59, 136*n*67